South University Library
Richmond Campus
2151 Old Brick Road
Glen Allen, Va 23060

MAR 2 7 2018

# FORGET ME NOT

## The Neuroethical Case Against Memory Manipulation

Peter A. DePergola II

Cognitive Science and Psychology

Copyright © 2018 Vernon Press, an imprint of Vernon Art and Science Inc, on behalf of the author.

All rights reserved. No part of this publication may be reproduced, stored in a retrieval system, or transmitted in any form or by any means, electronic, mechanical, photocopying, recording, or otherwise, without the prior permission of Vernon Art and Science Inc.

www.vernonpress.com

*In the Americas:*
Vernon Press
1000 N West Street,
Suite 1200, Wilmington,
Delaware 19801
United States

*In the rest of the world:*
Vernon Press
C/Sancti Espiritu 17,
Malaga, 29006
Spain

Cognitive Science and Psychology

Library of Congress Control Number: 2017962800

ISBN: 978-1-62273-364-4

Product and company names mentioned in this work are the trademarks of their respective owners. While every care has been taken in preparing this work, neither the authors nor Vernon Art and Science Inc. may be held responsible for any loss or damage caused or alleged to be caused directly or indirectly by the information contained in it.

Cover designed by Vernon Press, using elements from pixabay.com/Tumisu .

*For Laura and Mary, who give me every reason to remember*

"There is something quite definite I have to say, and I have it so much upon my conscience that … I dare not die without having uttered it. For the instant I die and so leave this world … the question will be put to me: 'Hast thou uttered the definite message *quite definitely?*' And if I have not done so, what then?"

- Søren Kierkegaard, *Papers and Journals: A Selection*


"You think you have [memory], but it has you. …"

- John Irving, *A Prayer for Owen Meany*


"There was a long hard time when I kept far from me the remembrance of what I had thrown away when I was quite ignorant of its worth."

- Charles Dickens, *Great Expectations*


"Memory is the scribe of the soul."

- Aristotle, *On Memory*


"Thence entered I the recesses of my memory, those manifold and spacious chambers, wonderfully furnished with innumerable stores; and I considered, and stood aghast. …"

- Augustine of Hippo, *Confessions*, Bk. XIII


"Gatsby believed in the green light, the orgastic future that year by year recedes before us. It eluded us then, but that's no matter – tomorrow we will run faster, stretch out our arms farther. … And one fine morning — So we beat on, boats against the current, borne back ceaselessly into the past.

- F. Scott Fitzgerald, *The Great Gatsby*

# Table of Contents

| | |
|---|---|
| ABSTRACT | xi |
| ACKNOWLEDGMENTS | xv |
| FOREWORD: NEUROETHICS & MEMORY MANIPULATION | xvii |
| 1. INTRODUCTION: THE FALSE HOPE OF DELIBERATE FORGETTING | 1 |
|     1.1. Status Quaestionis | 2 |
|     1.2. Argument and Method | 2 |
|     1.3. Clarification of Terms | 3 |
|     1.4. Notes | 4 |
| 2. THE NEUROSCIENCE OF MM: PRACTICES AND POSSIBILITIES, PROPONENTS, AND PROBLEMS | 7 |
|     2.1. Practices and Possibilities of MM | 7 |
|         2.1.1. Beta-Adrenergic Receptor-Blocking Pharmacologicals | 8 |
|         2.1.2. Electroconvulsive Therapy (ECT) | 10 |
|         2.1.3. False Memory Creation (FMC) | 12 |
|         2.1.4. Deep Brain Stimulation (DBS) | 14 |
|     2.2. Proponents of Limited-Use Memory Manipulation (LUMM) | 16 |
|         2.2.1. LUMM for Post-Traumatic Stress Disorder (PTSD) | 16 |
|             2.2.1.1. The Neurobiology of PTSD | 17 |
|             2.2.1.2. The Case in Favor of LUMM for PTSD | 19 |

| | |
|---|---|
| 2.2.2. LUMM for Substance Addiction | 22 |
|     2.2.2.1. The Neurobiology of Substance Addiction | 23 |
|     2.2.2.2. The Case in Favor of LUMM for Substance Addiction | 26 |
| 2.3. Problems of MM: A Preliminary Response to LUMM Proponents | 29 |
|     2.3.1. Biomedicalization and the Codification of New Diseases | 29 |
|     2.3.2. The Myth of Global Autonomy Loss | 32 |
| 2.4. Conclusion | 35 |
| 2.5. Notes | 36 |
| **3. THE CASE AGAINST EPISODIC DISINTEGRATION: THE MORAL SIGNIFICANCE OF AUTOBIOGRAPHICAL MEMORY FOR ETHICAL DECISION MAKING** | **47** |
| 3.1. Autobiography, Memory, and Judgment | 47 |
|     3.1.1. The Metaphysics of Autobiographical Memory | 48 |
|     3.1.2. Autobiographical Memory as Pillar of Moral Judgment | 51 |
| 3.2. Autobiographical Memory and Rationality | 53 |
|     3.2.1. Imaging and Socialization of Autobiographical Memory | 54 |
|     3.2.2. The Functional and Developmental Ontology of Rational Autobiographical Memory | 57 |
|     3.2.3. The Phenomenology of Autobiographical Reasoning | 61 |
|     3.2.4. Autobiographical Meta-Memory and Rational Prospection | 63 |
| 3.3. Autobiographical Memory and the Narrative of Human Emotion | 66 |
|     3.3.1. The Empirical Effects of Emotion on Autobiographical Memory | 67 |
|     3.3.2. The Contextual Effects of Emotion on Autobiographical Memory | 70 |

3.3.3. The Cognitive Effects of Autobiographical Memory on Empathic Intentionality . . . . . . 73

3.3.4. Emotional Memory Specificity and Autobiographical Narrative Centrality . . . . . . 76

3.4. Autobiographical Memory and the Emotional Nature of Rational Ethical Decision Making . . . . . . 78

3.4.1. Integrated Affective Foresight and Autobiographical Context Prediction . . . . . . 79

3.4.2. The Affective Structure of Autobiographical Moral Thought . . . . . . 82

3.4.3. The Primacy of Autobiographical Memory in Emotionally Rational Choice . . . . . . 85

3.4.4. Autobiographical Prospection and Rational Ethical Decision Making . . . . . . 88

3.5. Conclusion . . . . . . 91

3.6. Notes . . . . . . 92

4. THE CASE AGAINST AFFECTIVE DEGENERATION: THE MORAL SIGNIFICANCE OF EMOTIONAL RATIONALITY FOR ETHICAL DECISION MAKING . . . . . . 107

4.1. Emotion, Reason, and Choice . . . . . . 107

4.1.1. Emotion and Reason in Ancient Greek Philosophy . . . . . . 108

4.1.2. The Emotional and Rational Tenets of Choice . . . . . . 110

4.2. Emotion and Rationality . . . . . . 114

4.2.1. The Functional Ontology of Rational Emotion . . . . . . 114

4.2.2. The Epistemology of Rational Emotion . . . . . . 117

4.2.3. The Phenomenology of Rational Affective Awareness . . . . . . 120

4.2.4. Emotional Rationality as Necessary Condition for Evaluative Judgment . . . . . . 124

4.3. Emotional Rationality and Morality . . . . . . 127

4.3.1. The Cognitive Dimensions of Emotionally Rational Moral Intuition . . . . . . 128

    4.3.2. The Psychology of Emotionally Rational
    Moral Value     131

    4.3.3. The Rationality of Affective Moral
    Objectivity     134

    4.3.4. Emotional Rationality as Canon
    of Moral Motivation     137

4.4. Emotional Rationality and Ethical Decision Making     141

    4.4.1. The Cognitive Neuroscience of
    Emotionally Rational Ethical Decision Making     141

    4.4.2. Emotional Rationality as Ethical Insight
    and Practical Reason     144

    4.4.3. Emotionally Rational Moral Reflection
    as Ethical Problem Solving     147

    4.4.4. The Normative Logic of Emotionally
    Rational Ethical Decision Making     150

4.5. Conclusion     154

4.6. Notes     156

5. THE CASE AGAINST NARRATIVE DECAYING: THE MORAL SIGNIFICANCE OF NARRATIVE IDENTITY FOR ETHICAL DECISION MAKING     167

5.1. Narrative, Identity, and Development     167

    5.1.1. The Evolution of Narrative in History     168

    5.1.2. The Ontology of Narrative Identity     171

    5.1.3. Narrative Identity and Moral Development     175

5.2. Narrative Identity as Product of Autobiographical Memory and Emotional Rationality     179

    5.2.1. The Autobiographical and Emotionally
    Rational Ontology of Narrative Identity     179

    5.2.2. Narrative Identity as Moral Education,
    Moral Methodology, and Moral Discourse     183

    5.2.3. Narrative Identity as Ground and Object
    of Normative Ethical Principles     185

    5.2.4. Rigor in Narrative Judgment and Ethical
    Justification     188

5.3. The Requisite Unpredictability of Narrative Identity ... 191

    5.3.1. The Cognitive Correlates of Narrative Values ... 192

    5.3.2. The Variable and Creative Evolution of Narrative, Identity, and Agency ... 195

    5.3.3. The Developmental and Redemptive Metaphysics of Narrative Identity ... 199

    5.3.4. The Narrative Objectivity of Subjective Identity ... 203

5.4. The Threat of Manipulation to Narrative, Authenticity, and Ethical Decision Making ... 207

    5.4.1. Manipulation as Splintering of the Narrative Self ... 207

    5.4.2. Manipulation as Corrosive of Narrative Authenticity ... 211

    5.4.3. Manipulation as Destructive of Narrative Self-Control, Narrative Growth, and Narrative Responsibility ... 215

    5.4.4. Manipulation as Disabling of Narrative-Based Ethical Decision Making ... 218

5.5. Conclusion ... 222

5.6. Notes ... 224

6. CONCLUSION: THE TERMINAL NORMLESSNESS OF MEMORY MANIPULATION ... 237

6.1. The Case Against MM: The Normative Demands of Proportionate Reason ... 237

6.2. Implications ... 242

6.3. Notes ... 243

BIBLIOGRAPHY ... 245

INDEX ... 275

# ABSTRACT

An increasingly blurred understanding of the moral significance of accurate and authentic autobiographical memory for an adequate apprehension of self, other, and community suggests a critical need to explore the inter-relationships shared between autobiographical memory, emotional rationality, and narrative identity in light of the contemporary possibilities of memory manipulation (MM), particularly as it bears on ethical decision making. Grounding its thesis in four evidential effects – namely, (i) MM disintegrates autobiographical memory, (ii) the disintegration of autobiographical memory degenerates emotional rationality, (iii) the degeneration of emotional rationality decays narrative identity, and (iv) the decay of narrative identity disables one to seek, identify, and act on the good – the book argues that MM cannot be justified as a morally licit practice insofar as it disables one to seek, identify, and act on the good.

# ACKNOWLEDGMENTS

Writing this book required years of reflection on the moral significance of memory. Now that it is finished, I want to thank those who helped to create so many of the most beautiful and formative memories I have.

I am eternally indebted, above all, to my inspiring, supportive, and courageous wife, Laura, without whose love, friendship, and motivation this book would not have been possible. I am similarly indebted to my heroic mother, Mary, for her life-changing introduction to the world of clinical bioethics so many years ago.

My profoundest admiration and respect extend to my formidable mentor, Dr. Gerard Magill, whose rigorous attention and meticulous intellect served as my faithful and unwavering guide throughout the composition of an earlier dissertation version and, subsequently, its transformation into the present manuscript. Like the ropes that bound Odysseus to the ship's mast as he sailed past the Siren voices, Gerry's methods and expectations were the wise constraints that made me free. For that and much more, I am indebted to him beyond expression.

Steadfast thanks are due also for the encouragement of my supportive and talented colleagues at Baystate Health, the College of Our Lady of the Elms, University of Massachusetts Medical School, Tufts University School of Medicine, and Sacred Heart University.

I owe reverence and gratitude, finally, to my loyal patrons, Saints Augustine and Thomas Aquinas, whose intercession aided in pointing out the beginning, directing the progress, and helping in the completion of this work.

# FOREWORD: NEUROETHICS & MEMORY MANIPULATION

Gerard Magill, Ph.D., Vernon F. Gallagher Chair,
Center for Healthcare Ethics, Duquesne University

A bold stance that challenges received wisdom is a difficult endeavor for the best of writers. The author sets out on an exciting pathway to write an imaginative and original analysis of the relation between memory, emotion, identity and ethical decision-making. Being the first major book on this extraordinary subtle interaction, the reader not only evidences brinkmanship on the cutting edge of neuroscience but also offers a tour de force on moral normativity. In this astute study on the ethics of memory manipulation, there is a breathtaking view of over fifty years of memory research interpreted through the philosophical lens of normative ethics.

The outcome is dramatic and strategically significant. The drama accompanies a rigorous ethical defense of an absolutist position against memory manipulation, such as for patients suffering from PTSD or addiction. Hopefully, those who may be surprised at or opposed to such a counter-cultural stance will recognize the cautious and temperate analysis that carefully and precisely presents both sides of the ethical debate on memory manipulation. The strategic significance of the argument emerges from a persistent consideration of the evidence by a skilled researcher who started out in favor of memory manipulation only to be led by the data to oppose it, and strenuously so. Here is a master ethicist at work, balancing each side, presenting fair and accurate perspectives of the debate, perceptively seeing difficulties that have gone unnoticed, and pivoting to a conclusion that was as startling to the author as it may be to the reader. The outcome is stark: memory manipulation cannot be justified as an ethical medical practice – the noble goals to support and treat patients whose suffering elicits heart-wrenching compassion cannot be justified by what should be construed as immoral means of treatment. It could have been much more palatable for the author to have argued in favor of the mainstream defense of memory manipulation in the medical profession, as was his initial instinct at the start of the project. Yet, the study demonstrates how complex the topic is to grapple with intellectually and professionally. Fortunately, the clarity of writing and amenable manner of arguing present an approach that will be accessible to health professionals, researchers and scholars, as well as teachers and students.

To appreciate the contribution of this challenging and novel work, it can help to situate its charge within the broader context of bioethics and neuroscience, typically referred to as neuroethics. Over recent years, there has been a substantive increase in published works on neuroethics. They can be broadly categorized as follows. The appearance of several handbooks of neuroethics highlights the increasing appeal of this field both to professional ethicists and to a general audience insofar as these texts cover a wide variety of topics with scientific, medical, ethical, legal, and social implications, such as related to consciousness, intention and responsibility, or to aging and dementia.[1] Interestingly, the advances in neuroethics are so expansive that already the impact upon higher education policy is under consideration.[2] Not surprisingly, much of the recent literature deals with interdisciplinary issues about treatment and practical case studies in neuroscience, of which there is an abundance.[3] Also, the reciprocal impact of neuroethics discourse on theology and philosophy receives considerable scrutiny.[4] And the significance of neuroscience for morality itself presents captivating insights about the genesis and function of processes in ethical discernment, not only regarding the role of emotions and reason but also regarding the impact of moral cognition upon social conduct.[5] Predictably, the technological developments in neuroscience, including discoveries in nanotechnology and the neuroethics of biomarkers around breakthroughs in the human genome,[6] promise a fascinating future with the anticipation of exciting new social frontiers and ethical challenges.[7]

On such an expansive landscape of neuroethics this book makes its mark, and it is a very distinctive one to behold. The creativity and perspicacity of the analysis on the ethics of opposing memory manipulation is woven together with a treasure trove of classical writers, from ancient times onwards, whose time-tested insights shed light on this new topic that most could never have imagined. The list of giants is extensive, including Plato and Aristotle, Hume and Descartes, Dostoyevsky and Shakespeare, Freud and Heidegger, Husserl and William James, Kant and Locke, Nietzsche and Nussbaum, Pascal and Rawls, Ricoeur and Rorty, Sartre and Scheler, Charles Taylor and Wittgenstein, MacIntyre and McCormick.

The conversation with these greats enlightens and bolsters the analysis to yield a rich tapestry of scholarship that projects a clear-minded message to establish the significance of the work: memory manipulation cannot be morally justified and therefore its practice must cease – even the best of good ends do not justify immoral means. Taking bad, haunting memories away in an immoral manner simply compounds the original physical, emotional, mental harm with a moral harm that disintegrates identity, compromises values, and constitutes a morally corrosive approach to a profoundly human need. But such a bold stance does not bear any callousness for the many patients in dire need of treatment that memory manipulation has sought to address. Rather, the clarity of the ethical posture necessarily challenges professionals and society to up the game so to speak to provide urgent medical and social support for those who may see some relief in memory manipulation. The author faces up steadfastly to the implications of rejecting memory manipulation: there is a dire need to radically increase investment in and support for mental healthcare, especially in this fast-developing field of neuroscience where ethics has so much to contribute.

**Notes**

1. See, for example, Matthew Rizzo, Matthew, et al., eds., *The Wiley Handbook on the Ageing Mind and Brain* (Wiley-Blackwell, 2018); L. Syd M. Johnson, Karen S. Rommelfanger, eds., *The Routledge Handbook of Neuroethics* (Routledge, 2017); Jens Clausen, Neil Levy, *Handbook of Neuroethics* (New York: Springer, 2014); Judy Illes, Barbara J. Sahakian, eds., *The Oxford Handbook of Neuroethics* (New York: Oxford University Press, 2013); Robert H. Blank, *Intervention in the Brain: Politics, Policy and Ethics* (Cambridge, MA: MIT Press, 2013); John Bickle, ed., *The Oxford Handbook of Philosophy and Neuroscience* (New York: Oxford University Press, 2013).

2. See, Dana Lee Baker, Brandon Leonard, *Neuroethics in Higher Education Policy* (Palgrave MacMillan, 2016).

3. See, for example, Robert J. Sternberg, Susan T. Fiske, *Ethical Challenges in the Behavioral and Brain Sciences. Case Studies and Commentaries* (New York: Cambridge University Press, 2015); Anjan Chaterjee, Martha J. Farah, eds., *Neuroethics in Practice* (New York: Oxford University Press, 2013); Elis-

abeth Hildt, Andreas G. Franke, eds., *Cognitive Enhancement: An Interdisciplinary Perspective* (New York: Springer, 2013); Eric Racine, *Pragmatic Neuroethics: Improving Treatment and Understanding of the Mind-Brain* (Cambridge, MA: MIT Press, 2010).

4. See, for example, Geran F. Dodson, *Free Will, Neuroethics, Psychology and Theology* (Wilmington, Delaware: Vernon Press, 2017); Neil Messer, *Theological Neuroethics: Christian Ethics Meets the Science of the Human Brain* (New York: T&T Clark, Bloomsbury, 2017); E. Fuller Torrey, *Evolving Brains, Emerging Gods. Early Humans and the Origins of Religion* (New York: Columbia University Press, 2017); Thomas M. Crisp, et al., *Neuroscience and the Soul. The Human Person, in Philosophy, Science and Theology* (Eerdmans, 2016); Nasda Gligorov, *Neuroethics and the Scientific Revision of Common Sense* (New York: Springer, 2016); Tibor Solymosi, John R. Shook, eds., *Neuroscience, Neurophilosophy and Pragmatism: Brains at Work with the World* (Palgrave MacMillen, 2014); Charles T. Wolfe, ed., *Brain Theory: Essays in Critical Neurophilosophy* (Palgrave MacMillan, 2014); James J. Giordano, Bert Gordijn, eds., *Scientific and Philosophical Perspectives in Neuroethics* (New York: Cambridge University Press, 2010); Andrew B. Newberg, *Principles of Neurotheology* (Routledge, 2010).

5. See, S. Matthew Liao, ed., *Moral Brains: The Neuroscience of Morality* (New York: Oxford University Press, 2016).

6. See, Matthew L. Baum, The *Neuroethics of Biomarkers. What the Development of Bioprediction Means for Moral Responsibility, Justice, and the Nature of Mental Disorder* (New York: Oxford University Press, 2017); Sean Hays, et al., eds., *Nanotechnology, the Brain, and the Future* (New York: Springer, 2013).

7. See, for example, Judy Illes, ed., *Neuroethics: Anticipating the Future* (New York: Oxford University Press, 2017); Eric Racine, John Aspler, eds., *Debates about Neuroethics: Perspectives on its Development, Focus, and Future* (New York: Springer, 2017); Andrea Lavazza, ed., *Frontiers in Neuroethics* (England: Cambridge Scholars Publishing, 2016).

# 1.
# INTRODUCTION:
# THE FALSE HOPE OF DELIBERATE FORGETTING

Perhaps foremost among the controversial possibilities of cognitive neuroscience is the capacity to interrupt the reconsolidation of autobiographical memories,[1] thereby "dampening"[2] the acuteness of emotions associated with painful, and often perpetually traumatic, experiences of past events.[3] The use of pharmacologicals to extinguish the sting of traumatic memories is but one of the current pathways being explored in the name of benevolent therapy. Others include the implantation of false memories to alter or replace those considered unsavory, and targeted electroconvulsive and surgical techniques to delete specific recollections of the past.[4] Amid a field of gray matter, this book takes an absolutist position, and argues that any technique employed to directly and intentionally erase episodic memory poses grave neuroethical threats to the human condition that cannot be justified within a normative moral calculus.[5]

In its steadfast cultivation of increasingly effective and permanent memory manipulation techniques, modern neuroscience ominously mirrors the Homeric tale of Odysseus's unanticipated visit to the land of Lotus-eaters, whose inhabitants repeatedly consume the flowery plant to render themselves forgetful of everything once known, blissfully careless of all.[6] As the epic goes, Odysseus's shipmates, desiring to share in the euphoria of the natives, follow suit in consumption of the plant. Yet they receive only a blurry, amoral reality in return. While the plant successfully eradicates the episodic memories of the shipmates, it also expunges the positive and beneficial desire to engage in human relationships, thereby decorticating them of any corresponding sense of moral responsibility. The scene concludes with Odysseus dragging his shipmates, against their will, back into the boat as they wail in rebellion, pleading for one more taste of self-annihilation.

## 1.1. Status Quaestionis

Until now, the debate over memory manipulation (MM) has been riddled with fallacious and polarizing rhetoric.[7] For some, MM paves the already-too-tempting road to a brave new world in which human beings are gods of their own.[8] For others, the concepts of augmentation and perfection are mutually exclusive,[9] and manipulation has little to do,[10] in fact, with the desire to become god-like.[11] A third approach, heretofore uncharted in the literature, grounds the framework of the arguments posited in this book. It concerns the individual and collective relationships shared between autobiographical memory, emotional rationality, and narrative identity in the context of MM. The interconnectedness of these concepts is indispensable to any productive discussion over the ethics of manipulation, yet each respective interconnection remains underdeveloped. This book will argue that, with scrupulous examination, MM proves autobiographically, emotionally, narratively, and ultimately morally counterproductive to its intended end, however independently noble it may be.

An increasingly blurred understanding of the moral significance of accurate and authentic autobiographical memory for an adequate apprehension of self, other, and community suggests a critical need to explore the inter-relationships shared between autobiographical memory, emotional rationality, and narrative identity in light of the contemporary possibilities of MM, particularly as it bears on ethical decision making. To be sure, the issues of immediate import to the conversation over the general moral licitness of cognitive manipulation are manifold, and any singular analysis of topics, no matter how sweeping, will unavoidably fall short of adequacy. This book thus aims to address the foregoing three, each of which substantiate and build off of the others.

## 1.2. Argument and Method

The argument pursued herein grounds itself in four evidential effects: (i) MM disintegrates autobiographical memory; (ii) the disintegration of autobiographical memory degenerates emotional rationality; (iii) the degeneration of emotional rationality decays narrative identity; and (iv) the decay of narrative identity disables one to seek, identify, and act on the good. Drawing from this logic chain, the book endeavors to examine the respective relationships shared between autobiographical memory, emotional rationality, and narrative identity to demonstrate its argument that MM cannot be justified as a morally licit practice insofar as it disables one to seek, identify, and act on the good.

To secure the justification of its thesis, the book moves in five parts. Chapter 2 addresses the neuroscience of MM, including a specific analysis of (i) the practices and possibilities of MM; (ii) the proponents of limited-use memory manipulation (LUMM); and (iii) the problems of MM. Chapter 3 posits the case against episodic disintegration, including a specific analysis of (i) autobiography, memory, and judgment; (ii) autobiographical memory and rationality; (iii) autobiographical memory and the narrative of human emotion; and (iv) autobiographical memory and the nature of rational ethical decision making. Chapter 4 posits the case against affective degeneration, including a specific analysis of (i) emotion, reason, and choice; (ii) emotion and rationality; (iii) emotional rationality and morality; and (iv) emotional rationality and ethical decision making. Chapter 5 posits the case against experiential decaying, including a specific analysis of (i) narrative, identity, and development; (ii) narrative identity as a product of autobiographical memory and emotional rationality; (iii) the requisite unpredictability of narrative identity; and (iv) the threat of manipulation to narrative, authenticity, and ethical decision making. Finally, by way of conclusion, Chapter 6 posits the normative case against MM, including a specific analysis of the objective demands of proportionate reason and the corresponding implications for the future of memory research and treatment.

## 1.3. Clarification of Terms

Before progressing, it is important to clarify the philosophical meaning and usage of rationality as employed in this book. In the chapters to follow, the terms "rationality" (and its variations) and "emotional rationality" (and its variations) are used in two distinct senses, both of which derive from Max Weber's rationalization processes in history.[12] Regarding the former, the term "rationality" (and its variations) is used to indicate strictly practical rationality – the capacity to view and judge worldly activity in relation to one's pragmatic and egoistic interests, and to thereby make intelligent moral decisions in accordance with the facts of reality. Instead of implying patterns of action that manipulate daily routines on behalf of an absolute value system, practical rationality accepts given realities and calculates the most prudent means of dealing with the difficulties they present. Pragmatic action is ascendant, and practical ends are attained through the careful weighing, balancing, and calculating the most adequate means to achieve them. Hence, practical rationality exists as a manifestation of one's capacity for means-end rational action.[13]

Regarding the latter, the term "emotional rationality" (and its variations) is used to indicate strictly substantive rationality – the capacity to directly order actions into value patterns, and to thereby make intelligent moral decisions in accordance with one's life narrative. Unlike the purely means-end calculation of practical rationality, substantive rationality moves in relation to a past, present, and potential "value postulate."[14] In this way, substantive rationality exists as a manifestation of one's inherent capacity for value-rational action. In all cases, this type of rationality is considered valid insofar as it serves as a unique standard against which the flow of unending empirical events may be selected, measured, and judged.[15] Therefore, substantive rationality and the rationalization processes on which it is based always exist in reference to ultimate points of view, each of which implies an identifiable configuration of values that determine the direction of a potentially ensuing rationalization process. This direction depends on one's responsibility to certain ultimate values and the systemization of one's actions to conform to them. These values acquire "rationality" from their status as consistent value postulates. Hence, "rationality" and potential rationalization processes within a given arena must ultimately refer back to these value postulates.[16]

### 1.4. Notes

1. Turhan Canli and Zenab Amin, "Neuroimaging of Emotion and Personality: Ethical Considerations," in *Neuroethics: An Introduction with Readings*, ed. Martha J. Farah (Cambridge, MA: The MIT Press, 2010), 147-54.

2. Terms used to describe the general practice of memory manipulation are manifold – spanning from "dampening," to "bunting," to "erasing" – and invoke depictions of slightly altered forms of one and the same pharmacological pursuit.

3. These possibilities span from the augmentation of executive function, appetite, libido, sleep, and mood to the maturation and development of idiosyncratic personality. See Martha J. Farah, Judy Illes, Robert Cook-Deegan, Howard Gardner, Eric Kandel, Patricia King, Erik Parens, Barbara Sahakian, and Paul Root Wolpe, "Neurocognitive Enhancement: What Can We Do and What Should We Do?" in *Neuroethics: An Introduction with Readings*, ed. Martha J. Farah (Cambridge, MA: The MIT Press, 2010), 30-41.

4. At the time this book is written, the most current technique involves the selective erasure of forms of long-term synaptic plasticity that underlies different manifestations of episodic memory. See Jiangyuan Hu, Larissa

Ferguson, Kerry Adler, Carole A. Farah, Margaret H. Hastings, Wayne S. Sossin, and Samuel Schacher, "Selective Erasure of Distinct Forms of Long-Term Synaptic Plasticity Underlying Different Forms of Memory in the Same Postsynaptic Neuron," *Current Biology* 12 (July 2017): 1-12.

5. Erik Parens, "The Ethics of Memory Blunting: Some Initial Thoughts," *Frontiers in Behavioral Neuroscience* 4 (December 2010): 1-2; see especially p. 1.

6. President's Council on Bioethics, *Beyond Therapy: Biotechnology and the Pursuit of Happiness* (New York: HarperCollins Publishers, 2003), 205-13; see especially p. 207.

7. One cause of the current impasse is the idea that to augment something – here, to blunt or eliminate emotional memories tied to traumatic or undesirable events – de facto denotes the pursuit of self-mastery or perfection. See President's Council on Bioethics, *Beyond Therapy*.

8. This is the concern, for example, of the 2003 President's Council on Bioethics. See President's Council on Bioethics, *Beyond Therapy*.

9. This is the position, for example, of Allen Buchanan. See Allen Buchanan, *Beyond Humanity? The Ethics of Biomedical Enhancement* (New York: Oxford University Press, 2011).

10. To be clear, Buchanan does not delve into the ethics of memory manipulation, only neurocognitive enhancement. Thus, the term "manipulation" is here employed in its broadest – and most positive – sense.

11. These positions have been termed "anti-enhancement" and "anti-anti-enhancement," respectively. See Buchanan, *Beyond Humanity*.

12. Stephen Kalberg, "Max Weber's Types of Rationality: Cornerstones for the Analysis of Rationalization Processes in History," *The American Journal of Sociology* 85, no. 5 (March 1980): 1145-79.

13. Kalberg, "Max Weber's Types of Rationality," 1145-79; see especially pp. 1151-52.

14. Beyond representing a single value, a value postulate implies entire clusters of values that vary in comprehensiveness, consistency, and content. See Kalberg, "Max Weber's Types of Rationality," 1145-79; see especially pp. 1155-57.

15. Kalberg, "Max Weber's Types of Rationality," 1145-79; see especially pp. 1155-57.

16. Kalberg, "Max Weber's Types of Rationality," 1145-79; see especially pp. 1155-57.

# 2.

# THE NEUROSCIENCE OF MM: PRACTICES AND POSSIBILITIES, PROPONENTS, AND PROBLEMS

Before turning to the book's four-tiered logic chain, it is imperative, first, to explore the contemporary neuroscience of MM, both in its current practices and possibilities and what is thereby anticipated vis-à-vis its future developmental trajectory. Against this backdrop, the chapter then moves to identify and unpack the arguments and justifications of the strongest and most resistant opponents of the argument posited herein, namely, those in favor of limited-use memory manipulation (LUMM) as a last resort for individuals whose psychosocial condition poses a disproportionate and immediate threat to their overall health, well-being, and safety. The chapter concludes by illuminating the problems inherent to what it terms the "neuroethical astigmatism" of LUMM proponents, thereupon offering a preliminary retort.

## 2.1. Practices and Possibilities of MM

Recent advances in neuropharmacology, electroconvulsive therapy (ECT), false memory creation (FMC), and deep brain stimulation (DBS) have enabled researchers to arbitrate within systems of emotional, semantic, and working memory.[1] The contemporary capacities of cognitive manipulation thus seem limitless. This book focuses exclusively on those pharmaceutical and stimulation techniques that prevent or erase,[2] in the name of therapy, autobiographical memory considered "emotionally pathological." The use of pharmacologicals, ECT, and FMC and DBS to enhance emotional rationality related to autobiographical memory is an entirely separate issue whose moral significance and justification transcend the present scope. Memory enhancement remains, however, of pertinent interest to the broader conversation over manipulation.[3] Nevertheless, the range and methods explored herein possess a focus related to procedures of partial blunting and total erasure with the intention to bypass unnecessary psychological debilitation at the hand of unpleasant memories.[4]

### 2.1.1. Beta-Adrenergic Receptor-Blocking Pharmacologicals

Adrenaline, a stress hormone closely related to its counterpart, cortisol, is released when the adrenal gland receives a signal from the amygdala alerting it to a perceived threat.[5] As part of the limbic system, the amygdala regulates emotions and, as such, is among the most primitive parts of the brain, playing a critical role in the capacity to survive. Among other effects, adrenaline serves to embed non-conscious emotional memories of threatening events in the amygdala. However, if emotional memory is stored too densely in the amygdala, it can produce a disproportionately heightened fear response to external events. Since emotional memories stored in the amygdala transcend conscious control, they are often difficult to modulate, regulate, or expunge and can therefore detrimentally influence the nature and content of belief, affect, and various states of consciousness. Events erroneously perceived as stressors or threats can trigger chronic fear responses that put the brain, body, and mind on a constant state of alert. This describes both the pathology and the pathophysiology of some manifestations of depression, anxiety, and, most acutely, the emotionally disturbing "flashback" memories of traumatic events that characterize PTSD.[6]

The primary pharmacological agent examined for treatment of PTSD is propranolol, a beta-adrenergic receptor antagonist.[7] Propranolol suppresses noradrenergic activation by blocking beta 1 and beta 2 adrenoreceptors, located in target areas of the peripheral sympathetic nervous system, as well as in various brain regions, including the amygdala. Multiple studies have indicated that single doses of propranolol influence emotional processing and reduce physiological markers of acute arousal following exposure to emotionally evocative stimuli.[8] Additionally, fMRI studies have demonstrated that propranolol effects a reduction in amygdala responses to highly emotional pictures or emotional facial expressions.[9] Due to its ability to reduce heart rate and inhibit arterial vasoconstriction, beta-blockers have been administered for decades to treat hypertension and other cardiovascular diseases.[10] However, its inhibitory effect on the physiological aspects of emotional stimulation also makes it an effective and thus widely employed treatment for stress, acute anxiety, and performance anxiety. By virtue of its capacity to reduce emotional memory consolidation, propranolol has been proposed as a promising safeguard against PTSD if effectively administered prior to or immediately following a traumatic event.[11] Since the beta-adrenergic system is involved not only with response and memory formation, but also with the conditioning of emotional responses tied to memory, propranolol serves to both blunt memory formation and disassociate memory from an emotional response thereto.[12]

The first paper reporting experimental results of propranolol use in modulating memory and emotion in humans was published by Larry Cahill and colleagues in 1994.[13] Subjects received either propranolol or placebo one hour before exposure to emotional arousal or neutral stimulus. While no differences were recorded between groups for neutral stimuli, recall after emotional arousal was significantly higher in the placebo group. Other early study results proved somewhat contradictory: one recognized no difference in recall response,[14] while others indicated pronounced effectiveness.[15] Research on the effect of propranolol immediately following trauma has revealed that fewer individuals treated within six hours after the traumatic event develop PTSD than those receiving a placebo.[16] More recently, research has shifted to examine propranolol use during memory reconsolidation after the development of PTSD.[17] Responses in subjects who received the drug were significantly lower on all physiological measurements, and subsequent studies have shown similar results.[18]

The 2012 study of Robin Menzies found that in addition to fewer re-experiences of the event and diminished emotional distress, its thirty-six subjects reported a marked reduction in both the quality and quantity of their traumatic memories.[19] Building on his 2009 study – which demonstrated that treatment of five of six PTSD patients with propranolol effected the fragmentation of memory, difficulty accessing it, minimal or absent distress when thinking about it, and a feeling of emotional detachment from it[20] – Menzies concluded that propranolol diminishes the very integrity of traumatic memories, resulting in a degree of amnesia for the traumatic event. A press report based on the work of Menzie and colleagues in Canada suggests the effects can be genuinely profound in some patients. As one patient reports, "Before, I couldn't keep this thing away. Now, I can't find it. ... It's like you put a bomb under that memory. ... When I do think of it, it doesn't upset me. It's like a sad scene from a movie, not part of my life."[21] Nevertheless, while propranolol has enjoyed a safe history of use for other purposes, it does not ipso facto follow that it will translate safely to PTSD treatment, which poses different risks, including likely collateral effects on other (positive) memory systems.[22]

### 2.1.2. Electroconvulsive Therapy (ECT)

ECT is administered by delivering electricity directly to the brain through scalp electrodes to induce a generalized tonic-clonic seizure.[23] Modern ECT consists of delivering constant current (800-900 milliamperes) rectangular pulses with alternating polarity, which induces postictal disorientation, as well as both anterograde and retrograde amnesia.[24] The high current amplitude employed in conventional ECT devices exposes the entire brain to a suprathreshold stimulation that far exceeds its corresponding neural activation threshold. Pulse widths in the 0.5-2.0 milliseconds range are considered "brief," and widths less than 0.5 milliseconds are considered "ultra brief." The practice of ECT dates back to the 1930s, and its contemporary application is considered the standard par excellence of brain stimulation therapy for severe major depressive disorder. Approximately 1.5 million individuals receive ECT worldwide each year, with numbers increasing annually, and the reported clinical efficacy of the practice is unparalleled, with a remission rate of seventy percent.[25] Sigmund Freud originally reported that ECT produced admirable results in his *Studies in Hysteria*,[26] but later came to reject it as pretentious. Over a century later, ECT is surprisingly modernized, extending into mainstream medical treatments for schizophrenia, mania, and catatonia.[27]

While remarkably effective in the treatment of a wide range of psychiatric disorders, the use of ECT has been limited by its weighty and often intolerable cognitive side effects. For both patients and families, the retrograde amnesia induced by ECT is most bothersome of all.[28] ECT leads to the release of a substantial number of compounds within the central and peripheral nervous systems. Together with the significant preclinical evidence demonstrating the involvement of the cholinergic system in the amnestic effects of electrically induced seizures, the utility of procholinergic compounds remains significant to the etiology of the adverse cognitive side effects inherent to ECT. Several other potential chemical messengers have also been implicated in the cognitive side effects of ECT. Late attention has focused on the glutaminergic system and the n-methyl-d-aspartate (NMDA) receptor. In theory, the excitotoxicity induced by NMDA receptor activation following seizures may be causally related to the adverse effects of ECT. Other human data has focused on glucocorticoids (given the relationship between seizures, depression, and the hippocampus) and hypercortisolemia (with its link to postictal confusion).[29]

Recently, ECT applications have been used to intentionally alter and destroy memories. The study of Marijn Kroes and colleagues investigated whether strategically timed ECT bursts could disrupt subjects' disturbing episodic memories.[30] Forty-two patients were randomly assigned to one of three study groups and learned two emotionally averse slideshow stories, each consisting of eleven scenes accompanied by an auditory narrative. Memory for one of the two stories was reactivated a week later by presenting a partial version of the first slide and testing memory of the outcome. Immediately following memory reactivation, participants in groups A and B were anesthetized and received ECT, while participants in group C did not. As predicted, Kroes and colleagues found ECT-evoked disruption of memory for the reactivated story twenty-four hours post-ECT in group A, but not on immediate memory testing in group B. Taken together, the study concluded that results from groups A and B are consistent with a view that emotional episodic memories are impaired by a single ECT session in a time-dependent fashion insofar as they have been previously reactivated.[31] These results have peaked the interest of many in the PTSD research community.[32]

The 2008 study of Louisa Fraser and colleagues focus on a vital yet frequently overlooked neuroethical issue in the present context: the profoundly negative effect of ECT on autobiographical memory.[33] Memory impairment is now the primary concern surrounding the use of ECT, and growing evidence attests to the long-term (rather than immediate) consequences of ECT on retrograde and anterograde memory. More concerning still, ECT can have a particularly detrimental effect on personal memories, and studies have shown that autobiographical memory complaints can persist for up to three years after treatment.[34] The distinction between semantic – consisting of general, decontextualized information about one's past – and episodic – comprised of particular events situated in space and time – components is a largely accepted view of autobiographical memory.[35] Richard Weiner and colleagues reported one of few studies that show persistence in autobiographical memory deficits, assessed by objective criteria, up to at least six months after ECT.[36] Patients were randomly assigned to either sine wave or brief pulse ECT, with electrodes placed either bilaterally or unilaterally. Three days after completing a course of ECT, all but the brief pulse unilateral ECT groups exhibited significant impairment in the recall of personal memories. At six months of follow-up, significant differences between unilateral and bilateral groups were observed, with poorer results in the bilateral group. Regrettably, this study demonstrates the deleterious effects of ECT on both immediate and long-term autobiographical memory.[37]

### 2.1.3. False Memory Creation (FMC)

Human memory is subject to a multitude of errors, including source misattributions, distortions, and the creation of false memories.[38] Whereas false memory syndrome refers to the recovery of memories of traumatic events that did not take place, recovered memory therapy is based on the concept of regression. Psychoanalytic theory suggests that individuals often banish traumatic experiences from consciousness because they are too terrifying to examine. However, there is no scientific evidence to demonstrate that traumatic events are repressed.[39] On the contrary, the global prevalence of PTSD substantiates the notion that it is exceedingly difficult to leave highly disturbing events in the past. The concept of cognitive repression is a theory, not part of scientific psychiatry, and seventy years of efforts to study repression have yielded neither supporting nor persuasive evidence. While memories are difficult to repress, pseudomemories are frequently created from suggestion, social contagion, hypnosis, misdiagnosis, and the misapplication of regressive therapies. The study of Ira Hyman and F. James Billings Jr., for instance, found that approximately twenty-five percent of college students interviewed created false childhood memories.[40] While it is possible to for individuals to forget traumatic experiences and later remember them, considerable evidence suggests that many "recovered" memories were created as a result of suggestion or other psychological techniques.[41] Recent studies indicate that false memories can be planted by way of innocuous suggestions, and that these memories can play a role in shaping individual's subsequent attitudes and preferences.[42]

Historically, drug aversion therapies have included electroconvulsive techniques,[43] satiation,[44] and chemical aversant pairings with the undesired behavior.[45] Ethical concerns, as well as a lack of controlled scientific studies in these areas, have led to the demise of traditional aversion techniques. Despite the demise, it is important to note that some drug programs still include aversion therapy into their methodologies.[46] Current (acceptable) pharmacological treatments for substance addiction include the administration of inhibitory agents (e.g., Disulfiram) that act by blocking the breakdown of acetaldehyde, the chemical believed to contribute to withdrawal symptoms. The interaction of Disulfiram with any amount of the addictive substance enhances unpleasant physical symptoms,[47] deterring the individual from subsequent use. Today, an alternate approach to curbing substance addiction can be found in the false memory literature.[48] Results from a handful of studies have indicated that adopting a false memory as part of one's personal autobiography can affect an individual's current and future preferences related to that memory.[49]

To date, only one study has experimentally examined whether early substance-related memories would be prone to memory manipulation.[50] Seema Clifasefi and colleagues suggested to their trial participants that they had become sick during their early teenage years (prior to age sixteen) after exposure to a particular drug,[51] and examined whether it would increase confidence that the suggested event occurred and, if so, show a decrease in preference for the specific drug mentioned. Overall, experimental participants who received a false substance-related suggestion exhibited a significant increase in confidence (compared to controls) that the event did occur in the past. More importantly, individuals who received a false suggestion that they had become sick from a particular drug showed a trend toward diminished preference for that drug in a follow-up preference rating task.[52] The findings of Clifasefi and colleagues are consistent with the drug expectancy literature, which indicates that positive drug expectancies are associated with increased and risky drug-related behavior, whereas negative drug expectancies are associated with decreased drug-related behavior.[53]

To be sure, the 2013 study by Clifasefi and colleagues is not without limitations, five of which are especially worthy of note.[54] First, it is important to illuminate that only 19.6 percent of experimental subjects developed a memory or belief that the suggested alcohol memory occurred. Second, the fact that differences existed between experimental and control participants at baseline for alcohol preferences does not preclude the possibility that preference findings are due to regression to the mean. Third, the data suggests that those who reported a younger first drinking experience were more likely to adopt the false suggestion. Fourth, it can be argued that because early onset drinkers are known to be at higher risk for subsequent problems, these individuals may ultimately be remembering true events from their past. Finally, another argument can be made about early age-of-onset drinking that these individuals might show different cognitive vulnerabilities than their late age-of-onset counterparts. Beyond these limitations, using techniques to elicit false memories raises significant neuroethical issues. Questions concerning what happens to individuals upon learning that they have received an intervention based on deception, the appropriateness of creating fabricated memories to prevent addictive behavior prior to the onset of symptoms, and whether the ends fundamentally justify the means must be considered carefully before translating false memory creation into a clinical intervention to treat substance abuse.[55]

### 2.1.4. Deep Brain Stimulation (DBS)

DBS is a surgical procedure in which an electrode is implanted in one or more (deep) brain areas (such as the basal ganglia) and connected to a device that generates high-frequency electrical stimulation (130-180 hertz).[56] The impulse-generating device is subsequently implanted subcutaneously, typically within chest-range and carried by the patient. In stereotactic surgery, the electrodes are implanted in the brain after drilling four holes into the cranium. The patients are often kept conscious during the procedure so as to respond adequately to neurological testing. This is done to ensure that the various electrodes are located properly and produce the desired effect upon triggering.[57] First developed in the 1990s, this invasive method traditionally aimed to ameliorate symptoms associated with movement disorders, especially in advanced Parkinson's disease. However, DBS is now more frequently employed for diverse experimental applications such as psychiatric disorders, including depression, weight reduction, cluster headaches, epilepsy, substance addiction, post-traumatic stress, obsessive-compulsive disorder, intractable pain, and minimally conscious states. Today, over 30,000 implants have been reported worldwide, and most have gleaned impressive benefits and quality of life improvements.[58]

The use of electrical stimulation as a therapeutic tool has been known since antiquity.[59] The original electrical medium was the torpedo fish, and the main indication was pain. In the eighteenth and nineteenth centuries, "electrotherapy," including "electropuncture," enjoyed immense popularity and were prescribed for the treatment of various illnesses, including epilepsy, paralysis, chorea, deafness, blindness, rheumatism, and glandular enlargement, as well as for artificial respiration and resuscitation. The multipurpose use of electricity as a therapy to treat an array of ailments is strangely echoed in the manifold applications suggested of modern electrical brain stimulation. Over a quarter century has elapsed since the seminal publication of Alim-Louis Benabid and colleagues on chronic thalamic electrical stimulation for tremor in 1987.[60] DBS has kindled the imagination of many, resulting in thousands of scientific and non-scientific publications that span from the poetic depiction of DBS as a tool to "sing the mind electric" to the vacillating suggestion of using DBS to improve morality, treat antisocial behavior, or obliterate unsavory memories.[61]

Recent research on the promise of DBS to treat substance addiction raises especially complex neuroethical questions. The NAc has a pivotal role in the pathogenesis of addiction and is an important element in the mesocorticolimbic reward circuit. It is immediately involved in establishing the reward of drugs of abuse. Numerous researchers contend that dysregulation of the neurophysiological processes involved in creating the quality or intensity of rewarding experiences contributes to addiction.[62] For those reasons, the NAc is an ideal target for DBS, and early studies have shown promise.[63] For one individual who received DBS to alleviate anxiety and depression, stimulation in the NAc had the unintended consequence of improving the individual's comorbid alcohol dependence.[64] Spurred by this incidental finding,[65] Ulf Müller and colleagues initiated a pilot study of bilateral DBS of the NAc in five subjects with severe alcohol addiction, each of whom reported a total and permanent remission of their craving for alcohol following the procedure.[66] This result was repeated in 2011 by Jens Kuhn and colleagues,[67] and multiple cases since have noted the success of DBS on the NAc of individuals with heroin addiction.[68]

A notable 2006 article by M. Schüpbach and colleagues interprets a study conducted by psychologists, neurologists, and psychiatrists between 2000 and 2003 of twenty-nine subjects with Parkinson's disease treated with DBS.[69] The employed research technique was a repetitive, transparent, and unstructured interview for the purpose of qualitatively assessing the impact of subthalamic nucleus stimulation (STN) – a form of DBS – on participants' personal, marital, and socio-professional life. Despite having undergone a "successful surgery," eight patients (28%) reported debilitating cognitive side effects and a subsequent inability to resume a meaningful life. One of those eight patients – a thirty-eight-year-old woman – reported that, prior to STN, she was dynamic in spite of her motor disability, finding meaning in the effort to overcome the disease. Six months after the operation, she reported a seventy-five percent improvement and was able to cease her antiparkinson medication. However, after eighteen months of stimulation, the journalist, wife, and mother of one was no longer able to work, having lost inspiration and passion, both in her profession and life generally. As she wrote: "Now I feel like a machine; I've lost my passion. I don't recognize myself anymore."[70] She reported further that she lost interest in family, was easily exhausted, and experienced a loss of existential vitality leading to the interruption of professional activity. In brief, the loss of exuberance and life goals occurred despite major improvement to her motor symptoms by STN.[71] Tragically, experiences of alienation and the inauthenticity are hardly uncommon to post-DBS phenomenology;[72] hence, threats to personal identity from DBS remain real.[73]

## 2.2. Proponents of Limited-Use Memory Manipulation (LUMM)

As suggested above, the most formidable opponents of the arguments forwarded in this book are proponents of LUMM, who contend that the practice may be employed as a last resort for individuals whose psychosocial condition poses a disproportionate and immediate threat to their overall health, well-being, and safety. Since the strongest and most prominent examples of such conditions frequently center around PTSD and substance addition, each condition will be taken up in turn and explored as circumstances within which LUMM can be justified according to the logic of LUMM proponents.

### 2.2.1. LUMM for Post-Traumatic Stress Disorder (PTSD)

Proponents of LUMM for PTSD argue that is it morally reasonable, as a last resort, for individuals at risk of severe manifestations of the condition to be offered prophylaxis against the condition after enduring exceptionally traumatic and vastly disproportionate circumstances, such as the horrors of brutal rape or the recovery of fellow comrades' bodies.[74] Inasmuch as death associated with war is considered morally acceptable in particular circumstances, and the risk of death in war is typically outweighed by any benefits to be gleaned, proponents suggest that helping those who executed a military agenda, risked death and dishonor, and subsequently suffer from a condition associated with their service should be considered an ethical obligation. If it is reasonable, on the one hand, to ask individuals to engage in life-threatening and emotionally distressing activities, then it seems wrong, on the other hand, to deny them therapeutic medications that may significantly reduce their considerable risk of developing PTSD. Moreover, proponents note that objections to the preventative use of beta-blocking pharmacologicals overlook ethical questions about post-trauma debriefing that, they maintain, has little to no effect and, at worst, increases the risk of PTSD.[75] As Wayne Hall and Adrian Carter maintain, when compared with countless hours of psychological intervention of uncertain efficacy, a seven-day course of a low-toxicity drug seems trivial.[76]

### 2.2.1.1. The Neurobiology of PTSD

Psychological trauma often results from witnessing events that are perceived as life threatening or injurious to self or others.[77] Such experiences, which frequently evoke intense fear, horror, and helplessness, can lead to the development of PTSD. The condition was originally thought to represent a normative response (at the extreme end of the response continuum) to severe trauma or stress. However, it has become clear that the idiosyncratic response of an individual to trauma depends not only on stressor characteristics, but also on factors specific to the individual. For the majority of human beings, the psychological trauma induced by the experience of profound threat is acute and transient. Psychological trauma is typically characterized by phenomena that can be grouped into three domains: (i) reminders of exposure (including flashbacks, intrusive thoughts, and nightmares), (ii) activation (including hyperarousal, insomnia, agitation, irritability, impulsivity, and anger), and (iii) deactivation (including numbing, avoidance, withdrawal, confusion, derealization, dissociation, and depression). Self-limiting by definition, these reactions generally effect minimal impairment over time. For a (significant) minority of the population, however, the psychological trauma brought about by the experience of profound threat leads to a longer-term syndrome that has been defined, validated, and termed "PTSD" in the clinical literature. PTSD is accompanied by devastating functional impairment characterized by the presence of signs and symptoms in the three primary domains mentioned above.[78]

Contemporary neuroimaging has identified and confirmed characteristic changes in brain structure and function in individuals with PTSD.[79] Altered brain regions include the hippocampus, amygdala, anterior cingulate, insula, and orbitofrontal region. Together, these form a neural circuit that mediates adaptation to stress and fear conditioning. Changes in these circuits have been postulated to share a direct link to the development of PTSD. A hallmark feature of PTSD is reduced hippocampal volume. The hippocampus is implicated to control stress responses, declarative memory, and contextual aspects of fear conditioning. In fMRI studies, small hippocampal volumes were associated with trauma severity and memory impairments.[80] The functional role of the amygdala, which mediates stress responses and emotional learning, is also involved in the pathophysiology of PTSD. Given that increased amygdala activity has been linked to genetic traits that moderate PTSD, increased amygdala activity may represent a neurobiological risk factor for developing PTSD. The medial PFC comprises the ACC, subcallosal cortex, and medial frontal gyrus. The medial PFC exerts inhibitory control over stress responses and emotional reactivity through its connections with the amygdala, and mediates extinction of conditioned fear through active inhibition of acquired fear responses. Individuals with PTSD exhibit decreased volumes of the frontal cortex, including reduced ACC volume, which has been similarly correlated with the severity of PTSD symptoms.[81]

The neurobiological concerns observed in individuals with PTSD are numerous and likely reflect an enduring dysregulation of multiple stress-mediating systems that occur as a result of psychological assault.[82] These pathophysiological disturbances occur in individuals with genetic, epigenetic, and experiential predispositions when exposed to extreme conditions, and presumably signify an indelible sensory imprint of maladaptively processed experience that effects an imbalanced degree of emotional import and releases (or restrains) behavioral reactions that aim to defend against future trauma via activation (or deactivation) in a losing effort to secure equilibrium. Hence, a lack of baseline cortisol at the time of psychological trauma may facilitate overactivation in the central corticotropin-releasing hormone - norepinephrine (CRH-NE) cascade, resulting in prolonged and enhanced stress responses. This increased stress responsiveness may be further accented by inadequate regulatory effects of gamma-aminobutyric acid (GABA), serotonin, and neuropeptide Y (NPY). Additionally, altered norepinephrine and stress hormone activity may be involved in processes of learning and extinction, both of which are abnormal in PTSD. For instance, norepinephrine enhances the encoding of fear memories and glucocorticoids block the retrieval of emotional memories. The constellation of elevated noradrenergic activity and relative hypocortisolism may lead to the encoding of traumatic memories and the lack of memory retrieval inhibition, both of which presumably trigger the re-experience of phenomena in PTSD.[83]

Additionally, a malfunctioning hippocampus may account for some cognitive symptoms of PTSD, including declarative memory deficits.[84] Since the hippocampus is critical for context conditioning, an impaired hippocampus may facilitate generalization of learned fear in contexts unrelated to previous traumatic exposure and so impair the ability to discern between safe and unsafe stimuli. In combination with exaggerated amygdalar responses associated with PTSD, a limited capacity to discriminate threats may promote paranoia, hypervigilance, behavioral activation, exaggerated stress responses, and acquisition of additional fear associations. Disrupted PFC function may then serve to further exacerbate PTSD pathology as a result of deficient suppression of stress responses, fear associations, and extinction. To be sure, some neurobiological findings in patients with PTSD are controversial and require additional examination. Moreover, there are a number of understudied yet significant topics, including factors that impact resilience and vulnerability. For instance, stress-protective neurobiological factors such as activity in oxytocin and NPY-containing circuits could, in principle, be altered to promote resilience. Hence, there exists a general need for molecular biology to further explore PTSD to identify interactions between dispositional factors (both genetic and epigenetic) and trauma exposure to understand PTSD risk, gauge illness course, and predict treatment response. The effects of trauma on neurotrophic factors (in the hippocampus), neural plasticity (central nervous system [CNS]-wide), circuit remodeling (myelination patterns), and gene expression must be assessed in detail across illness duration. While difficult, such studies will necessitate accessing, assaying, and following populations at risk for exposure to trauma before exposure occurs.[85]

### 2.2.1.2. The Case in Favor of LUMM for PTSD

PTSD is a growing cause of human suffering that affects approximately one-third of all individuals exposed to major trauma.[86] Conventional psychological and pharmacological treatments for PTSD are often expensive, time-consuming, and of modest efficacy. On this basis, LUMM proponents argue that propranolol may be used, in extreme cases, to reduce the severity of psychological reactions to trauma and thereby reduce the risks of developing PTSD. While reasonable concerns have been raised about the use of drugs to alter memory, many (i) are based on wildly exaggerated and unrealistic scenarios that ignore the restricted and fleeting action of propranolol in affecting memory, (ii) underplay the utterly debilitating impact that PTSD has on those who suffer from it, and (iii) fail to acknowledge fully the extent to which other drugs – such as alcohol – are already used for this purpose. Anterograde amnesia is a well-known side effect of alcohol, as well as benzodiazepines available by prescription, such as Valium and Halcion, and illegally obtained benzodiazepines, such as Rohypnol. Unlike these drugs, propranolol has a retrograde amnesic effect, offering greater potential to ameliorate traumatic memories from the recent past.[87] Henry and colleagues offer a scathing critique of the ethical concerns forwarded by the 2003 President's Council on Bioethics (PCB) about the prophylactic and dampening use of propranolol. The authors comment that the PCB's concerns involve a series of speculative harms – for instance, that criminals may consume beta-blockers to reduce painful memories of their crimes – that fail to provide concrete reasons to oppose trials to assess the safety and effectiveness of propranolol. Moreover, the PCB also fails, in their judgment, to make a persuasive case for prescribing the clinical use of propranolol if clinical trials indicate its effectiveness.[88]

Wayne Hall and Adrian Carter expand the arguments of Henry and colleagues to articulate more broadly what is at stake, thereby forwarding the strongest (available) case in favor of LUMM for PTSD.[89] The authors offer a consequentialist argument in favor of using propranolol, namely, that it may be employed it to reduce the need for PTSD sufferers to use a more hazardous drug (e.g., alcohol) to treat their symptoms. In high doses, alcohol reduces anxiety and recall of emotionally traumatic memories, but chronic use for these purposes can quickly lead to dependence, a disorder that significantly reduces the chance of recovering from PTSD and has enormous health consequences, both individually and socially.[90] Following Henry and colleagues, the authors similarly reject the argument of the 2003 PCB that propranolol may be used by criminals to reduce regretful memories of their crimes, arguing instead that psychopaths cannot express interest in reducing the sting of such memories insofar as they do not possess the emotional capacity for regret.[91] Nonetheless, in the unlikely event that criminals used propranolol to numb their conscience, a positive outcome might include reduced alcohol abuse,[92] improved public order, and reduced burden on the families of criminals.[93] Hall and Carter further suggest that concerns about propranolol being used by the military to prevent soldiers from developing painful memories of war crimes and atrocities do not reflect the pharmacological properties of propranolol, which serves to attenuate reactions to trauma rather than procure global amnesia of events and conscience. Somewhat ironically, they mention that atrocities such as those at Srebrenica, Vietnam, and in World War II did not depend on the use of beta-adrenergic antagonists,[94] suggesting that the psychology of war appears sufficient to account for such acts attracting strong and justified societal opprobrium in the unlikely event that perpetrators of atrocities use propranolol for these means.[95]

In response to the medico-legal argument that damages payouts may be reduced by the effects of propranolol, Hall and Carter comment that this merely signifies the perverse incentives in the legal system rather than a compelling argument against the use of propranolol by victims of traumatic crimes to reduce the severity of PTSD.[96] This concern also seems exaggerated, they note, inasmuch as criminal actions that traumatize can be corroborated in ways that do not depend on the memory of the victim or the severity of the PTSD symptoms subsequently developed. For instance, no legal system would acquit a rapist on grounds that the victim did not develop PTSD. Nevertheless, some studies suggest that propranolol may actually improve recall of memories that are impaired by trauma.[97] Further, Hall and Carter remark that bioethicists who object to the preventative use of propranolol overlook moral questions about the genuine efficacy post-trauma debriefing, which is purported to reduce the risk of PTSD. According to Richard Bryant, contemporary evidence suggests that, at best, debriefing has no effect and, at worst, increases the risk of PTSD.[98] Compared to countless hours of psychological intervention of uncertain efficacy, the authors argue that a seven-day course of a low-toxicity drug seems trivial. Moreover, the use of propranolol to prevent the consolidation of traumatic episodic memories seems a risk worth taking in order to avoid a thirty-three percent chance of spending months undergoing psychotherapy and pharmacotherapy to treat PTSD and the common complications of alcohol and other drug dependence.[99]

While Hall and Carter sympathize with the concerns of Henry and colleagues regarding the potential for over-promotion of drugs to treat PTSD,[100] they make two observational points in reply.[101] First, they argue that propranolol is already off patent, which makes it exceedingly unlikely to be promoted by any drug company. Second, they argue that while it is plausible that the production of new drugs with similar effects may be promoted in this way in the United States – where direct-to-consumer advertising of pharmaceuticals is allowed and there are few regulatory limits to prevent superfluous promotion – this possibility simply denotes the need for improved regulation of the pharmaceutical industry rather than a robust argument against the use of propranolol per se. In conclusion, they reiterate that most of the commonly raised ethical objections to the use of memory dampening drugs, including propranolol, overstate the possible negative consequences of its use and run the risk of hindering a promising advance in the prevention of PTSD that may significantly reduce the need for PTSD suffers to turn to more harmful drugs, such as alcohol. Moreover, the authors reinforce that conventional arguments against the use of propranolol fail to provide cogent reasons for either preventing a trial of its safety and efficacy or for preventing its clinical use once proven to be safe and effective. For Hall and Carter, then, the criticisms of the PCB should be recognized only as a form of scare-mongering: a hazard of bioethical analyses that is the product of "worst-casing" the potential harms of new biotechnologies, often as a result of exaggerating their effectiveness.[102]

### 2.2.2. LUMM for Substance Addiction

Building on arguments in favor of LUMM for PTSD, proponents of LUMM for substance addiction argue that it is morally permissible, as a last resort, for individuals whose psychosocial condition poses a disproportionate and immediate threat to their overall health, well-being, and safety to be offered relief in the form of surgical or psychological memory editing. Impaired control over voluntary behavior is a marked feature in emerging neurobiological explanations of substance addiction, in clinical and diagnostic accounts, and in debates about addiction nosology.[103] Hence, drug cravings can manifest as such irresistible and powerful psychological forces that someone with an addiction is not capable, at certain times, of acting fully autonomously when the decision involves denying the persistence of cravings. An addict might be excessively subservient to the individual who supplies him with drugs, or with money for drugs, and therefore have his autonomy compromised by the rule of another. However, if the addict's autonomy is compromised in this way, it marks a consequence of an initial loss of autonomy that is characteristic of addiction.[104] It follows, therefore, that such a loss of autonomy undercuts the addict's ability to pursue his own goals.[105]

### 2.2.2.1. The Neurobiology of Substance Addiction

Communication in the brain is facilitated by neurotransmitters that are released from neurons at synapses where they interact as bonds with protein complexes, called receptors, on the surface of other cells, predominantly at the postsynaptic membrane.[106] The binding of a neurotransmitter to a receptor transduces a chemical signal that transfers activity-dependent information. The neurotransmitters can either be taken back up by the cell for future use by transporters or degraded and removed from the system. In the brain, pathways are complex integrative systems that contain numerous neurons or nuclei that relay information throughout a circuit and can be acted upon by other neurotransmitter systems that also integrate with that region. While addictive substances have diverse pharmacological profiles, their acute actions converge primarily on the mesocorticolimbic dopaminergic system. This pathway originates in the ventral tegmental area (VTA) and projects to the NAc, striatum, forebrain, and PFC. The PFC coordinates cognitive processes and actions aimed at an internal goal, while the NAc is believed to integrate information, effect an appropriate response, and control the motivational value of stimuli and reward enforcement. Immediately after initial exposure to a drug, extracellular levels of accumbal dopamine increase. Some enhance dopamine release from the presynaptic terminals as a consequence of increased neuronal activity in the VTA (e.g., alcohol, nicotine, opiates, and cannabis) while others inhibit the presynaptic uptake by the dopamine transporter in the NAc (e.g., cocaine and amphetamines).[107] Addictive substances produce a larger dopamine release that is maintained for longer than that of natural rewards. If exposure to the drug persists, there may be a loss of homeostatic regulation: a progressive increase in basal levels of dopamine is accompanied by a reduction in the lesser response to the drug, resulting in the appearance of tolerance to the drug.[108]

During acute withdrawal, dopamine rebounds to below basal levels so re-exposure to the drug or a drug-related cue is often sufficient to increase dopamine again.[109] This dopamine response has been hypothesized to contribute to addictive relapse, working on serotonergic, noradrenergic, glutamatergic, and GABAergic systems. While dopamine release may modulate the acute rewarding effects of an addictive substance, it does not solely mediate drug-seeking behaviors and persistent drug taking. Exposure to addictive drugs can have either a direct or indirect effect on numerous neurotransmitter systems. Unlike dopamine, which facilitates the response to initial drug use, these additional neurotransmitter systems play a greater role in mediating persistent drug use, contributing to the inability to terminate drug use and the likelihood of relapse after a period of abstinence. Glutamatergic inputs from the PFC, amygdala, hippocampus, and other brain regions modulate activity in the NAc either directly or indirectly by their influence on the VTA. Like to dopamine, initial exposure to a psychostimulant increases extracellular levels of glutamate in the NAc, PFC, and, to a lesser extent, the VTA. Unlike dopamine, however, this response increases the sensitivity of the receptors that bind glutamate to the effects of subsequent exposures to lower doses of the particular drug. This leads to reduced extracellular glutamate levels and, hence, decreased glutamate-driven activity over time. Upon reexposure to the drug or drug-associated cue, there is enhanced synaptic glutamate release that drives continued drug-seeking behaviors. Such dysregulation of the glutamatergic system is sufficient to alter drug-induced behaviors, even in light of normal dopaminergic responses procured in the NAc.[110]

Imbalance in the glutamatergic regulation of corticostriatal transmission has been termed the "glutamate hypothesis" of addiction, which suggests its cardinal role in mediating relapse.[111] This hypothesis is supported by studies demonstrating that the reinstatement of drug-seeking behaviors can be prevented using the procysteine drug N-acetylcysteine (NAC).[112] NAC increases glutathione synthesis, which restores glutamatergic signaling. Treatment with NAC is also able to restore prefrontal-driven long-term potentiation and long-term depression in the NAc, which are typically impaired during acute withdrawal. The therapeutic potential of NAC has now been trialed in preclinical human studies, where it has successfully reduced the desire to use drugs of abuse.[113] Astrocytes express the sodium-dependent glutamate transporter (GLT1), which is responsible for removing over ninety percent of glutamate from the extracellular space. Overexpression of GLT1 in the PFC and the NAc during extinction training is sufficient to inhibit cue-induced reinstatement to drug self-administration by suppressing the excess extracellular glutamate that normally occurs upon re-exposure to a drug. Beyond relapse, imbalances in glutamatergic transmission have been hypothesized to mediate responses to drugs including self-administration, reward learning, extinction, and behavioral sensitization that, in animal subjects, is manifested by increased psychomotor activity. In the NAc, the modulation of glutamatergic inputs onto medium spiny GABAergic neurons expressing $D_1$ dopamine receptors play a vital role in the development of sensitization to drugs. Thus, an allostatic shift – which marks an adaptive effort in a regulatory system in response to a chronic deviation from "normal," thereby establishing a new set-point – toward augmented glutamatergic function may contribute to the transition from controlled drug use to a compulsive and uncontrolled drug-dependent state and the high incidence of relapse.[114]

There are vast numbers of neuropeptides and corresponding receptors present in pathways that mediate addiction. The role of corticotropin-releasing factor (CRF) is highlighted as an example of the intricate part that neuropeptides play in mediating addictive behaviors.[115] Stress, either in the environment or due to substance withdrawal, can induce drug craving, which leads to relapse. The system mediating stress responses incorporates the hypothalamic pituitary axis (HPA) and extrahypothalamic regions (such as the extended amygdala). CRF is a neuropeptide that is responsible for activating the HPA, where it plays a mediating role in hormonal, autonomic, emotional, and behavioral responses to stress. Initial exposure to a drug engages the HPA, but this response becomes blunted with repeated exposures via feedback systems in response to circulating hormones. CRF-mediated actions on addictive behaviors depend on their interplay at extrahypothalamic sites.[116] These extrahypothalamic regions become sensitized to CRF after repeated exposure to substance abuse. During withdrawal, these regions become engaged and hyperactive, thereby increasing local CRF levels and perpetuating negative states of stress. While stress is sufficient to increase CRF levels in the VTA, it is neuroadaptive changes induced by prior drug abuse that enable the CRF to control local glutamate release, subsequently activating the dopaminergic system and perpetuating stress-induced relapse to drug-seeking behaviors. There remains debate about the particular sites of action for CRF beyond the HPA. CRF acts primarily through either CRF1 or CRF2 receptors, both of which are widely distributed throughout the brain. CRF1 receptors have been hypothesized to play a significant role in addiction sensitization and relapse. One study suggests that a CRF1 receptor antagonist was sufficient to decrease the reinstatement of drug-seeking behavior in a previously abstinent rodent that was given cocaine,[117] although more recent studies support the role of CRF1 receptors in active drug taking.[118] Chronic inhibition of CRF1 receptors is also sufficient to induce long-term adaptations to the dopaminergic system, including reducing the density of dopaminergic projections in the striatum and increasing dopamine receptor expression in a subtype-specific manner. Comparatively, stress-induced reinstatement to addictive substances can be prevented by infusions of a CRF2 receptor antagonist into the VTA. This most likely indicates inhibition of glutamate and dopamine release, even through CRF1 receptors are dominant in this region.[119]

### 2.2.2.2. The Case in Favor of LUMM for Substance Addiction

Addictive behaviors clearly undermine individual and population health, and exact a significant economic cost on global societies.[120] Clinicians, researchers, policy makers, and society at large are therefore eager to implement effective policies and programs to reduce the medical and economic burdens of addiction. Treatment is one important response to these burdens. Addiction treatment programs have traditionally engendered the view that patients are sufficiently impaired and concerned by their addictions to seek help voluntarily. However, the case-mix has shifted dramatically over time, and mandatory treatment pathways are becoming increasingly entrenched in addiction treatment programs and policies around the world. These pathways include legal mandates from the criminal justice system, formal mandates from employers and social assistance agencies, and informal mandates (e.g., threats, ultimatums, interventions, etc.) issued by family and friends, all compelling individuals with addiction to seek treatment. Mandated treatment policies and programs have been viewed as cost-effective and rehabilitative adjuncts to voluntary treatment and, on this basis, justifiable public health measures similar to seatbelt laws or mass immunization programs. The rationale for mandatory addiction treatment has recently been broadened to underscore findings from neuroscience research. Evidence of impairment in decision making and behavioral control in individuals with histories of substance abuse has been used to argue that individuals with such neurocognitive affliction are not capable of informed consent.[121]

Some scholars have expanded this argument by proposing that mandated addiction treatment should be used to restore patient autonomy and, to this end, can be justified according to a fundamentally humanitarian moral calculus.[122]

Impaired control over voluntary behavior is a marked feature in emerging neurobiological explanations of substance addiction, in clinical and diagnostic accounts, and in debates about addiction nosology.[123] There is growing evidence that chronic, sustained drug abuse is associated with neurocognitive changes and deficits, as revealed by neuroimaging studies[124] and neuropsychological testing.[125] Several studies propose that chronic exposure to drugs sets in motion neurobiological processes that result in overvaluing the reinforcing properties of a substance or behavior and an undervaluing of natural reinforcers (e.g., maintaining relationships, going to work, etc.).[126] These processes are associated with impaired voluntary control over one's behavior. Similarly, individuals experiencing addiction have neurological impairments that weaken their ability to make voluntary decisions in service of long-term outcomes. Despite cautionary assertions concerning the difficulty of making substantive generalizations or conclusions about the neuropsychological and neurobiological correlates of chronic drug use – due largely to that fact that findings are not always consistent in the nature or extent of deficits observed – results from neuroscientific studies have been used to argue that treatment is able to restore free will.[127] This suggests that drug cravings can manifest as such irresistible and powerful psychological forces that someone with an addiction is not capable, at certain times, of acting fully autonomously when the decision involves denying the persistence of cravings.[128]

Autonomy is a term with multiple meanings. In its maximal sense, autonomy means that human beings possess only the desires and beliefs they want to have and make choices uninfluenced by any factor they have not endorsed.[129] Certainly, if addiction threatens autonomy (as it seems to do), then it must be some less extravagant notion of autonomy that it undermines. In a minimal sense, autonomy is simply self-government. Just as autonomous nations are able to make major decisions of internal and external policy without undue interference from foreign powers, so autonomous persons are capable of governing themselves by setting their own short- and long-term ends and choosing the best means of achieving them. One obvious situation in which autonomy is compromised or lost is when the self is ruled by another. In the political domain, the loss of autonomy is almost exclusively described this way. The same kind of phenomenon can occur, more or less dramatically, in the substance addict as well. A slave, for instance, whose life is entirely in the hands of another, is a dramatic example of this phenomenon, while a dispositionally subservient person might represent a less dramatic instance of this partial loss of autonomy. An addict might be excessively subservient to the individual who supplies him with drugs, or with money for drugs, and therefore have his autonomy compromised by the rule of another. However, if the addict's autonomy is compromised in this way, it marks a consequence of an initial loss of autonomy that is characteristic of addiction. This initial loss of autonomy has left the addict vulnerable to this subservience, since it is the addiction that gives the individual who controls him undue influence.[130]

There need not be another party exercising undue influence over the addict to experience a weakening of autonomy. The individual who is able to supply his habit is unlikely to be at the control of another as the consequence of addiction.[131] It is sometimes postulated that addicts are controlled by the drugs they abuse. Carl Elliott, for instance, writes that the addict must go where addiction leads, because the addiction "holds the leash."[132] Elliott's imagery is, of course, a metaphor: an addiction cannot hold a leash, is not an agent, and has no desires or goals of its own. If addiction involves the loss of autonomy, then it must somehow undercut the addict's ability to pursue his own goals. Elliott's claim that addicts are in thrall to their addiction echoes a long tradition of theorizing about addiction, namely, that addiction exercises complete control over drug-seeking and consuming behavior – found in the writings of philosophers, psychologists, and clinicians. For Louis Charland, for instance, the addicted brain "has almost literally been hijacked by the drug";[133] for Alan Leshner, the initially voluntary behavior of drug-taking gradually transforms into involuntary drug-taking to the point where behavior is subsequently driven by compulsive cravings for the drug;[134] and for Harry Frankfurt, unwilling addicts struggle against their desires to no avail insofar as they are always "helplessly violated by their own desires."[135] For these authors, addiction is compulsive, which is to say that addicts are forced to act as they do by virtue of an irresistible desire. Desires are irresistible when they become powerful enough to overwhelm an individual's capacity to overcome or circumvent them. Thus, addiction is compulsive inasmuch as it produces desires that are so powerful that an addict cannot resist them.[136] This conception of how addiction functions dates back (at least) to William James, who commented that "'if a bottle of brandy stood at one hand and the pit of hell yawned at the other, and I were convinced that I should be pushed in as sure as I took one glass, I could not refrain.'"[137]

### 2.3. Problems of MM: A Preliminary Response to LUMM Proponents

Beta-blocking pharmacologicals, ECT, FMC, and DBS techniques used to block (i.e., blunt or dampen) or reverse (i.e., erase) the cognitive processes through which non-conscious recollections of past events deemed pathological and found to exacerbate PTSD and substance addiction are, as suggested above, currently offered as treatments for specific diseases of mentality. Intended as prudent therapies, these treatments are widely experimental in the context of targeted manipulation and therefore transcend the respective purposes for which they were originally designed. Due of the experimental nature of their implementation, the long-term effects of their novel application are widely unknown. While the potentially harmful neurocognitive and more general biological effects already suggest their restriction from general use, this book contends, for reasons beyond these implications, that even the most limited forms of MM are deeply problematic, and that arguments in its favor are indicative of an acute and far-reaching neuroethical astigmatism.

#### 2.3.1. Biomedicalization and the Codification of New Diseases

The primary problem and neuroethical astigmatism of LUMM as a treatment for PTSD concerns the likelihood for unsavory memories to become medicalized and subsequently codified as a new disease category. A lingering effect of contemporary biomedical technologies is the medicalization of what has heretofore been considered "normal" states of being.[138] Sociologists in the 1970s and 1980s defined medicalization as descriptive of at least two processes: first, placing what had previously been considered "normal" behavior under the medical gaze,[139] and second, taking something deemed by society as pathological and placing it under the jurisdiction of medicine.[140] In recent years, new processes of biomedicalization have expanded the diagnostic conditions of illness to include more symptoms and greater numbers of individuals. This expansion is exemplified by cases of clinical depression and bipolar disorder, and it is particularly evident in the extension of attention deficit hyperactivity disorder (ADHD) to include greater numbers of children and a growing adult population. The expansion of diagnoses is encouraged and promoted by pharmaceutical companies that produce drugs to treat disorders with the intention of codifying new disease categories. In turn, pharmaceutical companies sponsor disease awareness campaigns, advertise prescription drugs directly to consumers, and target clinicians at educational conferences and in medical offices to encourage them to prescribe their drugs. Sometimes referred to as "disease mongering,"[141] this newer process of medicalization allows pharmaceutical companies to capitalize on human suffering and exploit insecurities and unhappiness in order to increase drug sales.[142]

Propranolol in particular seems especially ripe for pharmaceutical rebranding.[143] A pharmaceutical company that wishes to manufacture and market a newer beta-blocker for the treatment of PTSD need only slightly alter its chemical composition to obtain a new patent and market the drug under a new name. It might, for example, promise fewer side effects, or longer-lasting effects than generic propranolol. The company responsible would then be able to brand the "new" (and likely more expensive) drug and market it with a new patent for the "new" ability to prevent PTSD. Granted this, various scenarios become possible. For instance, patients would be made aware of and offered the drug in the aftermath of a traumatic event. To sell more drugs, the company would specify a range of traumatic events for which its drug should be prescribed: rape, violent crimes, death of a loved one, and the like. Here, medicalization processes come into play. Trauma – its conception, parameters, and definition – is equal parts cultural and social, not medical. Yet the definition of trauma would be codified by the Food and Drug Administration (FDA) through its indications for use of the new drug, and the pharmaceutical company that manufactures it may continually broaden the scope of trauma in order to sell more of its product. Take, for example, a drug advertisement in which an individual is encouraged to ingest propranolol following an embarrassing or humiliating experience at the office. This quixotic yet sobering example provides a substantive reason to be concerned that a private company seeking to sell more drugs will promote an expanded set of PTSD causes, altering both a sense of the illness and interpretations of the experiences that may cause it.[144]

Moreover, the foregoing concern seems particularly acute in terms of employing the new drug as a prophylactic to trauma. Although the PCB has focused chiefly on the preventative uses of propranolol for military or emergency rescue teams,[145] the company producing a new drug for PTSD would presumably attempt to market directly to consumers.[146] Assuming the FDA approves the drug for this use, questions are inevitably raised over the breadth and depth of traumas for which the new drug is appropriate. This essentially social question would then become defined primarily by the pharmaceutical company. If the new drug is marketed as prophylactic, it would be advertised to consumers who may be exposed to trauma in the near future. It may eventually become tempting for all individuals to have the new drug on hand for consumption before or after trauma, idiosyncratically defined. Falling in line with methylphenidate for ADHD and selective serotonin reuptake inhibitors for depression diagnoses, propranolol may be positioned as another catalyst of "diagnostic bracket creep,"[147] in which the availability of a new drug encourages the expansion of a diagnostic category. This is complicated further by the added nebulous category of "prevention" rather than treatment where the potential for expansion is even greater. If modern history has demonstrated anything, it is that scientific breakthroughs are often double-edged swords. However, if the PCB's language of "evildoers" and "pain" that is "deserved" has resonance at all in high political circles, it has little utility in the scientific and rational evaluation of new medical technologies and their potential dangers.[148]

In addition to ethical qualms about biomedicalization and the codification of new diseases is the issue of capacity and, thus, informed consent.[149] It is hardly controversial to question the capacity of research subjects or medical patients to give informed consent in the immediate aftermath of severe psychic trauma. While victims of rape and witnesses to murder are generally assumed to have decisional capacity to accept diagnostic and forensic tests (as well as psychotherapeutic and psychopharmacologic interventions), the use of propranolol as a targeted method of manipulation would require healthcare professionals to accept a lower threshold of capacity. However, researchers or clinicians utilizing this method must take decisional capacity seriously if they wish maintain minimal treatment standards. If an individual is judged to be devoid of the ability to understand, evaluate, and reason about relevant information (whatever the cause), then this precludes the individual from free participation in PTSD research. No risk, however small, should be imposed in these circumstances. The prevention of PTSD with propranolol does not constitute a medical emergency as it has been traditionally defined – that is, when the consequence of withholding a particular treatment is that death will ensue or the patient's health will be substantially compromised. If, on the contrary, the prevention of PTSD were to become understood as an emergent circumstance (as defined above), then patients with capacity who refuse propranolol or whose surrogates consent for them would be physically forced or psychologically coerced into taking the drug against their will. In addition to this being an unjustifiable form of paternalism, such forceful and counterintuitive behavior would likely place an additional psychic burden on an already vulnerable person.[150]

### 2.3.2. The Myth of Global Autonomy Loss

The primary problem and neuroethical astigmatism of LUMM as a treatment for substance addiction concerns the myth that individuals with addiction suffer a global loss of autonomy that renders them incapable of acting freely. Notwithstanding its popular appeal, this characterization of addiction seems to be false.[151] While addiction undoubtedly produces powerful desires, there is ample data to suggest that it is not strong enough to overwhelm individuals in the aforementioned manner. Strictly speaking, the strength of a particular desire can be measured by examining the behavior of individuals who are subject to it. It is precisely this test for strength that underlies the claim above: proponents of the global loss of autonomy conception argue their position by highlighting the lengths to which addicts will go in order to procure drugs. Addicts will engage, they suggest, in degrading and risky activities, including stealing and lying. Moreover, addicts will spend time and effort not only in pursuit of drugs, but also in attempts to stop consuming them. This latter endeavor indicates that, irrespective of what else is true of them, addicts genuinely desire (on many occasions) to refrain from acting on their addiction. However, though proponents of the global autonomy loss myth are correct to hold that behavioral evidence unmistakably indicates that addicts have impaired autonomy, addiction behaviors do not fit the profile expected when subject to irresistible desires. Individuals with the capacity for voluntary action who are subject to the irresistible desire to achieve a particular goal will pursue it across a broad range of circumstances, realistic and unrealistic alike. Hence, only a countervailing incentive that is itself of comparable power can limit or prevent the behavior.[152]

To be sure, the fact that an addict might refrain from using a drug in front of law enforcement personnel is not evidence of a resistible desire; however, were the addict to refrain for much smaller incentives – for instance, in order to spend money on food (while not at risk of starvation), or in order to schedule it for a more convenient time – this would mark resistible compulsion.[153] Contemporary evidence patently demonstrates that addictive behavior is sensitive to incentives that are not extraordinary in nature, and that it is not therefore subject to irresistible desires. Joanne Neale has highlighted the effect of price on drug quantity consumed by addicts,[154] and Herbert Fingarette reports that alcoholics exhibit sensitivity to cost even after a priming drink.[155] Moreover, when a powerful reason to abstain is personally accepted and support is steadfastly provided throughout the withdrawal process, many addicts succeed in overcoming their addiction. New mothers, for instance, are frequently able to conquer their addiction in order to better care for their child.[156] Gene Heyman has emphasized that addicts can be treated through the constructive of positive and negative behavioral incentives.[157] Heyman draws heavily on the work of Stephen Higgins and colleagues, who have successfully used rewards (in the form of vouchers) in the treatment of cocaine addiction.[158] In a series of experiments, vouchers were paid to addicts in return for clear urine tests, with the value of each voucher increasing over time if the participant remained abstinent. The value of the vouchers did not exceed twelve dollars in United States currency, and was sometimes significantly lower than this figure. As Heyman notes, this is considerably less than the subjects were routinely spending on cocaine, yet the treatment modality was effective in encouraging the majority to abstain.[159]

The foregoing indications suggest that addicts are not subject to irresistible desires that entail a global loss of autonomy. Some evidence suggests that individuals with addiction may not be subject to desires to use drugs at all, at least on one understanding of the nature of desire, according to which human beings requisitely have positive attitudes toward desired objects.[160] Drugs may apparently be "wanted" – that is, they may possess a high incentive salience – without being "liked" at all.[161] David Balfour has identified the neural basis for this dissociation between the causal strength of a desire and the liking of its object as a consequence of the effects of dopamine on different regions of the NAc.[162] One region is involved in the subjective feelings of reward associated with the drug while the other confers incentive salience on the stimulus independently of its being pleasurable. Heyman utilizes Balfour's study as a basis for claiming that addiction is a "disorder of choice."[163] By this phrase, he indicates that (i) addiction is a syndrome in which choice is disordered, but also that (ii) addiction is a syndrome in which dysfunctional behavior is chosen. At least prima facie, this conclusion seems to imply that individuals ought to treat addictive behaviors in the same way as other voluntary actions and hence to regard them as freely chosen and morally irresponsible, worthy of condemnation and punishment. Although Heyman's line of thought is alluring, this text rejects it as acutely shortsighted. Autonomy manifests in degrees: it is not an all-or-nothing phenomenon. An individual may be capable of choice and suffer from diminished autonomy. While Heyman is technically correct to hold that addicts choose to act as they do, he fails to recognize how severely impaired their autonomy to choose is. Addicts need not be in thrall to anyone else, but it is clear that they fail to adequately govern themselves. Addicts experience great difficulty in imposing their will on themselves, not in the manner that myth proponents imagine (i.e., because they feel forced to act, against their will, by overwhelming desires), but because although they may identify with their moment-to-moment choices, they cannot effectively pursue future plans and projects.[164]

Despite the fact that some individuals are more vulnerable to addiction than others (as suggested by the high heritability of substance abuse disorders), modern neuroscience has produced a substantial corpus of material on changes in the brain that together suggest that the discount curves of addicts alter as a consequence of the chronic use of addictive substances. There is evidence that stimuli associated with substances to which an individual is addicted are highly motivating in ways that bypass capacities for conscious control. The motivational salience of a cue for the consumption of any good seems to be encoded as, or caused by, a surge in dopamine from the VTA. As individuals habituate toward a reward, this dopamine signal tends to attenuate. This attenuation fails to occur with regard to drugs of addiction, which may explain why their motivational salience increases even while the degree to which individuals prefer the drug tends to fall. Dopamine causes a heightened focus on predictors of reward and primes the motor system for action, leading to judgments that are difficult to revise and behavior that is difficult to inhibit. While these mechanisms cause judgments and behavior that would be strenuous for a well-functioning person to inhibit, addiction causes neuroadaptations that weaken the efficacy of the frontal mechanisms that regulate behavior. These neuroadaptations explain why addicts who sincerely wish to abstain from drug use nevertheless find it extremely difficult to prevent positive responses to drug-related cues. These neuroadaptations also explain the behavioral inconsistencies characteristic of addiction. Work in social psychology has demonstrated the existence of what may be a separate pathway whereby addicts find themselves oscillating between preferring abstention and preferring consumption.[165] Research on this phenomenon, known as "ego depletion," suggests that cognitive resources that individuals use to assess their options and inhibit prepotent responses are depletable. Utilizing these faculties leaves fewer available for subsequent self-control tasks and, hence, makes such tasks additionally cumbersome. In turn, ego depletion gives rise to the oscillation in preferences observed in addiction: when self-control resources are plentiful, the individual judges that abstention is best; when these resources are depleted, the individual experiences a judgment shift in favor of consumption.[166]

## 2.4. Conclusion

In order to adequately preface the book's logic chain pursued in the chapters to follow, it was necessary, first, to explore the contemporary neuroscience of MM. In its current forms, MM is successfully practiced via the prescription of beta-adrenergic receptor-blocking pharmacologicals to intentionally and permanently dampen the emotional processing (and thereby reduce the physiological markers) of episodic memories related to unsavory stimuli; via strategically timed ECT bursts to intentionally and permanently alter disturbing episodic memories; via FMC to intentionally and permanently replace long-term unsavory episodic memories with something considered more socially and emotionally appropriate; and via DBS to intentionally and permanently annihilate episodic memories linked to unhealthy and socially unacceptable behaviors. And, insofar as researchers continue to identify ways to arbitrate within systems of emotional, semantic, and working memory, the future trajectory of MM therefore seems limitless.

Against this backdrop, the chapter addressed, second, the arguments of the most formidable opponents to the arguments posited in this book, namely, proponents of LUMM. Typically centered around the examples of PTSD and substance addition, LUMM proponents argue that it is morally permissible, as a last resort, for individuals whose psychosocial condition poses a disproportionate and immediate threat to their overall health, well-being, and safety to be offered relief in the form of pharmacological, electrostimulant, hypnopsychological, or surgical memory editing. While thoughtfully formulated and compassionately intended, the chapter progressed, thirdly, to identify two fundamental problems indicative of a deep-seated neuroethical astigmatism within the arguments of LUMM proponents, namely, the lack of foresight regarding the likelihood for unsavory memories to become biomedicalized and subsequently codified as a new disease category, on the one hand, and the non-scientific myth of global autonomy loss, on the other hand.

At least two conclusions can be drawn from this chapter. First, in its current forms and practices, MM directly intends and successfully effects the disintegration – literally, the "scattering into disparate parts" a once cohesive whole – of targeted episodic (i.e., autobiographical) memories via direct (ECT, FMC, and DBS) and indirect (beta-antagonist receptor-blocking pharmacologicals) means. And second, arguments in favor of even the most limited forms of MM have heretofore been guilty of shortsightedness, unscientific origin, and, hence, logical underdevelopment. What follows, therefore, is a clear vision of the moral significance and indispensability of autobiographical memory – even in its most traumatic manifestations – and the dire consequences of snuffing it out.

### 2.5. Notes

1. For a superb neuroethical perspective of pharmaceutical cognitive enhancement, see Sharon Morein-Zamir and Barbara J. Sahakian, "Pharmaceutical Cognitive Enhancement," in *The Oxford Handbook of Neuroethics*, ed. Judy Illes and Barbara J. Sahakian (New York: Oxford University Press, 2011), 229-44.

2. The pharmacologicals typically used to dampen or erase memory are psychotropic in nature.

3. Walter Glannon, "Psychopharmacology and Memory," *Journal of Medical Ethics* 32, no. 2 (2006): 74-78; see especially p. 74.

4. See also Farah et al., "What Can We Do and What Should We Do?" 30-41.

5. Glannon, "Psychopharmacology and Memory," 74-78; see especially p. 74.

6. Glannon, "Psychopharmacology and Memory," 74-78; see especially p. 74.

7. Elsie Donovan, "Propranolol Use in the Prevention and Treatment of Posttraumatic Stress Disorder in Military Veterans," *Perspectives in Biology and Medicine* 53, no. 1 (Winter 2010): 61-74; see especially p. 64.

8. Sylvia Terbeck, Guy Kahane, Sarah McTavish, Julian Savulescu, Neil Levy, Miles Hewstone, and Philip J. Cowen, "Beta Adrenergic Blockade Reduces Utilitarian Judgement [sic]," *Biological Psychology* 92 (2013): 323-28; see especially p. 323.

9. See R. Hurlemann, H. Walter, A. K. Rehme, J. Kukolja, S. C. Santoro, C. Schmidt, K. Schnell, F. Musshoff, C. Keysers, W. Maier, K. M. Kendrick, and O. A. Onur, "Human Amygdala Reactivity is Diminished by the β-noradrenergic Antagonist Propranolol," *Psychological Medicine* 27 (2010): 1-10.

10. Donovan, "Propranolol Use," 61-74; see especially p. 64.

11. If administered prior to a traumatic event, propranolol would influence memory formation, acquisition, and encoding. If administered immediately after a traumatic event, propranolol would influence memory response and formation. If administered at a still later point, propranolol would influence memory recall, retrieval, and reconsolidation. See Donovan, "Propranolol Use," 61-74; see especially p. 64.

12. See Donovan, "Propranolol Use," 61-74; see especially p. 64.

13. Larry Cahill, Bruce Prins, Michael Weber, and James L. McGaugh, "β-Adrenergic Activation and Memory for Emotional Events," *Nature* 371 (October 1994): 702-04.

14. See, for instance, Anda H. van Stegeren, Walter Everaerd, and Louis J. Gooren, "The Effect of Beta-Adrenergic Blockade after Encoding on Memory of an Emotional Event," *Psychopharmacology* 163, no. 2 (September 2002): 202-12.

15. Christian Grillon, Jeremy Cordova, Charles Andrew Morgan III, Dennis S. Charney, and Michael Davis, "Effects of the Beta-blocker Propranolol on Cued and Contextual Fear Conditioning in Humans," *Psychopharmacology* 175, no. 3 (September 2004): 342-52.

16. Roger K. Pitman, Kathy M. Sanders, Randall M. Zusman, Anna R. Healy, Farah Cheema, Natasha B. Lasko, Larry Cahill, and Scott P. Orr, "Pilot Study of Secondary Prevention of Posttraumatic Stress Disorder with Propranolol," *Biological Psychiatry* 51 (2002): 189-92.

17. Alain Brunet, Scott P. Orr, Jacques Tremblay, Kate Robertson, Karim Nader, and Roger K. Pitman, "Effect of Post-Retrieval Propranolol on Psychophysiologic Responding During Subsequent Script-Driven Traumatic Imagery in Post-Traumatic Stress Disorder," *Journal of Psychiatric Research* 42, no. 6 (May 2008): 503-06.

18. Merel Kindt, Marieke Soeter, and Bram Vervliet, "Beyond Extinction: Erasing Human Fear Responses and Preventing the Return of Fear," *Nature Neuroscience* 12 (2009): 266-58.

19. Robin Menzies, "Propranolol, Traumatic Memories, and Amnesia: A Study of 36 Cases," *The Journal of Clinical Psychiatry* 73, no. 1 (January 2012): 129-30.

20. Robin Menzies, "Propranolol Treatment of Traumatic Memories," *Advances in Psychiatric Treatment* 15 (2009): 159-60.

21. Hannah Scissons, "Psychiatrist Studies Treatment for Traumatic Memories," *The StarPhoenix,* 27 January 2010.

22. Jennifer A. Chandler, Alexandra Mogyoros, Tristana Martin Rubio, and Eric Racine, "Another Look at the Legal and Ethical Consequences of Pharmacological Memory Dampening: The Case of Sexual Assault," *Journal of Law, Medicine & Ethics* 41, no. 4 (Winter 2013): 859-71; see especially p. 863.

23. Zhi-De Deng, Shawn M. McClintock, Nicodemus E. Oey, Bruce Luber, and Sarah H. Lisanby, "Neuromodulation for Mood and Memory: From the Engineering Bench to the Patient Bedside," *Current Opinion in Neurobiology* 30 (2015): 38-43; see especially p. 38.

24. Mitchell S. Nobler and Harold A. Sackeim, "Neurobiological Correlates of the Cognitive Side Effects of Electroconvulsive Therapy," *Journal of ECT* 24, no. 1 (March 2008): 40-45; see especially p. 40.

25. Deng et al., "Neuromodulation for Mood and Memory," 38-43; see especially pp. 38-39.

26. Sigmund Freud and Joseph Breuer, *Studies in Hysteria*, trans. Nicola Luckhurst (New York: Penguin Books, 2004).

27. Cody C. Delistraty, "The Ethics of Erasing Bad Memories," *The Atlantic*, 15 May 2014.

28. Nobler and Sackeim, "Neurobiological Correlates," 40-45; see especially p. 40.

29. Nobler and Sackeim, "Neurobiological Correlates," 40-45; see especially p. 41.

30. Marijn C. W. Kroes, Indira Tendolkar, Guido A. van Wingen, Jeroen A. van Waarde, Bryan A. Strange, and Guillén Fernández, "An Electroconvulsive Therapy Procedure Impairs Reconsolidation of Episodic Memories in Humans," *Nature Neuroscience* 17, no. 2 (February 2014): 204-08; see especially p. 204.

31. Alternatively, the control group (group C), which followed the same protocol as group A but did not receive anesthetics and ECT, showed improved memory for the reactivated versus non-reactivated story. See Kroes et al., "An Electroconvulsive Therapy," 204-08; see especially p. 204.

32. See, for instance, Maximilian Gahr, Carlos Schönfeldt-Lecuona, Manfred Spitzer, and Heiko Graf, "Electroconvulsive Therapy and Posttraumatic Stress Disorder: First Experience With Conversation-Based Reactivation of Traumatic Memory Contents and Subsequent ECT-mediated Impairment of Reconsolidation," *Journal of Neuropsychiatry and Clinical Neurosciences* 26, no. 3 (Summer 2014): E38-39.

33. Louisa M. Fraser, Ronan E. O'Carroll, and Klaus P. Ebmeier, "The Effect of Electroconvulsive Therapy on Autobiographical Memory: A Systematic Review," *Journal of ECT* 24, no. 1 (March 2008): 10-17.

34. Fraser et al., "The Effect of Electroconvulsive Therapy," 10-17; see especially p. 11.

35. Maria Semkovska and Declan M. McLoughlin, "Measuring Retrograde Autobiographical Amnesia Following Electroconvulsive Therapy," *Journal of ECT* 29, no. 2 (June 2013): 127-133; see especially p. 127.

36. Richard D. Weiner, Helen J. Rogers, Jonathan R. T. Davidson, and Larry R. Squire, "Effects of Stimulus Parameters on Cognitive Side Effects," *Annals of the New York Academy of Sciences* 462 (March 1986): 315-25.

37. Fraser et al., "The Effect of Electroconvulsive Therapy," 10-17; see especially p. 14.

38. Jason Arndt, "The Role of Memory Activation in Creating False Memories of Encoding Context," *Journal of Experimental Psychology* 36, no. 1 (2010): 66-79.

39. Leo Sher, "Memory Creation and the Treatment of Psychiatric Disorders," *Medical Hypotheses* 54, no. 4 (2000): 628-29; see especially p. 628.

40. Ira E. Hyman and F. James Billings Jr., "Individual Differences and the Creation of False Childhood Memories," *Memory* 6, no. 1 (1998): 1-20.

41. Sher, "Memory Creation," 628-29; see especially p. 628.

42. See Seema L. Clifasefi, Daniel M. Bernstein, Antonia Mantonakis, and Elizabeth F. Loftus, "'Queasy Does It': False Alcohol Beliefs and Memories May Lead to Diminished Alcohol Preferences," *Acta Psychologica* 143 (2013): 14-19; see especially p. 14.

43. Where an electric shock is used as a negative stimulus pairing when the individual is engaging in thoughts, urges, or behaviors related to the substance the individual wishes to avoid. See Clifasefi et al., "'Queasy Does It,' 14-19; see especially p. 14.

44. A technique primarily used with cigarette smokers whereby they smoke a large number of cigarettes in a short period of time to induce nicotine toxicity. See Clifasefi et al., "'Queasy Does It,'" 14-19; see especially p. 14.

45. Administering a repugnant smell or taste, or even an intravenous pharmacological agent, to induce sickness. See Clifasefi et al., "'Queasy Does It,'" 14-19; see especially p. 14.

46. See Clifasefi et al., "'Queasy Does It,'" 14-19; see especially p. 14.

47. Throbbing headache, nausea, vomiting, and weakness, for instance. See Clifasefi et al., "'Queasy Does It,'" 14-19; see especially p. 14.

48. Over the past twenty years, the literature on false memory has suggested the possibility of having individuals imagine an event that purportedly happened in their past through innocuous suggestions and later coming to feel confident that the event occurred. See Clifasefi et al., "'Queasy Does It,'" 14-19; see especially p. 14.

49. Clifasefi et al., "'Queasy Does It,'" 14-19; see especially p. 15.

50. See Clifasefi et al., "'Queasy Does It,'" 14-19.

51. The study of Clifasefi and colleagues focuses specifically on alcohol as the drug of choice; however, this text extends their application and scope to other addictive drugs.

52. Taken together, false memories about becoming sick from a specific drug in one's young adulthood appears to have implications for an individual's current and future associations with that drug. See Clifasefi et al., "'Queasy Does It,'" 14-19.

53. Clifasefi et al., "'Queasy Does It,'" 14-19.

54. Clifasefi et al., "'Queasy Does It,'" 14-19.

55. Clifasefi et al., "'Queasy Does It,'" 14-19; see especially p. 19.

56. See Michael B. Henderson, Alan I. Green, Perry S. Bradford, David T. Chau, David W. Roberts, and James C. Leiter, "Deep Brain Stimulation of the Nucleus Accumbens Reduces Alcohol Intake in Alcohol-Preferring Rats," *Neurosurgical Focus* 29, no. 2 (August 2010): 1-7; see especially p. 1.

57. Marcus Unterrainer and Fuat S. Oduncu, "The Ethics of Deep Brain Stimulation (DBS)," *Medicine, Health Care and Philosophy* [Epub ahead of print] (January 2015): 1-11; see especially p. 2.

58. Unterrainer and Oduncu, "The Ethics of Deep Brain Stimulation," 1-11; see especially p. 1.

59. Marwan Hariz, Patric Blomstedt, and Ludvic Zrinzo, "Future of Brain Stimulation: New Targets, New Indications, New Technology," *Movement Disorders* 28, no. 13 (2013): 1784-92; see especially p. 1784.

60. A. L. Benabid, P. Pollak, A. Louveau, S. Henry, and J. de Rougemont, "Combined (Thalamotomy and Stimulation) Stereotactic Surgery of the VIM Thalamic Nucleus for Bilateral Parkinson Disease," *Applied Neurophysiology* 50, nos. 1-6 (1987): 344-46.

61. Hariz et al., "Future of Brain Stimulation," 1784-92; see especially p. 1785.

62. Henderson et al., "Deep Brain Stimulation," 1-7; see especially p. 1.

63. DBS in the NAc has selectively blocked the return of psychostimulant use and reduced morphine-induced place preference. See Henderson et al., "Deep Brain Stimulation," 1-7; see especially p. 1.

64. Jens Kuhn, Doris Lenartz, Wolfgang Huff, SunHee Lee, Athanasios Koulousakis, Joachim Klosterkötter, and Volker Sturm, "Remission of Alcohol Dependency Following Deep Brain Stimulation of the Nucleus Accumbens: Valuable Therapeutic Implications?" *Journal of Neurology, Neurosurgery & Psychiatry* 78 (2007): 1152-53.

65. Ulf J. Müller, Jürgen Voges, Johann Steiner, Imke Galazky, Hans-Jochen Heinze, Michaela Möller, Jared Pisapia, Casey Halpern, Arthur Caplan, Bernhard Bogerts, and Jens Kuhn, "Deep Brain Stimulation of the Nucleus Accumbens for the Treatment of Addiction," *Annals of the New York Academy of Sciences* 1282 (2013): 119-28; see especially p. 123.

66. U. J. Müller, V. Sturm, J. Voges, H. J. Heinze, I. Galazky, M. Heldmann, H. Scheich, and B. Bogerts, "Successful Treatment of Chronic Resistant Alcoholism by Deep Brain Stimulation of Nucleus Accumbens: First Experience with Three Cases," *Pharmacopsychiatry* 42, no. 6 (November 2009): 288-91.

67. Jens Kuhn, Theo O. J. Gründler, Robert Bauer, Wolfgang Huff, Adrian G. Fischer, Doris Lenartz, Mohammad Maarouf, Christian Bührle, Joachim Klosterkötter, Markus Ullsperger, and Volker Sturm, "Successful Deep Brain Stimulation of the Nucleus Accumbens in Severe Alcohol Dependence is Associated with Changed Performance Monitoring," *Addiction Biology* 16, no. 4 (October 2011): 620-23.

68. See, for instance, Hongyu Zhou, Jiwen Xu, and Jiyao Jiang, "Deep Brain Stimulation of the Nucleus Accumbens on Heroin-Seeking Behaviors: A Case Report," *Biological Psychiatry* 69, no. 11 (2011): e41-42.

69. M. Schüpbach, M. Gargiulo, M. L. Welter, L. Mallet, C. Béhar, J. L. Houeto, D. Maltête, V. Mesnage, and Y. Agid, "Neurosurgery in Parkinson Disease: A Distressed Mind in a Repaired Body?" *Neurology* 66, no. 12 (June 2006): 1811-16.

70. Shüpbach et al., "Neurosurgery in Parkinson Disease," 1811-16; see especially p. 1812.

71. Felicitas Kraemer, "Me, Myself and My Brain Implant: Deep Brain Stimulation Raises Questions of Personal Authenticity and Alienation," *Neuroethics* 6 (2013): 483-97; see especially p. 489.

72. See Karsten Witt, Jens Kuhn, Lars Timmermann, Mateusz Zurowski and Christiane Woopen, "Deep Brain Stimulation and the Search for Identity," *Neuroethics* 6 (2013): 499-511.

73. See Françoise Baylis, "'I Am Who I Am': On the Perceived Threats to Personal Identity from Deep Brain Stimulation," *Neuroethics* 6 (2013): 513-26.

74. See Donovan, "Propranolol Use," 61-74.

75. See R. A. Bryant, "Early Interventions Following Psychological Trauma," *CNS Spectrum* 7, no. 9 (2002): 650-54.

76. See Wayne Hall and Adrian Carter, "Debunking Alarmist Objections to the Pharmacological Prevention of PTSD," *American Journal of Bioethics* 7, no. 9 (2007): 23-24; see especially p. 24.

77. Jonathan E. Sherin and Charles B. Nemeroff, "Post-Traumatic Stress Disorder: The Neurobiological Impact of Psychological Trauma," *Dialogues in Clinical Neuroscience* 13 (2011): 263-78; see especially p. 263.

78. Sherin and Nemeroff, "Post-Traumatic Stress Disorder," 263-78; see especially pp. 263-64.

79. Sherin and Nemeroff, "Post-Traumatic Stress Disorder," 263-78; see especially p. 270.

80. See, for instance, Sherin and Nemeroff, "Post-Traumatic Stress Disorder," 263-78; see especially p. 270.

81. Sherin and Nemeroff, "Post-Traumatic Stress Disorder," 263-78; see especially p. 271.

82. Sherin and Nemeroff, "Post-Traumatic Stress Disorder," 263-78; see especially p. 274.

83. Sherin and Nemeroff, "Post-Traumatic Stress Disorder," 263-78; see especially p. 274.

84. Sherin and Nemeroff, "Post-Traumatic Stress Disorder," 263-78; see especially p. 274.

85. Sherin and Nemeroff, "Post-Traumatic Stress Disorder," 263-78; see especially p. 274.

86. Hall and Carter, "Debunking Alarmist Objections," 23-24; see especially p. 23.

87. Adam J. Kolber, "Therapeutic Forgetting: The Legal and Ethical Implications of Memory Dampening," *Vanderbilt Law Review* 59, no. 5 (2006): 1561-1626; see especially pp. 1576-77.

88. Michael Henry, Jennifer R. Fishman, and Stuart J. Younger, "Propranolol and the Prevention of Post-Traumatic Stress Disorder: Is It Wrong to Erase the 'Sting' of Bad Memories?" *American Journal of Bioethics* 7, no. 9 (2007): 12-20.

89. Hall and Carter, "Debunking Alarmist Objections," 23-24.

90. Hall and Carter, "Debunking Alarmist Objections," 23-24; see especially p. 23.

91. Henry et al.,"Propranolol and the Prevention," 12-20; see especially pp. 16-17.

92. Alcohol is the overwhelming drug of choice for criminal offenders. See Hall and Carter, "Debunking Alarmist Objections," 23-24; see especially p. 24.

93. Hall and Carter, "Debunking Alarmist Objections," 23-24; see especially p. 24.

94. Ironically, this point may be used as a foundation on which their argument may be refuted.

95. Hall and Carter, "Debunking Alarmist Objections," 23-24; see especially p. 24.

96. Hall and Carter, "Debunking Alarmist Objections," 23-24; see especially p. 24.

97. See, for instance, B. A. Strange, R. Hurlemann, and R. J. Dolan, "An Emotion-Induced Retrograde Amnesia in Humans is Amygdala-and β-Adrenergic-Dependent," *Proceedings of the National Academy of Sciences of the United States of America* 100, no. 23 (November 2003): 13626-31.

98. See Bryant, "Early Interventions," 650-54.

99. Hall and Carter, "Debunking Alarmist Objections," 23-24; see especially p. 24.

100. See Henry et al., "Propranolol and the Prevention," 12-20.

101. Hall and Carter, "Debunking Alarmist Objections," 23-24; see especially p. 24.

102. Hall and Carter, "Debunking Alarmist Objections," 23-24; see especially p. 24.

103. T. Cameron Wild, Jody Wolfe, and Elaine Hyshka, "Consent and Coercion in Addiction Treatment," in *Addiction Neuroethics: The Ethics of Addiction Neuroscience Research and Treatment*, ed. Adrian Carter, Wayne Hall, and Judy Illes, (San Diego: Academic Press, 2012)," 153-74; see especially p. 155.

104. Wild et al., "Consent and Coercion," 153-74; see especially p. 155.

105. See Neil Levy, "Autonomy, Responsibility and the Oscillation of Preference," in *Addiction Neuroethics: The Ethics of Addiction Neuroscience Research and Treatment*, ed. Adrian Carter, Wayne Hall, and Judy Illes (San Diego: Academic Press, 2012), 139-51; see especially pp. 141-42.

106. Jhodie R. Duncan and Andrew J. Lawrence, "Molecular Neuroscience and Genetics," in *Addiction Neuroethics: The Ethics of Addiction Neuroscience Research and Treatment*, ed. Adrian Carter, Wayne Hall, and Judy Illes (San Diego: Academic Press, 2012), 27-54; see especially pp. 28-29.

107. Duncan and Lawrence, "Molecular Neuroscience and Genetics," 27-54; see especially p. 30.

108. Duncan and Lawrence, "Molecular Neuroscience and Genetics," 27-54; see especially pp. 29-31.

109. Duncan and Lawrence, "Molecular Neuroscience and Genetics," 27-54; see especially p. 31.

110. Duncan and Lawrence, "Molecular Neuroscience and Genetics," 27-54; see especially pp. 31-33.

111. Duncan and Lawrence, "Molecular Neuroscience and Genetics," 27-54; see especially p. 33.

112. See, for instance, Carmela M. Reichel, Khaled Moussawi, Phong H. Do, Peter W. Kalivas, and Ronald E. See, "Chronic N-Acetylcysteine during Abstinence or Extinction after Cocaine Self-Administration Produces Enduring Reductions in Drug Seeking," *The Journal of Pharmacology and Experimental Therapeutics* 337, no. 2 (2011): 487-93.

113. See, for instance, Kevin M. Gray, Noreen L. Watson, Matthew J. Carpenter, and Steven D. LaRowe, "N-Acetylcysteine (NAC) in Young Marijuana Users: An Open-Label Pilot Study," *American Journal of Addiction* 19, no. 2 (March 2010): 187-89.

114. Duncan and Lawrence, "Molecular Neuroscience and Genetics," 27-54; see especially p. 33.

115. Duncan and Lawrence, "Molecular Neuroscience and Genetics," 27-54; see especially pp. 31-41.

116. Duncan and Lawrence, "Molecular Neuroscience and Genetics," 27-54; see especially p. 41.

117. See Edmund Przegaliński, Małgorzata Filip, Małgorzata Frankowska, Magdalena Zaniewska, and Iwona Papla, "Effects of CP 154,526, A CRF1 Receptor Antagonist, On Behavioral Responses to Cocaine in Rats," *Neuropeptides* 39, no. 5 (October 2005): 525-33.

118. See, for instance, Sheila E. Specio, Sunmee Wee, Laura E. O'Dell, Benjamin Boutrel, Eric P. Zorrilla, and George F. Koob, "CRF1 Receptor Antagonists Attenuate Escalated Cocaine Self-Administration in Rats," *Psychopharmacology (Berl)* 196, no. 3 (February 2008): 473-82.

119. Duncan and Lawrence, "Molecular Neuroscience and Genetics," 27-54; see especially pp. 41-42.

120. Wild et al., "Consent and Coercion," 153-74; see especially pp. 153-54.

121. Wild et al., "Consent and Coercion," 153-74; see especially p. 154.

122. See Arthur Caplan, "Denying Autonomy in Order to Create It: The Paradox of Forcing Treatment on Addicts," *Addiction* 103, no. 12 (December 2008): 1919-21.

123. Wild et al., "Consent and Coercion," 153-74; see especially p. 155.

124. See, for instance, K. I. Bolla, D. A. Eldreth, E. D. London, K. A. Kiehl, M. Mouratidis, C. Contoreggi, J. A. Matochik, V. Kurian, J. L. Cadet, A. S. Kimes, F. R. Funderburk, and M. Ernst, "Orbitofrontal Cortex Dysfunction in Abstinent Cocaine Abusers Performing A Decision-Making Task," *NeuroImage* 19, no. 3 (July 2003): 1085-94.

125. See, for instance, Karen D. Ersche and Barbara J. Sahakian, "The Neuropsychology of Amphetamine and Opiate Dependence: Implications for Treatment," *Neuropsychology Review* 17, no. 3 (September 2007): 317-36.

126. See, for instance, Rita Z. Goldstein and Nora D. Volkow, "Drug Addiction and Its Underlying Neurobiological Basis: Neuroimaging Evidence for the Involvement of the Frontal Cortex," *American Journal of Psychiatry* 159, no. 10 (October 2002): 1642-52.

127. See Caplan, "Denying Autonomy," 1919-21.

128. Wild et al., "Consent and Coercion," 153-74; see especially p. 155.

129. Levy, "Autonomy, Responsibility, and the Oscillation of Preference," 139-51; see especially pp. 140-41.

130. Levy, "Autonomy, Responsibility, and the Oscillation of Preference," 139-51; see especially p. 141.

131. Levy, "Autonomy, Responsibility, and the Oscillation of Preference," 139-51; see especially p. 141.

132. See Carl Elliott, "Who Holds the Leash?" *American Journal of Bioethics* 2, no. 2 (Spring 2002): 48.

133. See Louis C. Charland, "Cynthia's Dilemma: Consenting to Heroin Prescription," *American Journal of Bioethics* 2, no. 2 (Spring 2002): 37-47; see especially p. 43.

134. See Alan Leshner, "Science-Based Views of Drug Addiction and Its Treatment," *Journal of the American Medical Association* 282 (1999): 1314-16.

135. Harry Frankfurt, "Freedom of the Will and the Concept of a Person," *Journal of Philosophy* 68 (1971): 5-20; see especially p. 12.

136. Levy, "Autonomy, Responsibility, and the Oscillation of Preference," 139-51; see especially p. 141.

137. See William James, *Principles of Psychology,* vols. 1-2 (Cambridge, MA: Harvard University Press, 1890); see especially vol. 2, ch. XXVI.

138. Henry et al., "Propranolol and the Prevention," 12-20; see especially pp. 17-18.

139. Talcott Parsons, "Definitions of Health and Disease in Light of American Values and Social Structures," in *Patients, Physicians and Illness,* ed. E. Gartley Jaco (New York: Free Press, 1979), 120-44.

140. Peter Conrad and Joseph W. Schneider, *Deviance and Medicalization: From Badness to Sickness* (St. Louis: The C. V. Mosby Company, 1980).

141. See, for instance, Ray Moynihan and David Henry, "The Fight Against Disease Mongering: Gathering Knowledge for Action," *PLoS Medicine* 3, no. 4 (2006): 425-28.

142. Henry et al., "Propranolol and the Prevention," 12-20; see especially pp. 17-18.

143. Henry et al., "Propranolol and the Prevention," 12-20; see especially p. 18.

144. See Henry et al., "Propranolol and the Prevention," 12-20; see especially p. 18.

145. See President's Council on Bioethics, *Beyond Therapy,* 214-33.

146. Henry et al., "Propranolol and the Prevention," 12-20; see especially p. 18.

147. Peter D. Kramer, *Listening to Prozac: A Psychiatrist Explores Antidepressant Drugs and the Remaking of the Self* (New York: Viking Press, 1993); see especially p. 15.

148. Henry et al., "Propranolol and the Prevention," 12-20; see especially p. 18.

149. Henry et al., "Propranolol and the Prevention," 12-20; see especially p. 15.

150. Henry et al., "Propranolol and the Prevention," 12-20; see especially p. 15.

151. Levy, "Autonomy, Responsibility, and the Oscillation of Preference," 139-51; see especially p. 142.

152. Levy, "Autonomy, Responsibility, and the Oscillation of Preference," 139-51; see especially p. 143.

153. Levy, "Autonomy, Responsibility, and the Oscillation of Preference," 139-51; see especially p. 143.

154. See Joanne Neale, *Drug Users in Society* (New York: Palgrave, 2002).

155. See Herbert Fingarette, *Heavy Drinking: The Myth of Alcoholism as a Disease* (Berkeley, CA: University of California Press, 1988).

156. See, for instance, Gene M. Heyman, *Addiction: A Disorder of Choice* (Cambridge, MA: Harvard University Press, 2009).

157. See Heyman, *Addiction*.

158. See, for instance, Stephen T. Higgins, Alan J. Budney, Warren K. Bickel, Florian E. Foerg, Robert Donham, and Gary J. Badger, "Incentives Improve Outcome in Outpatient Behavioral Treatment of Cocaine Dependence," *Archives of General Psychiatry* 51, no. 7 (1994): 568-76.

159. Levy, "Autonomy, Responsibility, and the Oscillation of Preference," 139-51; see especially pp. 143-44.

160. Levy, "Autonomy, Responsibility, and the Oscillation of Preference," 139-51; see especially p. 144.

161. See, for instance, Terry E. Robinson and Kent C. Berridge, "Addiction," *Annual Review of Psychology* 54 (2003): 25-53.

162. See David J. K. Balfour, "The Neurobiology of Tobacco Dependence: A Preclinical Perspective on the Role of the Dopamine Projections to the Nucleus Accumbens," *Nicotine & Tobacco Research* 6, no. 6 (2004): 899-912.

163. See Heyman, *Addiction*.

164. Levy, "Autonomy, Responsibility, and the Oscillation of Preference," 139-51; see especially p. 144.

165. See, for instance, Roy F. Baumeister, "Ego Depletion and Self-Control Failure: An Energy Model of the Self's Executive Function," *Self and Identity* 1, no. 2 (2002): 129-36.

166. Levy, "Autonomy, Responsibility, and the Oscillation of Preference," 139-51; see especially pp. 144-48.

# 3.
# THE CASE AGAINST EPISODIC DISINTEGRATION: THE MORAL SIGNIFICANCE OF AUTOBIOGRAPHICAL MEMORY FOR ETHICAL DECISION MAKING

In light of the foregoing chapter's conclusions – namely, that MM directly intends and successfully effects the disintegration of targeted episodic memories via direct and indirect means – this chapter aims to make the case against episodic disintegration by underscoring the moral significance of autobiographical memory for ethical decision making. To that end, it endeavors, first, to identify and unpack the interrelationships between subjective experience, autobiographical memory, and moral judgment. Second, it discusses the requisite dependence of subjective memory on objective reasoning. Third, it explores the correlation of autobiographical memory to the evolving narrative of human emotion. The chapter concludes by highlighting the necessity of autobiographical memory for emotionally rational ethical decision making.

## 3.1. Autobiography, Memory, and Judgment

The study of memory and cognitive learning arose from philosophical questions concerning the way individuals come to know themselves, others, things, and the world around them. Learning is the primary method by which one acquires knowledge, and remembering is the primary means by which one supports knowledge claims.[1] While the seventeenth to nineteenth centuries were marked by empiricist philosophers who speculated about the numerous factors that might affect the degree or strength of particular subjective associations,[2] it was philosophers writing in the twentieth century who first introduced to psychologists the distinction between episodic and semantic memory.[3] However, it was not until the "everyday movement" of the final two decades of the twentieth century when researchers first argued that attention should focus primarily on the ways in which individuals use autobiographical memory in their daily tasks.[4]

### 3.1.1. The Metaphysics of Autobiographical Memory

Autobiographical, or "episodic," memories enable individuals to create and maintain personal identity. By sharing past experiences and attending to the memories of others, individuals build and strengthen social relationships.[5] Paired with semantic self-knowledge, episodic memory is the primary database informing the self. Cognitive and neuropsychological research suggests episodic memories and semantic self-knowledge structures are stored in separate areas of the brain and serve diverse functional roles. Whereas semantic self-knowledge is typically applied at the personal level, autobiographical memories are used to inform behavior in particular circumstances. In this way, autobiographical memories possess an ancillary predictive utility for understanding meaningful life outcomes that transcend personal calculations referring to semantic self-knowledge, such as traditional questionnaire measures of personality.[6] Both positive and negative life experiences have an immediate impact on personal well-being, and memories of those experiences serve to guide choices, attitudes, and actions in life. Significant life experiences are encoded as autobiographical memories and remain associated with the cognitive-affective component experienced during the initial event.[7] Autobiographical memories remain attached to representations of how the initial event was experienced, often determined by the individual's goals during encoding. Although personal goals differ across situations, self-determination theory proposes that human beings possess an overarching desire for growth, which is expressed through the daily pursuit of three psychological needs: autonomy (the need to feel volitional and authentic in one's actions), competence (the need to feel effective and efficacious), and relatedness (the need to feel connected, and to care for and be cared for by others).[8]

Activation analyses of episodic memory encoding have pointed to a common element of left frontal and hippocampal regions, as well as specific regions in the temporal cortex and anterior cingulate.[9] Seed-correlation analysis between blood flow in the right hippocampus and all other brain areas during face encoding revealed that the right hippocampal region was strongly linked with a region in the anterior cingulate. Regarding retrieval, the activation analyses indicate that certain regions generally show decreased activity during episodic memory retrieval. This outcome has been suggested to reflect task-related inhibition from other brain regions. This inhibition of activity potentially reflects prevention of task-irrelevant processing and, if accurate, constitutes a significant aspect of episodic retrieval. Regarding regions that are specifically activated when retrieval is successful, a cryptic issue has concerned whether prefrontal brain regions are generally activated (reflecting "retrieval mode") or whether they are activated to a higher extent when retrieval is successful. Analyses of functional connectivity within the last two decades have gleaned that activity in the right prefrontal regions can reflect either retrieval mode or retrieval success, depending on the other regions to which it is functionally linked. This result is consistent with the notion that brain regions do not possess intrinsically singular functions, but that their functional role can vary across cognitive operations as a function of neural context. Finally, network analysis has been used to directly compare the neural interactions underlying episodic encoding and retrieval. Structural equation modeling of data from young subjects exhibited a shift from positive interactions involving the left prefrontal cortex during encoding to positive interactions involving the right prefrontal cortex during recall.[10]

Psychological writing on the subject of autobiographical memory began (at least) as early as William James.[11] Rather than considering memory as encompassing such acts as motor learning, habit formation, S-R strengthening, and the acquisition and use of knowledge, James viewed the nature of recollection as identical to what is now called episodic memory. The conscious recollection that accompanied memory retrieval was the defining characteristic of memory. As James writes: "Memory requires more than the mere dating of a fact in the past. It must be dated in *my* past. ... I must think that I directly experienced its occurrence."[12] Today, episodic memory is also defined by the nature of conscious awareness that accompanies retrieval, now known as autonoetic (self-knowing) awareness. The retrieval of episodic memory is not merely an objective account of what happened or was seen or heard. Rather, its contents are infused with the idiosyncratic perspectives, emotions, and thoughts of the individual doing the remembering. Hence, it necessarily involves the feeling that the present recollection is a re-experience of something that has happened before. Insofar as episodic recollection requisitely entails a conscious reexperience of a personal past, it is possible to conclude two theoretical propositions about autobiographical memory. First, autobiographical memory is critically different from all other varieties of memory and can be disassociated with them. There now exist at least three populations of subjects who show selective losses of episodic memory along with shared performance on other memory measures.[13] Second, autobiographical memory is closely related to other higher order mental achievements that are not traditionally considered acts of memory. This includes the autonoetic capacities to introspect upon present experiences and to anticipate future experiences through the imagination.[14]

In the exercise of episodic memory, individuals retrieve not only what events occurred (item memory), but also when they happened (temporal-order memory).[15] Temporal-order memory is thus a critical form of source memory and an integral and defining characteristic of autobiographical memory. In many circumstances, episodic memories are useful only to the extent that temporal-order information is simultaneously available – for instance, remembering where one left one's car keys today versus yesterday. Lesion and functional neuroimaging studies have shown that the PFC is a critical region for temporal-order memory, as is the medial temporal lobe (MTL.[16]) Nevertheless, the neural correlates of temporal-order memory, particularly as they relate to autobiographical events, are not well understood. Temporal-order memory involves both reconstruction and distance processes. Reconstruction processes are effortful operations that include receiving contextual details and using them to deduce the order of events. For example, when trying to determine whether, during a one-day trip to Catalonia, the visit to Montserrat occurred before or after lunch, one might remember the pleasant feelings of resting tired legs in a comfortable restaurant and conclude that the visit to the Benedictine monastery occurred before lunch. In contrast, distance processes are less effortful operations that rely on feelings associated with the strength of the memory trace. For example, one does not need to employ reconstruction processes to conclude that a clearly remembered trip to Paris occurred more recently than a vaguely remembered trip to Dublin. Although reconstruction and distance processes can be used to discern the temporal order of a common set of events, reconstruction processes are generally more effective for events that are close in time. Closeness in time benefits autobiographical reconstruction insofar as makes causal links more obvious.[17]

### 3.1.2. Autobiographical Memory as Pillar of Moral Judgment

Remembering and knowing are two subjective states of awareness associated with autobiographical memory judgment.[18] Remembering refers to the intimate experience of past events in personal history in which previous events and experiences are recreated with the awareness of reliving these events and experiences mentally. In this way, remembering entails cognitive time travel that engages one's innermost sense of self. Knowing refers to other experiences of the past in which one is impersonally aware of the knowledge possessed. Knowing indicates the general sense of familiarity individuals have with abstract knowledge, including the awareness of subjectively experienced events as objective facts. The notion that remembering and knowing could be studied in the laboratory was first suggested by Tulving, who proposed that two states of awareness reflect autonoetic and noetic consciousness – two forms of consciousness that respectively characterize episodic and semantic memory systems.[19] He reported experiments in which subjects were instructed to communicate their awareness states at the moment they recalled or recognized words previously encountered in a study list. Though Tulving used free-recall, cued-recall, and recognition tests, it was recognition memory that became the most commonly used remember/know paradigm, not least because recognition memory is most likely to be associated with experiences of knowing (and remembering), particularly when recognition is accompanied solely by feelings of familiarity. Moreover, the two states of awareness captured by remember/know responses were viewed, at the time, as additionally relevant to dual-component theories of recognition memory, which posited that recognition would be accomplished by one of two independent processes: recollection and familiarity.[20]

The cardinal premise underlying the use of remember/know responses is that the subjective states of awareness thereby measured cannot be dependably inferred from more conventional calculations of performance.[21] Since subjective mental experience cannot be deduced by purely objective measures of performance, subjective reports must be taken into account. Remember/know processes are not intended as introspective measures of underlying hypothetical constructs; their use also differs from classical introspection in that it only requires subjects to distinguish between kinds of mental experiences rather than report the details thereof.[22] Both remembering and knowing define general states of awareness that may be broken down into varieties of experiences. Just as remembering can be divided into more specific source monitoring judgments, so too may know responses be divided into additional response categories. The development of subsequent "guess" responses partly reflects particular concern about the interpretation of know responses – an issue that has plagued the remember/know paradigm. In Tulving's model, the default response of knowing remains open to abuse by subjects, who may use know responses to reflect various judgmental strategies that do not involve any awareness that items were selected from a study list. As a result, later studies have largely controlled for this by strongly discouraging guessing. However, allowing subjects to report guesses seems a better solution to this problem. Evidence suggests that it is guess responses, not know responses, which reflect various other judgmental strategies. These strategies appear to indicate awareness of the prevailing circumstances during the memory task, such as the general characteristics of the item or the frequency of previous responses.[23]

Most autobiographical memory judgments involve – or can be turned into – a choice between two elemental responses. This makes memory judgments ideally suited for studying retrieval and decision dynamics. Traditionally, such questions have been addressed using reaction time, but speed accuracy tradeoff, and an unwillingness of subjects to respond based on a bare minimum of information, limits the usefulness of the response time measure.[24] Insofar as making judgments requires inferring a continuous criterion from numerous attributes of the object in question, rule- and exemplar-based strategies may prove a clearer lens through which to view autobiographical memory as a pillar on which moral judgment rests. Rule-based strategies assume that individuals form hypotheses about the relationship between the cues and criterion and apply this knowledge to make judgments. Rule-based judgments have been chiefly captured with linear, additive models or cue abstraction models. Linear models describe individuals' judgments in a variety of tasks ranging from personal selection to medical diagnoses and have been found to match explicit judgment rules. Based on the lens model, linear models assume that individuals explicitly abstract a weight for each cue and then combine the weighted cue values additively. For instance, when judging the moral permissibility of palliative sedation for a particular patient, the clinicians involved would first determine the value of proportionate pain reduction and task demands that correspond with (minimally) adequate patient care. Thereafter, those clinicians would weigh the patient's diagnosis, prognosis, and pain intractability alongside the task demands of palliative sedation practice and combine this knowledge by adding the weighted cue values.[25]

By contrast, exemplar-based strategies rely on the retrieval of past experiences from long-term episodic memory.[26] Exemplar-based strategies assume that previously encountered objects are stored in memory along with their criterion values. To judge the new object (probe), all previous encountered objects (exemplars) and the associated criterion values are retrieved from memory. For instance, when judging the moral permissibility of palliative sedation for a particular patient, the clinicians involved would reflect on all the past instances in which palliative sedation was (and was not) appropriate. The more similar a retrieved exemplar is to the probe, the more it influences the final judgment. Accordingly, previous circumstances in which palliative sedation was morally permissible influence the attractiveness rating more than unrelated clinical experiences. Hence, exemplar-based strategies imply that individuals store concrete instances without abstracting any knowledge and engage in an associative similarity-based process during memory retrieval. Research suggests that individuals use both rule- and exemplar-based strategies, with strategy selection depending on task characteristics and individual differences. When individuals perform judgment tasks and receive feedback about the correct criterion, they tend to rely more on cue abstraction strategies if the criterion is a linear additive function of the cues (in linear tasks). However, individuals shift to exemplar-based strategies when (in multiplicative tasks) the judgment criterion is a non-linear function of the cues. In general, autobiographical memory judgments influence two aspects of strategy: execution and selection. Regarding the former, better episodic memory can enhance exemplar retrieval and thus lead to more accurate exemplar-based moral judgment. Regarding the latter, episodic memory abilities can fortify either the ability to choose the more accurate moral strategy or the preference for one moral strategy in particular.[27]

### 3.2. Autobiographical Memory and Rationality

The rational function of autobiographical memory is derived from its distributive property throughout manifold cortical systems.[28] While the understanding of cortical organization endures constant revision, two general observations can be gleaned from contemporary neuroimaging techniques. The first is that prefrontal brain regions are the most immediately involved in examined memory domains, including those immediately related to rational recollection.[29] These findings hint at the heterogeneity of the prefrontal cortex, and thus further exhibit the neurological complexity of rationality. The second involves the interaction between prefrontal and posterior brain regions during the encoding and retrieval of individual memories. This indicates that the posterior regions, which store and maintain information, are refreshed by frontal regions, which consequently mediate rehearsal processes of working memory indispensable to rational cognition.[30]

### 3.2.1. Imaging and Socialization of Autobiographical Memory

Over the past two decades, significant progress has been made in identifying the brain regions that support various facets of self-knowledge.[31] The neural basis of autobiographical memory in particular has been amply investigated and is now well categorized, with neuroimaging studies consistently demonstrating that the retrieval of autobiographical information recruits a specific set of frontal, parietal, and temporal regions. The tasks used in these studies typically consist of remembering specific events or retrieving semantic knowledge of facts about one's life. While it is likely that individuals occasionally engage in autobiographical reflection while executing such tasks, the brain regions specifically involved in this process have not been examined. Thus, while the neural substrates of autobiographical memory are well known, the brain regions that contribute to deriving a sense of self and identity from personal memories remain to be elucidated. Arnaud D'Argembeau and colleagues address this question by contrasting neural activity associated with two different ways of considering the same autobiographical memories.[32] Several days prior to fMRI scanning, study participants were asked to select a set of memories that were important in developing and sustaining their sense of self and identity. During scanning, participants were instructed to approach each of these memories in two distinct ways: in some trials, they were tasked with remembering the concrete content of the event (autobiographical remembering), whereas in others they were tasked with reflecting on the broader meaning and implications of their memory (autobiographical reasoning). Contrasting the neural activity associated with these two ways of approaching the same self-understanding enabled D'Argembeau and colleagues to identify that, relative to remembering, autobiographical reflection recruits a left-lateralized network involved in conceptual processing, including the dorsal medial PFC, inferior frontal gyrus, middle temporal gyrus, and angular gyrus.[33] These findings support the notion that autobiographical reflection and the construction of personal narratives go beyond mere remembering insofar as they require deriving meaning and value from past experiences.[34]

Reconstruction from memory relies not only on incomplete representations stored in episodic cognitive units, but also on categorical knowledge learned from past experiences.[35] Beginning with Francis Bartlett's seminal research in the 1930s,[36] it is evident that individuals use knowledge of cultural and social norms, as well as cognitive expectations of the surrounding environment, to facilitate performance across tasks. Long-term episodic memory encoding is known to rely on gist and meaning, including schemas, scripts, and frames. Over time, autobiographical memory becomes more abstract and information is retained at this higher abstract level of knowledge. However, the preferred (basic) cognitive level of abstraction for categorization has also proven to be a preferred level of retention in episodic memory. Hence, the relative contribution of categorical knowledge and how it operates on long-term episodic memory remains unclear. Perhaps the most striking effect of categorical knowledge is the average improvement in performance for stimuli associated with pre-experimental knowledge. In their study of categorical knowledge and episodic memory across domains, Pernille Hemmer and Kimele Persaud document that, within a continuous domain, study participants' recall was better for familiar objects (i.e., fruits and vegetables), even when participants did not remember studying the objects, compared to unfamiliar objects (i.e., random shapes). Similarly in the temporal domain, fifty percent of participants achieved perfect or near perfect performance in the reconstruction from memory of the true order of events for which they had categorical knowledge, whereas only five percent achieved equivalent performance in the reconstruction from memory of events for which they had no prior expectations. Taken together, these findings demonstrate that categorical knowledge exerts a strong influence on reconstructive autobiographical memory.[37]

Socialization of autobiographical memory implies that fundamental memory functions, skills, strategies, and practices are affected by social learning – that in some way and for some purposes, episodic memory is improved, generally in accord with dominant cultural values, through exposure to training of practices by socialization agents, parents, teachers, or others.[38] The assertion that socialization is significant to, and constitutes a unique aspect of, autobiographical memory is consistent with evidence that cognitive enculturation practices define other unique aspects of human cognitive functioning. The history of autobiographical memory in both oral and literate cultures indicates the significance of the socialization of memory vis-à-vis effective learning procedures for committing material to memory. David Rubin has documented the strategies and skills used to secure memory for oral materials, such as Greek epics and traditional ballads.[39] Throughout history and well into the twentieth century, superior memory has been considered the mark of superior intelligence. It was universally accepted, for instance, that memory needed to be trained, and the educational system in the West was predominantly designed to impart methods for increasing memory capacity and skills. For example, the method of rehearsal through imaginatively locating sequential materials to be recalled in spatially ordered locations originated with the ancient Greeks. Mary Carruthers reports the importance of memory to scholars in medieval Europe,[40] highlighting the extraordinary memory performances of scholastics, such as Thomas Aquinas, who were dependent upon memory for reproducing texts available only in single copies in widely dispersed monastery libraries. In China, from ancient times until the modern era, exams for entering civil service requires years of dedicated study that consisted of committing classic texts to memory.[41]

Children begin talking about the past almost as soon as they begin speaking, sometime between sixteen and twenty months of age.[42] However, children's ability to refer to a specific event that occurred in the remote past develops dramatically between two and three years of age, and this typically occurs through joint reminiscing between parents and children. Children's references to the past are framed by adults, usually parents, inasmuch as adults provide the majority of content and structure for the narrative. In the earliest conversations about the past, parents tell children about what happened and children confirm or repeat parental contributions. Research with parent-child dialogue about the past suggests that children learn the forms and functions of such dialogue through participating in parent-guided conversations. A substantial body of research has now established that there are consistent individual differences in the ways mothers reminisce with young children. Several longitudinal studies indicate that an elaborative maternal reminiscing style facilitates children's developing autobiographical skills.[43] Children not only learn to provide details in their episodic reports through parent-guided conversations, but also how to structure their autobiographical memories into organized narratives. There is also growing evidence that providing emotional evaluation of past events is learned through joint reminiscing. Emotional aspects of the past are particularly significant insofar as they move personal narratives from stories about what happened to stories about what happened to oneself. It is emotions that bind past events with self-concept to become part of autobiographical memory. Moreover, a great deal of research has established that, as adults, females' autobiographical narratives are longer, more detailed, and more emotionally laden than are males'. Related to this difference are cross-cultural discrepancies in autobiographical memories, particularly the finding that members of Western cultures have earlier first memories than their Eastern counterparts.[44] Although this research does not establish a direct link between early parent-child reminiscing and the construction of enduring, accessible autobiographical memories, the relations compliment the direction expected from socialization theory.[45]

### 3.2.2. The Functional and Developmental Ontology of Rational Autobiographical Memory

Rational autobiographical memory is a uniquely human form of memory that goes beyond recalling the details of an event to include memory of how the event occurred, what it means, and why it matters.[46] More than mere episodic recall, it abounds with thoughts, emotions, and evaluations about what happened, thereby providing explanatory frameworks replete with human intentions and motivations. Such memories comprise the story of one's life, rich in interactions and relationships, and in a very deep sense provide an enduring sense of self through narrative identity. Thus understood, rational autobiographical memory is socially and culturally mediated in (at least) two ways. First, it emerges within social interactions that center on the telling and retelling of significant life events. Second, it is tempered by the sociocultural models available for organizing and understanding human existence, including narrative genres and life scripts. In this sense, rational autobiographical memory is a socioculturally mediated skill. In modern industrialized societies, the ability to have and tell stories about one's life is crucial. As argued by Katherine Nelson, this skill has become increasingly important as human beings transition from traditional cultures, where individuals are defined in terms of their social relationships and societal and vocational roles, to more industrialized cultures, where individuals move in and out of multiple geographical locations, social relationships, and vocational roles across their lifetimes.[47] Whereas in traditional cultures individual lives gain coherence and consistency through stability of location, role, and relationships, in modern industrialized cultures individual lives gain coherence and consistency through an individual narrative that weaves these disparate parts together. Hence, in modern cultures, autobiographical narratives serve to create a sense of individual consistency and coherence across time.[48]

Although the ability to narrate single stories is typically ingrained by the end of childhood, a full life narrative involves the integration of multiple autobiographically significant memories into an overarching story that encompasses an entire life (to date), and this does not seem to develop fully until adolescence.[49] The life story may manifest in (at least) two ways. The first concerns the ability to integrate multiple episodes into an overarching, causally connected, coherent life narrative. While life narratives are produced, global coherence is generated by the overall temporal structure and the causal-motivational and thematic connections created between individual events, particularly the type of autobiographical reasoning that links events to one another. A second manifestation of the life story involves more partial autobiographical reasoning, and links personal experiences with other, distant biographically noteworthy experiences and facts, including the development of the self, and by these means attributes self-defining power to these memories. Thus, whereas in early childhood narratives of personal past express a sense of self through its situation in relevant contexts and attributing actions and responsibility to the self, the birth of the life story ties the personal past so tightly to identity that narrative and self are no longer separable. When recounting specific past experiences, narrators may claim that prior actions are atypical for them and may be explained or excused by reference to circumstance. When narrating one's life as a whole, however, the narrative may no longer be dismissed as irrelevant to the narrator's identity. Rather, the life story defines who the narrator claims to be. To be accepted as morally responsible persons, narrators must demonstrate an understanding of how their personality and values have developed, influenced by both the unsavory circumstances of life and by their own actions based on enduring values. Moreover, narrators are challenged to construct personal continuity across change, both for their own well-being and for being accepted as agents who assume responsibility for past actions. With the advent of the life story in early adolescence, the development of autobiographical remembering and narrating affiliates with the development of an understanding of personal identity.[50]

Various cognitive and socioemotional skills converge in adolescence to allow individuals to begin creating advanced autobiographical narratives and reasoning techniques.[51] Requisite cognitive skills include temporal understanding, causal and hypothetical reasoning, and textual interpretation. Temporal understanding involves both the ability to recreate temporal sequences of action within events and the more complex ability to sequence events across long periods of time. In Western cultures, for instance, it is typically not until ages six to eight years that children are able to accurately locate in time that events were more recent or distant. The ability to locate past events in terms of calendar time is only acquired fully around age twelve. A second set of cognitive skills that develop in adolescence (and is crucial for the formation of a life narrative) is causal reasoning. Adolescents develop cognitive skills that allow for more integrated, nuanced, and systematic reasoning about relations among events, including hypothetical events. For example, older children begin to construct internal and external causes for the temporally extended development of drawing and verbal skills, which can be combined in preadolescence. In early adolescence, human development is conceived of in terms of habits and attitudes that constitute individuals in terms of abstract psychological traits, which allow for the construction of personal identity across superficial change. Reasoning about psychological characteristics leads adolescents to construct a sense that underlying predispositions or tendencies are continuous over time, and any individual's moral behavior should be viewed as demonstrating this type of consistency. A third set of cognitive tools for constructing the life story involves the ability to summarize and interpret narrative texts, which require stepping back from the text and, for instance, extracting a moral from the story or reasoning about the intentions of the author(s). Moreover, epistemological development of thinking about how individuals come to know and how to be certain about what they know contributes to an emerging understanding that narrating one's life implies more than mere remembering, namely, a constructive and interpretative effort.[52]

As noted above, cultural frameworks influence personal memory and narratives. Culturally canonical biographies and life scripts provide the necessary infrastructure for organizing individual life narratives.[53] Cultural lenses also influence the interpretation of single events through what has been labeled "master narratives." Master narratives are programmatic representations that contain information about cultural norms that individuals should follow and use to position themselves while constructing and sharing an autobiographical narrative. Research on master narratives has focused less on autobiographical memory and narrative coherence and more on gender identity and positioning. Thorne and McLean have identified three master narratives in particular that disclose autobiographical approaches to trauma: the "John Wayne," "Florence Nightingale," and "Vulnerability" narratives, respectively.[54] Thorne and McLean investigated the extent to which these master narratives differ in their frequency by gender in a sample of trauma narratives collected from Euro-American adolescents. Results indicated that only the Florence Nightingale master narrative differed in frequency by gender, with more female participants following this structure.[55] Moreover, master narratives may provide evaluative frameworks for life narratives as well. Dan McAdams and colleagues present evidence that, in both emerging and middle-age Euro-American adults, individuals who tell more redemptive episodes in their life story score higher on measures of psychological well-being and generativity.[56] Nevertheless, there is limited research to date on the development of master narratives and their relation to autobiographical memory coherence. Future investigation would do well, therefore, to examine the effects of master narrative knowledge on both event narrative and overall autobiographical coherence. For instance, a more extensive list of prominent master narratives pertaining to autobiographical memories for experiences would provide a clearer account of the role of master narratives in constructing coherent autobiographical narratives. It is also critical to extend research to examine the types of master narratives that are prevalent in a wider variety of cultures. Finally, a better understanding of the quality and role of master narratives could illuminate the influences of sociocultural conventions on the ways in which autobiographical narratives are organized and rationally evaluated.[57]

### 3.2.3. The Phenomenology of Autobiographical Reasoning

Hermann Ebbinghaus laid the foundations for a psychology of memory in 1880 by substituting everyday remembering with an experimental paradigm that focused on the controlled learning of elements with identical value,[58] the correct recollection of which could be counted and related to the passing of time.[59] Perhaps the most explicit opponent to Ebbinghaus was his colleague at the University of Berlin, Wilhelm Dilthey, who proclaimed life stories as the model for the humanities (Geisteswissenschaften), which approach their subject through historical understanding.[60] Dilthey was not interested in memory per se, but in obtaining the most exhaustive account of phenomena that must be understood against the backdrop of history. Psychology began to leave Ebbinghaus's "memorizing trap" only a century later, moving in the direction of Dilthey by developing an interest in everyday remembering as opposed to memorizing laboratory material. This new object of psychological research was termed "autobiographical memory." Save for involuntary memories, everyday remembering is an active, reconstructive activity that involves at least some degree of argumentative linking or narrative reasoning. Biographically relevant events are contextualized by life, linked both chronologically and by arguments that underscore thematic or causal-motivational implications, similarities, or consequences. In 2000, Tilmann Habermas and Susan Bluck coined the term "autobiographical reasoning" for the activity of explicating the biographical relevance of memories.[61] Autobiographical reasoning creates links between remembered events and other distant parts of one's life, as well as to the self and its development. It thus refers to the remembering subject's life as the relevant frame of reference, thereby implying the life story. In this way, autobiographical reasoning requires cognitive effort transcendent of the mere act of remembering.[62]

Autobiographical reasoning is typically used to develop and secure a sense of personal persistence and self-continuity.[63] Personal persistence, in turn, is predicated on the maintenance of two conflicting autobiographical reasoning processes: one based on self-change, and the other on self-stability. The first process, which involves acknowledging and understanding the immense amount of change one has undergone throughout one's life, is necessary to avoid becoming overwhelmed with feelings of personal stagnation. The second process, which establishes a stable sense of self across time and context, is necessary to avoid becoming disconnected from one's personal history and possible future. Both processes are important insofar as failing to establish a sense of personal persistence can result in the experience of uselessness and isolation, thereby increasing the probability of engagement in self-destructive behavior. The relation between autobiographical reasoning and personal persistence is mediated by the construction of a coherent life story. A coherent story permits the mitigation of discrepancies and inconsistencies among temporally distinct self-concepts, and leads individuals to feel that, notwithstanding the numerous changes that inevitably attend to the passage of time, the self of the past set the stage for the self of the present, which in turn will develop into the self of the future. To cultivate this perception, however, the cavalcade of personal events that compromise the life story must be associated with the advancement of temporally distinct self-concepts. Autobiographical reasoning is required to draw a seamless connection between these discrete constructs. Hence, autobiographical reasoning represents the lifeblood of a coherent life narrative.[64]

Despite the foregoing, it is not the case that autobiographical reasoning is the necessary condition for thinking about one's past. In her review of autobiographical reflection, Ursula Staudinger draws an important distinction between reminiscence and life review.[65] Both refer to the recollection of a life event, but in the case of the latter, individuals move beyond mere recall to imbue past experience with explanation and evaluation. Most appropriately understood, therefore, autobiographical reasoning is a function of life review rather than reminiscence as such.[66] Contemporarily, narrative psychologists employ the term "autobiographical reasoning" to refer to processes including reporting an insight or lesson learned from a previous experience, evincing wisdom when reflecting on one's life, proffering a redemptive twist when describing personal hardships, and describing a stable personal characteristic within a self-narrative.[67] The vast majority of research on autobiographical reasoning has focused on processes of change as opposed to stability. As a result, there has been a great deal of variability in the operationalization of change-based reasoning processes. Research on reasoning processes that indicate a sense of self-change has typically emphasized causal self-event connections.[68] Such connections are present when the narrator describes an event as causing a change in personality. In contrast, autobiographical reasoning processes entailing a sense of stability have often been explored by considering illustrative self-event connections,[69] which are present when the narrator refers to a stable personal attribute and then illustrates an example in which this attribute contributed to an event's occurrence.[70]

Historical research has indicated that, when it comes to autobiographical reasoning, not all events are created equally.[71] Negative, disruptive, and challenging events, for instance, appear to require more reasoning than do expected or positive events. Parents begin to scaffold rudimentary forms of autobiographical reasoning at a young age by providing more explanation of negative than positive events.[72] Thus, reasoning about a cancer scare may bring peace, but reasoning about one's wedding day may not. In fact, as Sonja Lyubomirsky and colleagues suggest, trying to explicate positive events may undermine well-being insofar as various parts of the canonical life story (e.g., one's wedding day) do not require explanation.[73] Nevertheless, even within traumatic events there are limits to the efficacy of reasoning. Roy Baumeister and colleagues have argued that human beings distance themselves from events in which they harm others in order to maintain positive self-perceptions.[74] When harm is directed at the self, there are also limits to the efficacy of autobiographical reasoning. This is exemplified in the study of Robyn Fivush and Jessica Sales in which the authors examine how mother-child dyads address acute (visiting the emergency department for an asthma attack) and chronic (the ongoing conflict of controlling the child's asthma) stressors.[75] Somewhat surprisingly, children of mothers who helped them to develop especially detailed narratives of the acute stressor were less likely to cope effectively than children whose mothers constructed a less detailed narrative. In contrast, for narratives of chronic stressors, children whose mothers provided an emotional and explanatory framework for apprehending the stressor had the highest well-being. Thus, the event being reasoned about and the context of that reasoning – including having a good scaffolder, in this case – are as important to well-being as how one reasons about the event.[76]

### 3.2.4. Autobiographical Meta-Memory and Rational Prospection

In his search for the first principle, Descartes questioned all of what many scholars now call his basic object-level cognitive process,[77] allowing that he could imagine that things were different than they seemed or that they might not exist at all.[78] It was not thinking per se that was indisputable to Descartes, but rather thinking about thinking. What he could not imagine was that the individual engaged in such self-reflective processing did not exist. The reality of the individual who knows about knowing, then, was taken by Descartes as the irreducible core – the foundation upon which all other knowledge was and must be built. It seems, however, that he based his philosophy on a misnomer, namely, "cogito ergo sum." What Descartes should have said was "metacogito ergo sum." The capacity for self-reflection, including knowledge about one's own knowledge, skills, aptitudes, and memories, has long been considered fundamental to understanding what human beings are. Accordingly, Descartes was not the only philosopher to emphasize the import of metacognition. William James once commented on the "tip-of-the-tongue state," a phenomenon that has become a major topic in modern studies of metacognition.[79] Although little attention was paid to metacognition during the behaviorist era, the significance of the topic was acknowledged anew during the cognitive revolution.[80] Some fifty years ago, autobiographical metamemory was first advanced as a phenomenon amenable to detailed scientific study by J. T. Hart, who devised a system that came to be known as the recall, judge, recognize (RJR) paradigm.[81] According to the classic 1977 definition,[82] autobiographical metamemory refers to an individual's knowledge and awareness of autobiographical memory, or anything pertinent to information storage and retrieval.[83]

For (at least) the past half-century, researchers have made an admirable effort to successfully describe the phenomenology of autobiographical memory. However, theoretical and empirical efforts to integrate individual findings into a more complete picture of autobiographical metamemory are necessary.[84] In particular, the degree to which events are recollected (autobiographical reminiscence), believed to have occurred (autobiographical belief), and appraised vis-à-vis accuracy (autobiographical authenticity) are distinct components of autobiographical metamemory that require further investigation. Autobiographical reminiscence includes conscious awareness of remembering, re-experiencing perceptual details of the event, recognizing the spatial and temporal characteristics of the event, and novel appraisal of the event as it influences current emotion. This cumulative recollective experience results in a subjective feeling of re-experiencing the past that differentiates remembering from imagining. Autobiographical belief (or belief in occurrence) is the conviction that an event occurred to oneself in the past. While memories have long been recognized as including a sense of genuineness, less attention has been given to the degree that reminiscence and belief are dissociable, and a growing body of evidence supports such a distinction.[85] Interestingly, suggesting false events frequently results in reports that events occurred without accompanying recollection,[86] whereas studies of nonbelieved memories exhibits a strength of recollection that exceeds that of autobiographical belief.[87] Finally, autobiographical authenticity refers to whether the details of an event are being recollected in the way in which they actually occurred. Questions regarding belief in occurrence may be thought of as potential errors of commission: misremembering details by confusing aspects of one event with another, by incorporating post-event information, or constructing details of an event based on schematic information. In this way, autobiographical authenticity is the individual's functional assessment of the accuracy of what is recalled (independent of what might be considered objectively accurate).[88]

Rational prospection refers to the process of representing and planning for possible future states of the world in light of one's autobiographical memories.[89] Several recent reviews in human psychology, paired with an emerging debate in comparative psychology, have helped draw attention to the role of prospection in cognition.[90] Evidence in support of the critical role of episodic memory in prospective thinking comes in two forms, the first of which from neuroimaging studies. In their 2007 study, Donna Addis and colleagues scanned participants who were asked to imagine experiencing an event based on cued words and a specified time horizon and then to describe the event after being scanned.[91] The investigators found that the region associated with episodic memory (i.e., medial left pre-frontal parietal) was also active during imaging of past and future events. The second form of support is based on clinical studies. Prior to their 2002 study, Stanley Klein and colleagues hypothesized that patients with impaired episodic memory should also show impaired prospective thinking.[92] When the investigators compared participant "D. B.," a seventy-eight-year-old male with severe anterograde amnesia, with two healthy age-matched controls, D. B. showed impaired performance on recollective tasks as expected, but inaccurate responses to tests of prospective memory. However, when compared to healthy age-matched controls, clinical studies of participants with autism, schizophrenia, depression, and impulse-control problems also revealed impairments in prospective thinking. These findings are theoretically problematic insofar as not all of these groups simultaneously experience episodic memory impairments. If there exists any consistent pattern across studies, it is that participants have more difficulty generating positive examples of prospective events than their negative counterparts.[93]

With the exception of some neuroimaging evidence and work in prospective memory, most other claims concerning the necessary connection between the core cognitive requisites (i.e., episodic memory, prospective memory, hypothetical thinking, and conditional representations) and rational prospection exceed the evidence.[94] Moreover, no evidence to date suggests that these cognitive processes have a core functional basis in prospective thinking. An alternative conceptualization, proposed by Magda Osman, examines the role of planning and causal learning in prospection, both of which remain under-researched.[95] Osman's key claim is that goals are, by definition, prospects. By exploring the surrounding environment, individuals learn to discover ways of maintaining what they need and also extending their reach, both of which require representations of the future in the form of goals. The ability of individuals to construct autobiographical representations of the future and adapt to future outcomes is built on basic contingency learning mechanisms, which generate rational expectations and plans of action. These expectations and plans of actions are guided by the discovery and maintenance of goals. Hence, if goals are rational prospects, and contingency learning connects actions with intended outcomes guided by goals, then planning and causal learning are essential properties of rational prospection insofar as they enable individuals to represent how to act in the future. The strong claim being made here is that accurate autobiographical representations of the future will depend on the accuracy of causal autobiographical representations that support the planned actions needed to achieve a future goal. On this basis, it can be argued that the more individuals unpack their causal beliefs and the consequences of their actions in achieving future goals, the higher their judgment of likelihood will be vis-à-vis achieving the goals in the future.[96]

### 3.3. Autobiographical Memory and the Narrative of Human Emotion

The interconnected structures within the limbic system possess a pivotal emotional mechanism immediately related to autobiographical memory. These structures – which include hippocampal formation, fornix, mammillary bodies, the mammillothalamic tract, cingulate gyrus, and cingulum – confirm the existence of a uniform system whereby information is temporarily circulated and ultimately associated and synchronized with emotional and motivational subjective states prior to being transmitted into long-term storage areas.[97] Other limbic systems, such as the amygdalar and septal nuclei, have become regarded as still more intimately associated with emotional regulation. Nevertheless, both sets of mechanistic systems temper and tone the emotional consolidation of autobiographical memory through operations of information evaluation.[98] Hence, disorders of memory systems that control emotions render individuals incapable of rationalizing and evaluating information, the consequence of which is significantly reduced memory capacity. This much denotes the important nature of proper emotional embedding within neurocognitive memory circuits.[99]

### 3.3.1. The Empirical Effects of Emotion on Autobiographical Memory

The most autobiographically memorable events in life are typically those arousing significant emotion.[100] It is well documented that emotional materials can attract more attention or garner more elaborative encoding than neutral materials, and that such enhanced encoding can lead emotional materials to be better remembered than their neutral counterparts. Yet the reason why emotional memories are so resistant to forgetting is not yet fully understood. Despite this, several important empirical regularities have been observed in pursuit to understand the effects of emotion on episodic memory in humans, five of which will be explored herein. The first empirical observation is that that memory advantage for emotional materials increases over time.[101] Numerous laboratory experiments indicate that negative emotional materials are recalled and recognized better than neutral materials.[102] Although these effects may be partially attributed to enhanced encoding of emotion (compared to neutral items), several studies indicate that the effects of emotion are either absent or significantly reduced when memory is tested immediately, and tend to increase in magnitude after several hours.[103] To illustrate the delayed effects of emotion, Tali Sharot and Andrew Yonelinas presented study subjects with a mixture of negative and neutral images.[104] Images were divided into two lists that were studied one day apart. Immediately after exposure to the second list, participants completed a recognition memory test for the collection of studied images mixed with negative and neutral non-studied images. Overall recognition performance was subsequently assessed for the emotional and neural items that had just been studied as well as those studied one day prior. For items studied and tested on the same day, emotional and neutral items were recognized equally well. However, for items studied one day prior, a recognition memory advantage for emotional materials over neutral materials was present. Hence, emotional and neural memories were equally well encoded, but the advantage of emotion emerged only after a delay.[105] Similar effects have been observed in studies of recognition for words and visual images, as well as in tests of free recall,[106] with the delayed effects of emotion appearing as soon as two hours post examination.[107]

A second observation of the effect of emotion on episodic memory is that emotion impacts recollection rather than familiarity.[108] Recognition memory judgments can be based on either the recollection of qualitative information about a study event or on assessments of familiarity. Studies that have directly contrasted the contributions of recollection and familiarity to autobiographical memory have demonstrated that emotion impacts recollection but has little or no effect on familiarity.[109] Many of these studies have examined autobiographical memory under conditions in which the relative increase in recollection may reflect better encoding of emotional (rather than neutral) items, but others, such as Sharot and Yonelinas,[110] have found that recollection advantage for emotional materials is time-dependent. This suggests that not all forms of episodic memory benefit from emotion. Instead, emotion improves recollection, but does not benefit familiarity. Thus, any account of emotion and autobiographical memory needs to account for the observed selectivity of these effects.[111] A third observation of the effect of emotion on episodic memory is that that emotion impacts the recollection of items rather than contexts. That is, not all forms of recollection are increased by emotion. This observation is consistent with a growing body of research suggesting that emotion increases recollection for details that are intrinsic to the emotional item(s) or object(s) while having little (or even a negative) effect on the details that are extrinsic or contextual in nature.[112] These results further highlight the fact that the effects of emotion on episodic memory are selective, and demonstrates that emotion affects the ability to recollect aspects of the emotional item(s) rather than increasing all aspects of the emotional event, such as the contextual or peripheral features thereof. Thus, emotion does not simply enhance memory for emotional events, but selectively improves recollection for the emotional items(s) in the event (compared to other contextual details).[113]

A fourth observation of the effect of emotion on episodic memory is that it depends largely on the amygdala.[114] Although selective amygdala lesions are rare, several cases have reported that bilateral amygdala damage either eliminates or drastically reduces the normal advantage seen for emotional (compared to neutral) materials. In the 1995 study of Larry Cahill and colleagues, for example, subjects were presented with a set of slides along with an accompanying story that included neutral materials and negatively arousing materials. In a subsequent recognition test one week later, participants with selective bilateral amygdala damage performed well at recognizing the neutral slides but, unlike the controls, showed no advantage in recognizing the negative slides.[115] Additionally, another human study, conducted by Elizabeth Phelps and colleagues, revealed that in immediate tests, amygdala damage did not entirely eliminate the advantage of emotion but, unlike in controls, the advantage of emotion did not increase over time.[116] Neuroimaging results provide analogous evidence that the amygdala plays a critical role in producing the effects of emotion on episodic memory. Consistent with a role in recollection, amygdala activity during encoding scales with the vividness of subsequent memory, and its involvement during retrieval is associated with recollection rather than familiarity processes. Moreover, the amygdala is selectively involved in creating and retrieving emotional memories that carry item-specific details, but not necessarily other forms of contextual information. These lesion and imaging results demonstrate that the amygdala plays a central role in producing the advantage of emotion in autobiographical memory, and further verify that the episodic advantage is selective to the recollection of the emotional item(s) in particular (rather than influencing familiarity or recollection of contextual or background information).[117]

A final observation of the effect of emotion on episodic memory is that it does not depend on the hippocampus.[118] Several studies have investigated emotion and memory in participants with large MTL lesions that have included the amygdala, hippocampus, and surrounding perirhinal cortex, and found that these subjects exhibit reduced effects of emotion on memory.[119] Nevertheless, these impairments appear to be due to amygdala damage rather than damage elsewhere in the MTL. This observation is based on studies of patients with MTL damage (not including the amygdala) whose emotional memory advantage is not disrupted.[120] In contrast to the clear body of brain-imaging evidence linking the amygdala to emotional memory processes, neuroimaging findings have been more varied with regard to the role of other MTL regions. Some studies have found that MTL regions contribute comparably to neutral and emotional memory formation,[121] whereas others have found that the anterior hippocampus and rhinal cortex are more involved in emotional (rather than neutral) encoding.[122] Findings have been similarly varied with respect to the role of the MTL during emotional retrieval.[123] Although MTL activity (by itself) has been an inconsistent predictor of the emotional memory advantage, several studies have identified correlations between amygdala activity and other MTL regions during emotional (rather than neutral) encoding, including regions in the anterior hippocampus,[124] anterior parahippocampal gyrus,[125] and posterior parahippocampal gyrus.[126] In sum, these neural studies indicate that the hippocampus is not necessary for producing the delayed effects of emotion. While hippocampal damage reduced episodic memory, it reduced memory for both emotional and neutral items equally. Moreover, hippocampal activity is related to the encoding of both emotional and neutral materials. Nevertheless, some imaging studies suggest that the anterior hippocampus and perirhinal cortex may be more involved in emotional processes than posterior regions such as the parahippocampal cortex.[127]

### 3.3.2. The Contextual Effects of Emotion on Autobiographical Memory

Personal experiences are preserved as episodic memories, which are constituted by the events and contexts in which they took place.[128] These experiences are more accurately remembered when they are emotional (compared to non-emotional events). However, it remains unclear whether the corresponding valence – that is, the degree of pleasantness, ranging from positive to negative – might differently affect the subsequent retrieval of memory. Numerous studies of event-related potentials (ERPs) have attempted to establish whether events with positive or negative valence are more memorable. Some report that negative events are remembered better than positive events,[129] whereas others report converse results,[130] or even no difference between positive and negative events.[131] Nevertheless, few neurophysiological studies have investigated whether neutral events are more accurately remembered when they are encoded within emotional contexts. Other studies have examined the specific effects of negative sentences used as contexts for the recognition of neutral words, but report benefit when compared to words encoded in neutral sentences.[132] Except for a singular outlier,[133] which observed no differences for words presented over positive and negative scenes from the International Affective Positive System,[134] studies that have directly compared positive and negative contexts have reported higher recognition for neutral events encoded under positive contexts compared with negative and neutral contexts.[135]

The neural activity that obtains during the encoding of neutral events presented in emotional contexts has heretofore been rarely investigated.[136] Only one fMRI study has addressed this question, the results of which demonstrate that the amygdala exhibits greater activation for events encoded in a negative context, whereas the PFC is more active for events encoded in a positive context.[137] However, neural activity during the encoding of neutral events in emotional contexts has not been addressed using the ERP technique. The study of Markus Kiefer and colleagues examined subsequent memory effects (SMEs) for words that referred to positive and negative traits.[138] The words were encoded under two distinct emotional states (good and bad moods) evoked by watching various films. Thus, SMEs in this study were not analyzed using the conventional procedure of varying the emotional valence for each stimulus. SMEs between 500 and 600 milliseconds after stimulus onset were more positive in centro-parietal derivations and negative in frontotemporal electrode sites for individuals in a good mood compared to those in a bad mood. Hence, the study demonstrates that an emotional state influenced the neurophysiological responses that predicted ulterior autobiographical memory.[139]

Positive affect may vivify the richness of autobiographical memory retrieval for several reasons.[140] Relative to negative emotion, it has been suggested that position affect promotes relational cognitive processing, allowing for the activation of surrounding information in additional to the emotionally pertinent details.[141] Moreover, enhanced richness of positive autobiographical memory may be caused by increased elaboration and rehearsal of events that resonate with the generally positive self-schema that most individuals maintain over time.[142] Indeed, as Anne Rasmussen and Dorthe Bersten demonstrate, positive memories have more personal and social function than negative memories, suggesting that positive memories are often retrieved in order to amplify self-concept or to facilitate bonding with others.[143] However, regardless of arousal, retrieval of all positive events should satisfy these functions, indicating that emotional stimulation may not have a strong contextual effect on memory richness in positive autobiographical memories. Studies have also revealed an enhancing effect of negative valence on some contextual components of memory retrieval. Negative autobiographical narratives contain more central details than positive events.[144] Additionally, when comparing emotional reactions within a unified autobiographical narrative, ratings of vividness were related with feelings of anger and sadness, but not with happiness or surprise.[145] Lastly, when comparing positive and negative recollections of the same potent emotional event, negative autobiographical memories tend to be more contextually accurate and contain more detail-specific events than positive memories.[146]

The research detailed above illustrates that increased negative emotion enhances autobiographical memory retrieval in a way that is contextually unique from the effects of positive emotion.[147] Unlike positive emotion, negative affect promotes sophisticated processing of the particular details of an event.[148] This detail-oriented processing may occur because negative autobiographical memories signal danger and have a more direct function than do their positive counterparts. That is, negative events (to a greater extent than positive events) are used to help direct future behavior in order to avoid similar negative circumstances. It has been argued that high-arousal negative affect may have different effects on autobiographical memory as compared to low-arousal negative affect.[149] It is possible that high-arousal negative events may have increased directive function relative to low-arousal negative events, leading to a fortified relationship between arousal and episodic memory richness in negative events. Such a difference can help to explain why many individuals have such exceedingly vivid and elaborate representations of personal trauma (including flashbulb memories).[150] Although the mnemonic effects of valence and arousal have been studied in depth, it remains unclear whether the contextual effects of emotional arousal are equivalent for negative and positive autobiographical events. Previous studies have indicated that individuals with depression have a tendency to retrieve memories that are less specific than control participants.[151] While research has confirmed reduced contextual specificity in both individuals with emotional disorders and healthy young adults induced into a negative emotional state,[152] it has not examined contextual memory specificity in healthy young adults who tend to retrieve more negative autobiographical memories than average.[153]

### 3.3.3. The Cognitive Effects of Autobiographical Memory on Empathic Intentionality

Autobiographical memories provide human beings with a sense of past, but recently it has been argued that they provide much more, namely, the same social interaction variables as the personality quality of empathy.[154] As various scholars have argued, autobiographical memory is not just referenced to the self but is personally significant, concerned with episodes that have personal meaning.[155] Personal meaning manifests from emotions, motivation, and goals that are formed through interaction with others in the world. Thus understood, autobiographical memory is declarative, explicit memory for particular experiences in the past, recalled from the idiosyncratic perspective of the self in relation to others. While the respective literature on autobiographical memory and empathy have been separate, both have focused on similar influences. For instance, both emphasize the import of social interaction variables. Research on autobiographical memory has stressed the quality of parent-child interactions styles,[156] the quality of parent-child attachment,[157] and parental warmth.[158] Similarly, research on empathy has stressed parent-child interaction styles,[159] the quality of parent-child attachment,[160] and parental warmth.[161] Studies directly linking autobiographical memory and empathy are equal parts rare and heterogeneous. For example, Richard Harris and colleagues look at how dispositional empathy facilitates recall, using films as prompts for autobiographical memory.[162] More recently, Susan Bluck and colleagues empirically support an alternative point of view, claiming that autobiographical memory sharing can serve the function of eliciting empathy.[163] For this group, autobiographical memory constitutes a cardinal psychological process in empathy insofar as the use of one's past experience constructs models that allow one to understand the inner world of others. Moreover, Bluck and colleagues demonstrate how the communication of autobiographical memories to a listener can enhance the narrator's empathy toward that person.[164] The foregoing group of studies, which indicate how the evolution of both autobiographical memory and empathy have strong relationships to parallel social interactive factors, lend support to the thesis that these two concepts are inextricably linked.[165]

There is no universally accepted definition of empathy, nor is there consensus on what constitutes its general contents.[166] However, empathy has been traditionally understood as a deep-seated, genuine, affective resonance with the emotional state or condition of others.

Today, there is increasing consensus that empathy is a multifaceted process, composed of various parts. Mark Davis, for instance, describes empathy as a multidimensional phenomenon comprising emotional and cognitive components that can be described as a set of distinct but related interacting dimensions.[167] In particular, Davis identifies four primary dimensions of empathy: perspective-taking, fantasy, empathic concern, and personal distress. On the one hand, perspective-taking and fantasy refer to the cognitive aspects of empathy and represent two distinct antecedent forms of affect experienced in response to the emotions of others. Specifically, perspective-taking is defined as one's willingness to adopt the point of view and perspective of others, while fantasy describes the tendencies to identify and imagine oneself through the feelings and actions of fictional characters in movies, books, plays, and other forms of role taking. On the other hand, empathic concern and personal distress are purely and typically emotional. Specifically, empathic concern includes feelings of warmth, compassion, and sympathy for others; it aims to comprehend the emotional state of others and, as such, is other-focused. Oppositely, personal distress is self-focused and is related to feelings of anxiety, discomfort, fearfulness, and uncertainty in emotionally taxing situations or when observing negative emotions in others, which subsequently induces the desire to avoid contact with the distressed individual, if possible.[168]

The neuroscience of episodic memory and autobiographical imagination has predominantly evolved independently of the neuroscience of social cognition (generally) and empathy (particularly).[169] Although initial findings from brain-damaged amnesic patients suggest that brain regions supporting memory and imagination may not be essential to the completion of some mentalizing tasks,[170] work from various clinical populations and neuroimaging studies demonstrate that brain systems supporting memory and imagination plausibly shape empathy.[171] Demonstrating a link between autobiographical memory and empathic intentionality, Michael Lombardo and colleagues discovered that differences in self-referential memory between autistic patients and healthy controls disappeared when measures of empathy were included as a covariate.[172] Conversely, independence across these mental processes also exists. For instance, older adults exhibit diminished capacities to remember the past and imagine the future,[173] yet demonstrate preserved levels of empathic intentionality across the lifespan.[174] These behavioral findings underscore a sophisticated relationship between processes, indicating that episodic memory and autobiographical imagination may contribute to, but are distinct from, the capacity for empathy. While fMRI studies have provided initial evidence of the neural overlap between memory, imagination, and social cognition in the default network (i.e., the neurocognitive unit recruited for remembering and imagining specific personal experiences as well as understanding and empathizing with others),[175] the neural relationship of memory and imagination with empathic intentionality in action remains unexamined. Nevertheless, recent neuroimaging studies are beginning to uncover commonalities.[176]

Investigating the neural conditions that facilitate prosocial thoughts and behavior, Carrie Masten and colleagues discovered that regions of the default network were more strongly activated when subjects viewed social exclusion (e.g., a person in need) during a ball-tossing game compared to when they viewed social inclusion.[177] Greater activity was observed for the medial PFC and precuneus when subjects viewed social exclusion compared to inclusion. Further, after controlling for trait empathy levels in an investigative mediation analysis, only activity in the medial PFC accurately predicted prosocial behavior (e.g., consoling the excluded player outside of the scanner).[178] Investigating the neural substrates of an enhanced empathic intentionality toward ingroup members relative to outgroup members, Vani Mathur and colleagues determined that, across individuals, the difference in medial PFC activity when observing someone in need (e.g., in a natural disaster) for ingroup versus outgroup members predicts state empathy and willingness to donate money and time to help ingroup members.[179] One possible explanation for the involvement of the medial PFC across the foregoing studies on empathy comes from its role in self-referential processing.[180] Psychological research has historically demonstrated that subjects who view a person in need as similar to themselves display heightened empathic concern for the similar-to-self person in need.[181] Insofar as neuroimaging studies illustrate the medial PFC as being preferentially activated when mentalizing about similar and psychologically-close others,[182] medial PFC activity may support empathic intentionality to the extent that it reflects a perceived self-overlap, as greater medial PFC activity likely indicates an increase in perceived self-other similarity.[183] The preceding studies (typically) interpret activity within the default network as a proxy of mentalizing processing. However, inasmuch as similar regions are activated under conditions of autobiographical simulation,[184] this activity plausibly represents constructing vivid scenarios rather than simulating thoughts and feelings or shared component processes, such as self-referential processing.[185]

### 3.3.4. Emotional Memory Specificity and Autobiographical Narrative Centrality

Emotional memory specificity is operationally understood as the recall of a unique set of emotionally salient events confined to a single episode within a twenty-four hour period.[186] Whether about pleasure or pain, specific emotional memories are textured and autobiographically evocative reconstructions of personal history. Difficulty in activating specific emotional memories could lead to maladaptive functioning insofar as it disables individual sufferers from benefiting from the autobiographical information available through vivid and detailed affective memories. The re-experiencing of vivid emotional memories provides cognitive information about the probability of a desired goal outcome while simultaneously reminding the individual of what the affective experience of the outcome would be like. In similar fashion, one can employ the affective quality of memory to maintain a positive mood or repair a negative one. To secure healthy functioning, the capacity for emotional memory specificity must also be accompanied by the capacity to connect the memory narrative to adaptive self-structures in the long-term self. That is, individuals will be effective at pursuing goals and navigating challenges in life to the extent that they generate informative self-evident connections through the employment of memories, scripts, and the life story to explain, reveal, or cause change in the self.[187]

At each level of narrative identity, researchers have found that the capacity for meaning-making in response to an emotional experience is (generally) predictive of psychological health, well-being, and growth.[188] The work of Tali Boritz and colleagues examines memory specificity as it occurs in therapy transcripts of thirty-four clients undergoing brief emotion-focused and client-centered therapy for depression. Their findings demonstrate that memory specificity significantly increased over the course of the therapy, suggesting that the overgeneral (i.e., nonspecific) memory bias in depression may be subject to change by the result of treatment.[189] J. Mark Williams and colleagues argue that the increased visual and experimental imagery of specific memory is more likely to evoke the original affective content of the remembered event, thereby providing a rich narrative context within which to explore salient emotions.[190] Similarly, Leslie Greenberg and Lynne Angus suggest that the disclosure of emotionally salient, specific memories is of critical import for effective emotion-focused therapy for depression to the extent that emotions can only be understood – and have personal meaning – when they are situated within a narrative framework that identifies what is felt and about whom in relation to a specific need or issue.[191] As such, the differentiation of emotional experiences in the context of salient personal stories may be a key intervention strategy for narrative change and beneficial outcomes for clients in psychotherapy. Nevertheless, few studies to date have explicitly examined the relationship between personal autobiography and emotional memory specificity in actual therapy sessions.[192]

Autobiographical narrative centrality is the (final) medium through which objective events are experienced, interpreted, and expressed subjectively. The psychological study of narrative centrality is equal parts extensive and diverse.[193] In studies of personality, life stories are the foundation of major theoretical approaches. Early clinical psychotherapeutic therapies were often based on understanding narratives told by patients, and this continues today in the form of therapies that seek to modify or reinterpret central autobiographical narratives. Reshaping autobiographical narratives of negative or stressful events (to have diminished interpretations and implications) is a common strategy for comforting individuals in non-clinical settings and for reducing symptoms in clinical settings.[194] In acute PTSD, autobiographical narrative centrality assumes an additional role to the extent that narrative is used to evaluate the criteria and symptoms necessary to diagnose PTSD. A second (and very different) behavioral approach to understanding reactions to stressful events is the measurement of individual differences, including personality tests. In contrast to the autobiographical narratives of events, personalities traits are seen as a relatively stable attribute of a person, which is expressed in a wide range of cognition, emotion, and behavior, rather than being focused on a singular event. The autobiographical narrative centrality of single events, on the one hand, and personality traits, on the other hand, are fundamentally different insofar as narratives of particular events cannot be changed without restructuring the master narrative of other events and behaviors that directly impinge on personality traits. Several types of therapies known to reduce PTSD symptoms focus on changing the central structure of autobiographical narratives of the traumatic event(s). In contrast, personality – particularly higher-order dimensions such as negative affectivity – is nonspecific and relatively stable (though not fixed) across time and situations.[195]

Measures of autobiographical narrative centrality and personality traits also differ in the time course of how they are affected by traumatic events.[196] As suggested above, personality is hypothesized to be generally stable in adults and should change less over time than the narrative interpretation of an event. Previous studies indicate that negative emotionality – often measured by neuroticism – is not affected by detrimental events, such as divorce or unemployment, [197] or more traumatic events encountered in middle adulthood.[198] In contrast, the time course of autobiographical narrative centrality is more dynamic. The interpretation of whether a negative event is central to identity, is resolved, or is related to positive growth cannot exist apart from individual interpretation. In other words, autobiographical narrative centrality is contingent upon not only information, attitudes, and skills obtained prior to the onset of trauma, but also upon the changes within the narrative's evolution both during and after trauma. In the study of David Rubin and colleagues, reducing the negative evaluation of traumatic memories provided closure for individual sufferers and made the memories less narratively central, thereby reducing the frequency with which the memories were recalled.[199] Hence, the frequent recall of a negative event, rather than the valence of the event's sequelae, may be of ultimate import and help to explain why the post-traumatic growth inventory correlates positively with the PTSD checklist.[200] The distinction between measures pertaining to the individual and measures pertaining to narratives about an event therefore appears to be extremely useful for approaching the understanding of emotional responses to stressful events, the clearest of which is PTSD. However, the dichotomy may be further useful if one views disorders (such as depression) as related to specific events or the accumulation thereof. Moreover, although effects have not been demonstrated in clinical samples, the foregoing results are substantial enough to be of practical significance in non-clinical populations.[201]

### 3.4. Autobiographical Memory and the Emotional Nature of Rational Ethical Decision Making

Sound ethical decisions flow forth from both emotional and rational recollections of events in history. Two fundamental methods mark the query of human memory. The first, concerning tasks related to recall, stimulates associations related to regenerations of past memorable events. The second, concerning tasks related to autobiographical memory judgment, stimulates – by way of third-party participation – carbon-copy recollections of past events, the product of which renders subjects able to answer pointed questions pertaining to the memory in question.[202] These questions may apply to efforts to grade and categorize events, or to broader efforts to compare events in autobiographical history by dissecting relevant dimensions. Because both methods implicate the frontal lobes and are bound by the hippocampus, it follows that the work of autobiographical memory-judgment cognition in the rational apprehension of emotive values is critical to substantive ethical decision making.[203]

### 3.4.1. Integrated Affective Foresight and Autobiographical Context Prediction

The mental simulation of (potential) future episodes provides adaptive value insofar as it supports planning, problem solving, and the execution of prospective intentions.[204] A particular advantage of episodic simulations is their capacity to communicate the affective qualities that future events might contain. This cognitive experience, in turn, can induce motivational incentives for farsighted decisions. Moreover, by encoding imagined scenarios into long-term autobiographical memory, individuals can use their prior simulations as lenses through which to view real experiences. The VMPFC plays a critical role in mediating simulations of future affective episodes and, thereby, in contributing to integrated affective foresight. The VMPFC is part of a network that is consistently engaged during both the construction of potential future scenarios and the recollection of past autobiographical events. It is thus part of a sophisticated system that provides episodic details and constructive processes by which the details for the simulation of potential scenarios are recombined. The VMPFC exhibits bidirectional anatomical connections with other nodes of this simulation network, such as the hippocampal formation, as well as with several structures implicated in the processing of affective autobiographical information. Portions of the VMPFC feature a higher spine density and number of dendritic spines per cell than comparable cortical areas, making it particularly suitable for the integration of affective inputs. On this basis, Roland Benoit and colleagues hypothesize that the VMPFC supports processes that integrate knowledge related to elements that constitute a potential future episode to simulate the emergent and overall affective quality thereof.[205]

The hypothesis of Benoit and colleagues is grounded in two lines of evidence that associate VMPFC functioning with both mnemonic processes and the computation of subjective value.[206] On the one hand, this region contributes to superior memory for episodes whose elements entail pre experimental associations. The VMPFC seems to support this mnemonic benefit of prior knowledge through interactions with posterior cortical regions that are likely implicated representations of individual elements. If the VMPFC augments new autobiographical memories by supporting the integration of prior affective knowledge, it presumably supports a similar function during episodic future simulations. On the other hand, there is considerable evidence for a contribution of the VMPFC to the computation of emotional and subjective value.[207] Activation in the medial PFC and adjacent anterior cingulate is greater during the simulation of positive (rather than negative) experiences, and it is paired with the anticipated reward magnitude of imagined experiences. The VMPFC may support such value representations by serving as a hub that links information about the simulated episodic details with associated affective responses. Based on these lines of research, Benoit and colleagues designed a procedure that examined BOLD signal changes during episodic simulation as a function of both (i) the degree of knowledge about the constituting events, and (ii) the anticipated affective quality of the events.[208] This procedure allowed them to confirm four critical predictions: (i) greater VMPFC engagement exists during the simulation of (more) familiar elements; (ii) familiarity-dependent coupling exists between (more) content-specific cortical regions and the VMPFC; (iii) a stronger familiarity-dependent coupling predicts a greater mnemonic benefit of familiarity; and (iv) there exists a concrete VMPFC representation of the emergent affective quality of future episodes.[209] Together, these functional properties provide an objective basis for ascribing a key role to the VMPFC in mediating adaptive benefits of episodic simulations. By supporting the anticipation and retention of future autobiographical episodes and associated affective states, this region has the potential to augment future-oriented decisions, even for scenarios that have yet to be encountered.[210]

An emerging view among neuroscientists is that the brain has evolved primarily to allow organisms to accurately predict the outcomes of events and behaviors.[211] In particular, it has been suggested that organisms have been able to adapt to environments and societies of increasing complexity because the brain has evolved further complex neural circuitry that support the capacity to make dynamic and conditional decisions and autobiographical predictions. These neural clusters developed to retain information over diverse time-scales depending on the desired goal. Different brain areas generate and retain sequences of information, and this ability is accounted for by state-dependent changes in network dynamics, internally generated oscillatory activity, and dedicated "time cells." Hence, many elemental features of autobiographical prediction analyses occur automatically so that individuals (naturally) seek control of the outcomes to their behaviors. While the specific mechanism of a self-generated – and thus auto-regulated – autobiographical prediction system is unknown, Sheri Mizumori suggests that such a mechanism may (to some extent) mirror principles of self-regulation at the synaptic and neural circuit levels.[212] Eve Marder and Jean-Marc Goaillard contend that groups of neurons or neural networks sense changes in firing collectively to regulate experience-dependent population activity levels and patterns of activation.[213] In this way, homeostatic plasticity enables groups of neural circuits to find a balance between flexible and stable processing as needed to learn from experiences and to be responsible to future autobiographical inputs. The details of how cell networks or their connections engage in homeostatic regulation remains to be discovered. Nevertheless, while specific homeostatic neural plasticity mechanisms have not been used to account for complex learning, current theories of reinforcement-based and context-based learning and autobiographical memory commonly rely on the autoregulation of feedback loops.[214]

If applied to the autoregulation of episodic context prediction analyses, a homeostatic framework has the potential to impact the regulation of future autobiographical memories and decisions.[215] Such a framework includes variables that are monitored by sensors and subsequently regulated by controllers, thereby involving multiple, interactive, and hierarchically organized information loops. At the cellular or synaptic levels, homeostatic plasticity mechanisms regulate cell excitability around a neural activity set point such that neurons retain maximal responsivity to future inputs. This process enables neurons to achieve a balance between synaptic stability and flexibility. Hence, a prediction error, or mismatch signal, may result in higher or lower firing rates, at which time controller mechanisms engage to bring firing rates back to set level points. Of particular interest are mechanisms by which cortical memory may impact the threshold for signaling autobiographical prediction errors. One cortical area of significance is the PFC given (i) its intrinsic recurrent circuitry and detailed excitatory and inhibitory extrinsic connections with both the hippocampal/temporal lobe and reward valuation systems, and (ii) its role in attention and working memory. In other words, the PFC orchestrates and coordinates the level of neural excitability in different prediction error brain areas according to homeostatic principles and, in this way, bias the nature of the outputs of connected brain areas according to autobiographical interpretation and recent decisional outcomes. The PFC also possesses strong functional connections with other cortical memory areas (e.g., parietal and temporal cortices, respectively), and is thus strategically positioned to influence long-term autobiographical memory updating based on prediction error analyses. Consequently, the most recent episodic memories can be fed forward to the hippocampus for future autobiographical context evaluation.[216]

### 3.4.2. The Affective Structure of Autobiographical Moral Thought

The somatic marker hypothesis proposes that autobiographical decision making is a process that depends on emotion and that both the amygdala and OFC are parts of a neural circuit vital for moral judgment and ethical decision making.[217] Although both structures pair exteroceptive sensory data with interoceptive data concerning somatic/emotional states, they do so at different levels, thus making distinctive contributions to the process. Clinical observations have revealed that subjects with bilateral damage of the OFC and amygdala exercise poor judgment and decision making in the social realm.[218] Antoine Bechara and colleagues demonstrate this autobiographical decision-making impairment using neuropsychological tests to measure decision making through a set of gambling tasks – a paradigm employed to simulate real-life situations in terms of uncertainty, reward, and punishment.[219] To accomplish this, Bechara and colleagues investigate the performance of normal control subjects with demographic characteristics matched to a group of subjects with bilateral damage to the VMPFC and a separate group of subjects with bilateral damage to the amygdala. Significant observations included that normal subjects avoided the disadvantageous gambling tasks and preferred the proportionate ones, whereas subjects with bilateral damage to the VMPFC and amygdala, respectively, did not avoid – and in fact overwhelmingly preferred – the disadvantageous tasks.[220] These results suggest that the autobiographical performance profile of subjects with bilateral damage to either the VMPFC or amygdala is comparable to their real-life incapacity to decide advantageously.[221]

Given the importance of the ventromedial cortex and amygdala in both autobiographical moral thought and somatic state activation, the question is raised over the different roles, if any, that are played between the ventromedial cortex and the amygdala in emotional processing, somatic state activation, and decision making.[222] Somatic states can be effected from primary and secondary inducers, respectively. Primary inducers are stimuli that are innate or learned to be pleasurable or aversive. Once they manifest in the immediate environment, they automatically (and obligatorily) elicit a somatic response. Examples of primary inducers include the encounter of a fear object or a stimulus predictive thereof. Secondary inducers are entities generated by the recall of an autobiographical or hypothetical emotional event, including moral thoughts and episodic memories about the primary inducer that, when brought to working memory, elicit a somatic state. Examples of secondary inducers include the emotional response elicited by the autobiographical memory of encountering a fear object, an unsavory event, or the imagination of achieving a desired event. Evidence suggests that the amygdala is a critical substrate in the neural system necessary for the processing of primary inducers.[223] It pairs the features of primary inducers, which can be processed subliminally (via the thalamus) or explicitly (via early sensory and high-order association cortices), with the somatic state associated with the inducer. This somatic state is evoked by effector structures such as the hypothalamus, basal forebrain, ventral striatum, periaqueductal gray, and other brain-stem nuclei. Evidence also suggests that the ventromedial cortex is a trigger structure for somatic states from secondary inducers.[224] Once somatic states from primary inducers are themselves induced, signals from these somatic states are relayed to the brain. Representations of these signals can remain at the brain stem level, or can reach the insular and posterior cingulate cortices and be perceived as a feeling.[225]

Episodic moral sensitivity is a form of autobiographical moral thought that refers to the detection and interpretation of an episodic moral issue or situation, including awareness of how various agents may be affected by an action taken in response to the issue in question.[226] This interpretive awareness is the first component of ethical decision making inasmuch as it gives rise to the need to make a moral judgment, select a moral action, and process the components of moral behavior. Given its theoretical distinction from deliberative moral reasoning, understanding the process of episodic moral sensitivity is important to the extent that (i) individuals vary dramatically in their ability to detect moral issues, and (ii) moral failure often results from a lack of moral sensitivity. Defining the neural information processing related to autobiographical moral thought may inform these theories and their debate and provide useful inferences with regard to the bases of immoral actions. The relationship between brain states and states of autobiographical moral thought and behavior has been informed by neurological case studies. Such studies indicate that autobiographical moral thought and behavior are mediated in large part by the PFC. In vivo fMRI studies have provided an intricate model of the neural representations of moral cognition, implicating a distributed cortical network involving diverse frontal and posterior cortical areas comprised of the anterior medial PFC and OFC, posterior STS, and limbic regions (e.g., amygdala). In their novel neuroimaging study, Diana Robertson and colleagues endeavor to isolate the neural correlates of one specific component of autobiographical moral thought, namely, the ability to recognize and interpret moral issues – and compare neural processing related to moral sensitivity to two different forms of moral content represented by issues of justice and care.[227]

Robertson and colleagues hypothesized that the social and personal affective implications related to the interpretive autobiographical detection of moral issues are associated with distinct neural information processing events when compared to the same task for non-moral problems encountered in everyday decisions.[228] To test this possibility, they developed narratives that depicted moral and non-moral issues encountered by a hypothetical individual in a typical work setting. The results of the study implicated three brain areas as critical for episodic moral sensitivity: the medial PFC, the posterior cingulate, and the posterior STS. Notwithstanding remarkable variability in stimuli and tasks between studies, the findings of Robertson and colleagues are strikingly similar to results of prior fMRI studies of implicit and explicit autobiographical moral judgments that revealed primary neural activations in medial cortical regions and the posterior STS.[229] Regarding the medial PFC, Robertson and colleagues identified that greater activation of the medial PFC was associated with greater sensitivity to moral issues versus either neutral or nonmoral issues. The medial PFC has been implicated in diverse brain functions, including imitation, emotional response monitoring, error detection, evaluative judgments, and pain perception, each of which is linked to self-monitoring and self-reflective moral behavior. Regarding the posterior cingulate cortex, Robertson and colleagues observed that sensitivity to moral issues was associated with activation of the posterior cingulate cortex located on the dorsal bank of the cingulate gyrus. The posterior cingulate cortex has been implicated in seemingly diverse functions, including the experience of emotion, recall of emotional and autobiographical memories, reward processing, forgivability, and self-reference. Finally, regarding the posterior STS, Robertson and colleagues identified activation in the posterior STS during effort related to assessing the intentions of other individuals or violations of moral expectations. Its demonstrated role in social perception and maintaining behavioral representations are consistent with a role of the posterior STS in social perspective formation in service of episodic moral sensitivity.[230]

### 3.4.3. The Primacy of Autobiographical Memory in Emotionally Rational Choice

Despite many advances in the field of autobiographical memory, one area that has heretofore received little systematic investigation is the relationship between memory processes, value objects, and emotionally rational choice. To remedy this underdevelopment, Gregory Jones and Maryanne Martin investigate how individuals make emotionally rational choices among value objects that are defined in terms of different forms of autobiographical influence.[231] Two phases in each of three distinct experiments mark the method of Jones and Martin. First, in the recall phase, participants recalled a number of value objects on the basis of criteria such as usability, and were subsequently probed about value object characteristics such as emotional significance. Second, in the choice phase, participants ordered all recalled items in terms of their personal value. Jones and Martin hypothesized that analyses of these responses would allow two linked questions to be addressed: first, whether objects selected according to different autobiographical criteria within the recall phase differ systematically in value within the choice phase; and second, which value object properties (across all recalled objects) predominate in determining emotionally rational choice. In the first experiment, participants recalled a set of four objects. For three of these sets, the object was the one most valued by the individual participant due to its appearance, its usability, or its linkage to autobiographical memory. In addition to these three restrictive criteria, an additional object was selected with an unrestricted remit so as to encompass the whole range of participants' most valued objects. In the second experiment, the specification of the memory criterion was narrowed and the usability and appearance criteria subsumed within unrestricted selection. Additionally, the examples of valued objects provided were based on the most frequent responses in the relevant classes of the first experiment. For the third and final experiment, a procedural omission was introduced (during the recall phase) of any object examples for the different selection criteria. The three selection criteria of appearance, usability, and autobiographical memory linkage were subsequently investigated, with participants recalling two objects for each (rather than one).[232]

The outcomes of two sets of primary analyses within each of the three different experiments described above provided remarkably consistent evidence that, when individuals choose their most valued objects, the criterion used is that of linkage to autobiographical memory.[233] Appearance, usability, the opinion of others, and economic value all proved less valuable than linkage to episodic memory. At least three significant insights can be gleaned from the experiments of Jones and Martin. First, given the backward temporal reference inherent to autobiographical memory, it is unsurprising that objects selected on the basis of memory had (on average) been possessed significantly longer than those selected on the basis of appearance or usability. However, it did not appear that an object assumed greater value simply by reaching a threshold age. Second, it is a logical possibility that the apparent influence of autobiographical memory linkage on emotionally rational choice is mediated through economic value (i.e., enhanced memories might be associated with particularly expensive objects, with emotionally rational choice determined primarily by cost). To the contrary, however, it was found that the economic salience of objects selected on the basis of their linkage to autobiographical memory was extremely low. Indeed, across all objects there was a significant negative relation between memory salience and economic salience. Finally, the results suggest that the primacy of autobiographical memory linkage in determining emotionally rational choice of object is widely distributed among diverse groups of individuals. Similar relations among value objects selected according to diverse criteria were observed for both genders, different nationalities, and across participant lifespan.[234]

Context memory retrieval tasks often implicate the left ventrolateral PFC (LVPFC) during functional imaging.[235] Although this region has been conventionally linked to controlled semantic processing of materials, it may also play a more general role in selecting among competing episodic representations during demanding retrieval in emotionally rational choice. Hence, the LVPFC response during context memory retrieval may reflect either semantic processing of memoranda or adjudication of interfering autobiographical memories evoked by memoranda. To distinguish between these hypotheses, Sanghoon Han and colleagues contrast context and item memory retrieval tasks for meaningful and non-meaningful memoranda using fMRI.[236] Increased activation during context memory compared to item memory occurred exclusively for meaningful memory probes. In contrast, even demanding context retrieval for non-meaningful materials failed to engage the LVPFC. These data points demonstrate that the activation previously seen during episodic tasks likely reflects semantic processing of the probes during episodic retrieval in emotionally rational choice, not selection among competing elicited autobiographical representations. Posterior middle temporal gyrus and the body/head of the caudate demonstrated a similar selective response as the LVPFC, although resting-state functional connectivity analyses suggested that these two regions likely share disparate functional relationships with the LVPFC.[237]

The study of Han and colleagues yield two important findings regarding the primacy of autobiographical memory in emotionally rational choice. First, insofar as context memory-linked LVPFC activation does not reflect selection processes operating directly on interfering autobiographical memory representations, non-meaningful materials should have similarly demonstrated a clear context superior to item response (granted that context retrieval was especially demanding with non-meaningful materials). However, the differential context greater than item memory was only observed using meaningful materials. This indicates that the activation is likely an indirect response to challenging autobiographical retrieval, or episodic interference, and is consistent with the hypothesis that subjects respond to episodic interference by more selectively processing the semantic characteristics of the probes when it is possible to do so. In the cognitive literature, this is often referred to as "semantic elaboration," and it is presumably necessary for vanguarding the subset of semantic features that would likely have been the principal focus of processing during the original encoding experience. Second, insofar as the LVPFC is functionally connected with two regions that appear to be independent from one another, namely, the caudate head/body and left posterior middle temporal gyrus – all three regions appear to display a pattern of mean amplitude, and appear to do so for different functional reasons. This finding corroborates previous work linking the ventrolateral PFC and middle temporal gyrus during controlled semantic retrieval. The LVPFC, then, in conjunction with the middle temporal gyrus, appears to support a semantic elaboration strategy performed on the probes in response to interfering autobiographical representations, and caudate findings further suggest this strategy as an instrumental behavior motivated by the expectation that it will facilitate successful episodic retrieval outcomes in pursuit of emotionally rational choice.[238]

### 3.4.4. Autobiographical Prospection and Rational Ethical Decision Making

While it is widely acknowledged that memory and decision making are closely related, the neural processes linking episodic memory and decision making have only recently been the targets of systematic study.[239] One emerging line of research that can illuminate the particular relationship between autobiographical memory and rational ethical decision making centers on the roles of two particular forms of episodic memory: autobiographical prospection and autobiographical counterfactual reflection. Autobiographical prospection involves the construction of possible future personal episodes or scenarios, whereas autobiographical counterfactual reflection involves simulating alternative versions of outcomes of past personal episodes that could have happened but did not occur. In the past several years, there has been a flurry of research concerning autobiographical prospection, motivated in large part by the observation that a common core brain network is involved in both episodic memory and episodic future thinking. Though there has been less research concerning autobiographical counterfactual reflection, several recent papers have explored the phenomenon and its relationship to autobiographical memory.[240] Despite the scant amount of relevant studies that have been reported, it seems plausible that autobiographical prospection and autobiographical counterfactual reflection both engage cognitive regions that are similarly recruited when individuals remember specific past experiences from their ordinary lives. Generally, the overlap of this core network with the default network is consistent with theoretical perspectives that have emphasized the role of this network in supporting various forms of mental simulations. More precisely, the joint recruitment of the core network is consistent with the proposal that it supports processes that can be generally applied to construct episodes, irrespective of their actual historical occurrence.[241]

   Several recent behavioral studies have compared the cognitive properties of remembered past events and imagined future events.[242] These studies have revealed numerous similarities between the two, such as parallel responses to experimental manipulations that increase the availability or vividness of autobiographical details and reductions in the episodic specificity of remembered and imagined events in a variety of populations, including older adults, schizophrenics, depressed individuals, patients with PTSD, and amnesic patients. Despite these similarities in the cognitive properties of autobiographical memory and autobiographical prospection, differences have also been documented. Specifically, several studies have demonstrated that remembered past events are subjectively experienced as richer in sensory detail than imagined future events or imagined events in general.[243] Similarly, studies that have employed objective methods for characterizing the amount of episodic detail that participants provide when remembering or imagining have indicated greater levels of episodic detail in remembered past events than imagined future events.[244] Felipe De Brigard and Kelly Giovanello recently compared both subjective properties and objective features of remembered autobiographical events with autobiographical counterfactual reflections and reported evidence that remembered events were experienced as clearer and more detailed, and objectively contained more episodic details than either counterfactual or prospective simulations.[245] However, while the phenomenological features of autobiographical prospection and autobiographical counterfactual reflection did not differ from one another (in most respects), De Brigard and Giovanello found that participants experienced lower emotional intensity during autobiographical counterfactual reflection relative to both autobiographical past and future simulation, regardless of the simulated event's corresponding valence.[246]

The process of ethical decision making is often shrouded in uncertainty. To hedge this ambivalence, individuals tend to strategize either by envisioning possible scenarios that might occur as a result of future choice, or by simulating alternative scenarios that might have occurred as a result of having chosen differently in the past.[247] Both strategies are highly dependent on autobiographical memory, and as such are prone to memory-related biases. However, given the number of differences between autobiographical prospection and autobiographical counterfactual reflection, it is unclear how these related simulation processes influence rational actions and behaviors. Nevertheless, episodic simulations may have a particular impact on rational ethical decisions that have long-term consequences insofar as they allow for the "pre-feeling" of what it may be like to experience a specific future situation. To the degree that this emotional state (of the anticipated episode) is possible, it may motivate farsighted rational choices that would make it more likely to actually experience the simulated future event. Simulating the future episode enables individuals to bridge the gap between the moment of decision making and the moment of reward delivery, and this allows for an immediate experience of the anticipated event's affective impact. The experienced emotional state may subsequently increase the rational valuation of the imagined moral reward, and thus effectively attenuate its discounting. Taken together, this suggests that processes mediated by the core network can be utilized to imagine the future consequences of one's moral actions. Given that similar constructive processes are likely employed by autobiographical counterfactual reflection, it is reasonable to conclude that considering alternative outcomes of past events has the potential to influence one's future rational ethical decisions.[248]

A typical consequence of upward autobiographical counterfactual reflection is the feeling of regret, and there is some intriguing evidence for the impact of regret on rational ethical decisions.[249] In their study of the relationship between the OFC and regret, Nathalie Camille and colleagues asked participants to make repeated choices between two risky economic gambles, and assessed the emotional reactions to the outcome.[250] Unsurprisingly, participants were happier when their choice resulted in gain rather than loss. However, their affective experience was not only determined by the outcome of their choice, but also by an objective comparison with the outcome of the alternative, foregone option. That is, the same nominal "win" induced happiness when the rejected gamble would have led to a loss, but it actually induced unhappiness when the alternative gamble would have led to a greater result. Thus, the comparison between what had been and what could have been triggered the emotion of regret. Camille and colleagues further modeled participants' choices and demonstrated that their rational calculations were not only influenced by the ethical values at stake in the two gambles, but also by the avoidance of anticipated autobiographical regret. Critically, this was not the case for a group of participants with lesions including the OFC. They neither reported regret nor did their choices reveal the disposition to avoid regret. In the long-term, healthy participants accumulated greater "wins," indicating that the experience of autobiographical regret and its subsequent avoidance biased rational ethical decisions toward farsighted choices. In a follow-up fMRI study, Giorgio Coricelli and colleagues associated greater regret with enhanced activation in regions including the medial PFC and hippocampus.[251] Though participants in the foregoing studies may not have simulated elaborate autobiographical counterfactual reflections, the data is consistent with the possibility that a mechanism may have supported the effect of regret on rational ethical decisions in similar fashion to the one shown to effectively attenuate discounting via autobiographical prospection. Both "prefeeling" a possible future scenario and reminiscing about foregone past choices thus provide motivational incentives that foster rational ethical decisions.[252]

### 3.5. Conclusion

As the first link in the book's logic chain, this chapter endeavored to underscore the moral significance of autobiographical memory for ethical decision-making and to thereby make the case against direct and intentional episodic disintegration via MM. It suggested, first, that insofar as episodic memories enable individuals to create and maintain personal identity, episodic memory integration serves as the primary database informing the self of historical events, present relationships, future obligations, and aggregate meanings. As such, this complex autobiographical network – moving from objective events to subjective interpretations – serves, finally, as the fundamental locus of moral judgment. Second, the chapter addressed the mutual dependence of reason and memory. More than simple recall, the rational process of episodic reflection and prospection abounds with thoughts, emotions, and evaluations about events in history, thereby providing explanatory frameworks replete with intentions and motivations In turn, these memories secure an enduring sense of a storied and, hence, responsible self.

Third, the chapter addressed the intimate correlation between emotion and memory. To be sure, episodic memories provide a sense of historical past, but they also provide the same social interaction variables as the personality quality of empathy. This sense of personal meaning manifests from emotions, motivations, and desires that are formed through interaction with others in the world. Finally, the chapter addressed the interplay between memory, reason, emotion, and ethical decision making. Episodic recollections have a particular impact on rational ethical decisions insofar as they allow for the "pre-feeling" of specific future situations. The experienced emotional state subsequently increases the rational valuation of the imagined moral reward, thereby attenuating its discounting through a process by which future consequences of present actions are projected against past moral successes and failures.

Against this backdrop, at least three conclusions can be drawn. First, episodic memory serves as the fulcrum for moral judgments, bridging objective rationality with subjective values. The practice of MM effectively burns this bridge, annihilating the very means by which one is able to morally reflect, judge, and decide. Second, insofar as rational deliberation requires an episodic database against which to reflect, rationality cannot be fully (or usefully) separated from episodic memory. Hence, when episodic memory is erased through MM, so too is one's capacity for rationality. Finally, episodic memory has a function beyond mere historical data storage; it also serves to make life emotionally intelligible and, as such, "sensible." Since moral values are first apprehended by the emotions, the disintegration of episodic memory through MM also precludes the ability to make meaning out of what remains. Continuing in this vein of thought, the book now turns to critically examine the immediate consequence of disintegrating episodic memory: the degeneration of emotional rationality.

### 3.6. Notes

1. This is exemplified, for instance, when a court witness claims to "remember seeing Jones at the murder scene." See Gordon R. Bower, "A Brief History of Memory Research," in *The Oxford Handbook of Memory*, edited by Endel Tulving and Fergus I. M. Craik (New York: Oxford University Press, 2000), 3-32.

2. John Locke, John Stuart Mill, and Thomas Brown, for instance. See Bower, "A Brief History," 3-32; see especially pp. 3-4.

3. Henri Bergson, Bertrand Russell, and Endel Tulving, for instance. See Bower, "A Brief History," 3-32; see especially p. 22.

4. Bower, "A Brief History," 3-32; see especially pp. 26-27.

5. Francesca Gino and Sreedhari D. Desai, "Memory Lane and Morality: "How Childhood Memories Promote Prosocial Behavior," *Journal of Personality and Social Psychology* 102, no. 4 (April 2012): 743-58; see especially p. 743.

6. Frederick L. Philippe, Richard Koestner, Geneviève Beaulieu-Pelletier, and Serge Lecours, "The Role of Need Satisfaction as a Distinct and Basic Psychological Component of Autobiographical Memories: A Look at Well-Being," *Journal of Personality* 79, no. 5 (October 2011): 905-38; see especially pp. 909-10.

7. Philippe et al., "The Role of Need Satisfaction," 905-38; see especially p. 906.

8. Frederick L. Philippe, Richard Koestner, Serge Lecours, Geneviève Beaulieu-Pelletier, and Katy Bois, "The Role of Autobiographical Memory Networks in the Experience of Negative Emotions: How Our Remembered Past Elicits Our Current Feelings," *Emotion* 11, no. 6 (2011): 1279-90; see especially p. 1280.

9. Lars Nyberg and Roberto Cabeza, "Brain Imaging of Memory," in *The Oxford Handbook of Memory*, ed. Endel Tulving and Fergus I. M. Craik (New York: Oxford University Press, 2000), 501-519; see especially p. 510.

10. Nyberg and Cabeza, "Brain Imaging of Memory," 501-19; see especially p. 510.

11. Mark A. Wheeler, "Episodic Memory and Autonoetic Awareness," in *The Oxford Handbook of Memory*, ed. Endel Tulving and Fergus I. M. Craik (New York: Oxford University Press, 2000), 597-608; see especially p. 597.

12. James, *Principles of Psychology*, vols. 1-2; see especially vol. 2, p. 612.

13. Wheeler, "Episodic Memory," 597-608; see especially p. 598.

14. Wheeler, "Episodic Memory," 597-608; see especially pp. 597-98.

15. Peggy St. Jacques, David C. Rubin, Kevin S. LaBar, and Roberto Cabeza, "The Short and Long of It: Neural Correlates of Temporal-order Memory for Autobiographical Events," *Journal of Cognitive Neuroscience* 20, no. 7 (2008): 1327-41; see especially p. 1327.

16. St. Jacques et al., "The Short and Long," 1327-1341; see especially p. 1327.

17. St. Jacques et al., "The Short and Long," 1327-1341; see especially pp. 1327-28.

18. John M. Gardiner and Alan Richardson-Klavehn, "Remembering and Knowing," in *The Oxford Handbook of Memory*, ed. Endel Tulving and Fergus I. M. Craik (New York: Oxford University Press, 2000), 229-44; see especially p. 229.

19. See Endel Tulving, "Memory and Consciousness," *Canadian Psychology* 26 (1985): 1-12.

20. Gardiner and Richardson-Klavehn, "Remembering and Knowing," 229-44; see especially pp. 229-30.

21. Gardiner and Richardson-Klavehn, "Remembering and Knowing," 229-44; see especially p. 230.

22. Gardiner and Richardson-Klavehn, "Remembering and Knowing," 229-44; see especially p. 230.

23. Gardiner and Richardson-Klavehn, "Remembering and Knowing," 229-44; see especially p. 238.

24. See Douglas L. Hintzman, "Memory Judgments," in *The Oxford Handbook of Memory*, ed. Endel Tulving and Fergus I. M. Craik (New York: Oxford University Press, 2000), 165-77; see especially p. 172.

25. Janina A. Hoffmann, Bettina von Helversen, and Jörg Rieskamp, "Pillars of Judgment: How Memory Abilities Affect Performance in Rule-Based and Exemplar-Based Judgments," *Journal of Experimental Psychology: General* 143, no. 6 (2014): 2242-61; see especially p. 2243.

26. Hoffmann et al., "Pillars of Judgment," 2242-61; see especially p. 2243.

27. Hoffmann et al., "Pillars of Judgment," 2242-61; see especially pp. 2243-44.

28. Each cortical system – from those controlling data acquisition and analysis to semantic. episodic, and working memory – is defined by the functional contribution it makes to the whole. See Nyberg and Cabeza, "Brain Imaging of Memory," 501-19.

29. Some imaging has shown distinct engagement in regions within the prefrontal cortex for memory operation. See Nyberg and Cabeza, "Brain Imaging of Memory," 501-19; see especially p. 512.

30. Nyberg and Cabeza, "Brain Imaging of Memory," 501-19; see especially pp. 512-14.

31. Arnaud D'Argembeau, Helena Cassol, Christophe Phillips, Evelyne Balteau, Eric Salmon, and Martial Van der Linden, "Brain Creating Stories of Selves: The Neural Basis of Autobiographical Reasoning," *Social Cognitive and Affective Neuroscience* 9 (2014): 646-52; see especially p. 646.

32. See D'Argembeau et al., "Brain Creating Stories," 646-52.

33. Interestingly, the VMPFC – an area responsible for generating personal and affective meaning – was not consistently engaged across participants. See D'Argembeau et al., "Brain Creating Stories," 646-52; see especially p. 646.

34. D'Argembeau et al., "Brain Creating Stories," 646-52; see especially p. 646.

35. Pernille Hemmer and Kimele Persaud, "Interaction Between Categorical Knowledge and Episodic Memory Across Domains," *Frontiers in Psychology* 5 (June 2014): 1-5; see especially p. 1.

36. See Francis E. Bartlett, *Remembering: A Study in Experimental and Social Psychology* (Cambridge, UK: Cambridge University Press, 1932).

37. See Hemmer and Persaud, "Interaction Between Categorical Knowledge," 1-5; see especially p. 3.

38. Katherine Nelson and Robyn Fivush, "Socialization of Memory," in *The Oxford Handbook of Memory*, ed. Endel Tulving and Fergus I. M. Craik (New York: Oxford University Press, 2000), 283-95; see especially p. 283.

39. See David C. Rubin, *Memory in Oral Traditions* (New York: Oxford University Press, 1995).

40. See Mary J. Carruthers, *The Book of Memory: A Study of Memory in Medieval Culture* (Cambridge, UK: Cambridge University Press, 1990).

41. Nelson and Fivush, "Socialization of Memory," 283-95; see especially pp. 283-84.

42. Nelson and Fivush, "Socialization of Memory," 283-95; see especially p. 286.

43. For the most comprehensive of these studies, see Elaine Reese, Catherine A. Haden, and Robyn Fivush, "Mother-Child Conversations About the Past: Relationships of Style and Memory Over Time," *Cognitive Development* 8 (1993): 403-30.

44. Indeed, as Mary Mullen notes, the average age of earliest memory for many individuals of Asian descent is five or six years. See Mary K. Mullen, "Earliest Recollections of Childhood: A Demographic Analysis," *Cognition* 52 (1994): 55-79.

45. Nelson and Fivush, "Socialization of Memory," 283-95; see especially pp. 287-88.

46. Robyn Fivush, Tilmann Habermas, Theodore E. A. Waters, and Widaad Zaman, "The Making of Autobiographical Memory: Intersections of Culture, Narratives and Identity," *International Journal of Psychology* 46, no. 5 (2011): 321-45; see especially p. 322.

47. See Katherine Nelson, "Language and the Self: From the 'Experiencing I' to the 'Continuing Me,'" in *The Self in Time: Developmental Perspectives*, ed. Chris Moore and Karen Lemmon (Mahwah, NJ: Lawrence Erlbaum Associates, Inc., Publishers, 2001), 15-33.

48. Fivush et al., "The Making of Autobiographical Memory," 321-45; see especially pp. 322-23.

49. Fivush et al., "The Making of Autobiographical Memory," 321-45; see especially p. 328.

50. Fivush et al., "The Making of Autobiographical Memory," 321-45; see especially p. 328.

51. Fivush et al., "The Making of Autobiographical Memory," 321-45; see especially p. 329.

52. Fivush et al., "The Making of Autobiographical Memory," 321-45; see especially p. 329.

53. Fivush et al., "The Making of Autobiographical Memory," 321-45; see especially p. 334.

54. The John Wayne (JW) master narrative refers to the narrator's position of courage and stoic resolve during disproportionately negative experiences and expressing little or no negative emotion. The Florence Nightingale (FN) master narrative expresses negative emotions as a result of traumatic events; however, these emotions are not immediately linked to or followed by concern for others. Finally, the Vulnerability (V) master narrative expresses intensely negative emotions and feelings of helplessness as a consequence of the negative event. See Avril Thorne and Kate C. McLean, "Telling Traumatic Events in Adolescence: A Study of Master Narrative Positioning," in *Connecting Culture and Memory: The Development of an Autobiographical Self*, ed. Robyn Fivush and Catherine A. Haden (Mahwah, NJ: Lawrence Erlbaum Associates, 2003), 169-85.

55. Thorne and McLean, "Telling Traumatic Events," 169-85.

56. See Dan P. McAdams, Jeffrey Reynolds, Martha Lewis, Allison H. Patten, and Phillip J. Bowman, "When Bad Things Turn Good and Good Things Turn Bad: Sequences of Redemption and Contamination in Life Narrative and their Rela-

tion to Psychosocial Adaptation in Midlife Adults and in Students," *Personality and Social Psychology Bulletin* 27, no. 4 (April 2001): 472-83.

57. Fivush et al., "The Making of Autobiographical Memory," 321-45; see especially pp. 334-35.

58. See Hermann Ebbinghaus, Urmanuskript "Über das Gedächtnis" (Passau, Germany: Passavia Universitätsverlag, 1983).

59. Tilmann Habermas, "Autobiographical Reasoning: Arguing and Narrating from a Biographical Perspective," *New Directions for Child and Adolescent Development* 131 (Spring 2011): 1-17; see especially p. 1.

60. See Wilhelm Dilthey, "Über Eine Beschreibende und Zergliedernde Psychologie," in *Gesammelte Schriften*, ed. Wilhelm Dilthey, vol. 5 of *Die geistige Welt: Einleitung in die Philosophie des Lebens*, ed. Wilhelm Dilthey (Stuttgart, Göttingen: Teubner, Vandenhoeck & Ruprecht, 1895), 139-240.

61. See Tilmann Habermas and Susan Bluck, "Getting a Life: The Emergence of the Life Story in Adolescence," *Psychological Bulletin* 126, no. 5 (September 2000): 748-69.

62. Habermas, "Autobiographical Reasoning," 1-17; see especially pp. 2-3.

63. William L. Dunlop and Jessica L. Tracy, "The Autobiography of Addiction: Autobiographical Reasoning and Psychological Judgment in Abstinent Alcoholics," *Memory* 21, no 1 (2013): 64-78; see especially p. 65.

64. Dunlop and Tracy, "The Autobiography of Addiction," 64-78; see especially p. 65.

65. See Ursula M. Staudinger, "Life Reflection: A Social-Cognitive Analysis of Life Review," *Review of General Psychology* 5, no. 2 (June 2001): 148-60.

66. Dunlop and Tracy, "The Autobiography of Addiction," 64-78; see especially pp. 65-66.

67. Dunlop and Tracy, "The Autobiography of Addiction," 64-78; see especially p. 66.

68. See, for instance, M. Pasupathi, E. Mansour, and J. R. Brubaker, "Developing a Life Story: Constructing Relations Between Self and Experience in Autobiographical Narratives," *Human Development* 50 (2007): 85-110.

69. See Pasupathi et al., "Developing a Life Story," 85-110.

70. Dunlop and Tracy, "The Autobiography of Addiction," 64-78; see especially p. 66.

71. Kate C. McLean and Cade D. Mansfield, "To Reason or Not to Reason: Is Autobiographical Reasoning Always Beneficial?" *New Directions for Child and Adolescent Development* 131 (Spring 2011): 85-97; see especially p. 86.

72. See, for example, Amy Bird and Elaine Reese, "Emotional Reminiscing and the Development of an Autobiographical Self," *Developmental Psychology* 42, no. 4 (July 2006): 613-26.

73. See Sonja Lyubomirsky, Lorie Sousa, and Rene Dickerhoof, "The Costs and Benefits of Writing, Talking, and Thinking About Life's Triumphs and Defeats," *Journal of Personality and Social Psychology* 90, no. 4 (2006): 692-708.

74. See Roy F. Baumeister, Arlene Stillwell, and Sara R. Wotman, "Victim and Perpetrator Accounts of Interpersonal Conflict: Autobiographical Narratives

About Anger," *Journal of Personality and Social Psychology* 59, no. 5 (November 1990): 994-1005.

75. See Robyn Fivush and Jessica McDermott Sales, "Coping, Attachment, and Mother-Child Narratives of Stressful Events," *Merrill-Palmer Quarterly* 52, no. 1 (January 2006): 125-50.

76. McLean and Mansfield, "To Reason or Not to Reason," 85-97; see especially pp. 86-87.

77. Janet Metcalfe, "Metamemory," in *The Oxford Handbook of Memory*, ed. Endel Tulving and Fergus I. M. Craik (New York: Oxford University Press, 2000), 197-211; see especially p. 197.

78. See René Descartes, *The Philosophical Work of Descartes*, vols. 1-2, trans. Elizabeth S. Haldane and G. R. T. Ross (Cambridge, UK: Cambridge University Press, 1911).

79. See James, *Principles of Psychology*, vols. 1-2; see especially vol. 1, pp. 243-44.

80. Metcalfe, "Metamemory," 197-211; see especially pp. 197-98.

81. See J. T. Hart, "Memory and the Feeling-of-Knowing Experience," *Journal of Educational Psychology* 56, no. 4 (August 1965): 208-16.

82. John H. Flavell and Henry M. Wellman, "Metamemory," in *Perspectives on the Development of Memory and Cognition*, ed. Robert V. Kail and John W. Hagen (Hillsdale, NJ: Lawrence Erlbaum Associates, Inc., Publishers, 1977), 3-34.

83. Agnieszka Niedźwieńska, "Metamemory Knowledge and the Accuracy of Flashbulb Memories," *Memory* 12, no. 5 (2004): 603-13; see especially p. 603.

84. Alan Scoboria, Jennifer M. Talarico, and Lisa Pascal, "Metamemory Appraisals in Autobiographical Event Recall," *Cognition* 136 (2015): 337-49; see especially p. 337.

85. See, for instance, Giuliana Mazzoni, Alan Scoboria, and Lucy Harvey, "Nonbelieved Memories," *Psychological Science* 21 (August 2010): 1334-40.

86. See, for instance, Daniel M. Bernstein, Nicole L. M. Pernat, and Elizabeth F. Loftus, "The False Memory Diet: False Memories Alter Food Preferences," in *Handbook of Behavior, Food and Nutrition*, ed. Victor R. Preedy, Ronald Ross Watson, and Colin R. Martin (New York: Springer, 2011), 1645-63.

87. See, for instance, Giuliana Mazzoni, Andrew Clark, and Robert A. Nash, "Disowned Recollections: Denying True Experiences Undermines Belief in Occurrence But Not Judgments of Remembering," *Acta Psychologica* 145 (January 2014): 139-46.

88. Scoboria et al., "Metamemory Appraisals," 337-49; see especially pp. 338-39.

89. Magda Osman, "What are the Essential Cognitive Requirements for Prospection (Thinking About the Future)?" *Frontiers in Psychology* 5 (June 2014): 1-4; see especially p. 1.

90. See, for instance, Frans B. M. de Waal and Pier Francesco Ferrari, "Towards a Bottom-Up Perspective on Animal and Human Cognition," *Trends in Cognitive Sciences* 14, no. 5 (May 2010): 201-07.

91. See Donna Rose Addis, Alana T. Wong, and Daniel L. Schacter, "Remembering the Past and Imagining the Future: Common and Distinct Neural Substrates

During Event Construction and Elaboration," *Neuropsychologia* 45, no. 7 (April 2007): 1363-77.

92. See Stanley B. Klein, Judith Loftus, and John F. Kihlstrom, "Memory and Temporal Experience: The Effects of Episodic Memory Loss On An Amnesic Patient's Ability to Remember the Past and Imagine the Future," *Social Cognition* 20, no. 5 (2002): 353-79.

93. Osman, "What are the Essential Cognitive," 1-4; see especially p. 2.

94. Osman, "What are the Essential Cognitive," 1-4; see especially p. 2.

95. Osman, "What are the Essential Cognitive," 1-4; see especially p. 3.

96. Osman, "What are the Essential Cognitive," 1-4; see especially p. 3.

97. This system is known by neuroscientists as the "Papez circuit." See Hans J. Markowitsch, "Neuroanatomy of Memory," in *The Oxford Handbook of Memory*, ed. Endel Tulving and Fergus I. M. Craik (New York: Oxford University Press, 2000), 465-84; see especially p. 469.

98. These sets of mechanistic systems belong to the basolateral limbic circuit, which includes the mediodorsal nucleus. See Markowitsch, "Neuroanatomy of Memory," 465-84; see especially p. 469.

99. Markowitsch, "Neuroanatomy of Memory," 465-84; see especially pp. 469-70.

100. Andrew P. Yonelinas and Maureen Ritchey, "The Slow Forgetting of Emotional Episodic Memories: An Emotional Binding Account," *Trends in Cognitive Neuroscience* 19, no. 5 (May 2015): 259-67; see especially p. 259.

101. Yonelinas and Ritchey, "The Slow Forgetting," 259-67; see especially p. 260.

102. See, for instance, Kevin S. LaBar and Roberto Cabeza, "Cognitive Neuroscience of Emotional Memory," *Nature Reviews Neuroscience* 7, No. 1 (January 2006): 54-64.

103. See, for instance, Tali Sharot and Andrew P. Yonelinas, "Differential Time-Dependent Effects of Emotion on Recollective Experience and Memory for Contextual Information," *Cognition* 106 (2008): 538-47.

104. See Sharot and Yonelinas, "Differential Time-Dependent Effects," 538-47.

105. See Sharot and Yonelinas, "Differential Time-Dependent Effects," 538-47.

106. Yonelinas and Ritchey, "The Slow Forgetting," 259-67; see especially p. 260.

107. See, for instance, Tali Sharot, Mieke Verfaellie, Andrew P. Yonelinas, "How Emotion Strengthens the Recollective Experience: A Time-Dependent Hippocampal Process," *PLoS ONE* 2, no. 10 (2007): e1068.

108. Yonelinas and Ritchey, "The Slow Forgetting," 259-67; see especially p. 260.

109. See, for instance, Andrew M. McCullough and Andrew P. Yonelinas, "Cold-Pressor Stress After Learning Enhances Familiarity-Based Recognition Memory in Men," *Neurobiology of Learning and Memory* 106 (November 2013): 11-17.

110. See Sharot and Yonelinas, "Differential Time-Dependent Effects," 538-47.

111. Yonelinas and Ritchey, "The Slow Forgetting," 259-67; see especially p. 261.

112. See, for instance, Elizabeth A. Kensinger, "Negative Emotion Enhances Memory Accuracy: Behavioral and Neuroimaging Evidence," *Current Directions in Psychological Science* 16, no. 4 (2007): 213-18.

113. Yonelinas and Ritchey, "The Slow Forgetting," 259-67; see especially p. 261.

114. Yonelinas and Ritchey, "The Slow Forgetting," 259-67; see especially p. 261.

115. See Larry Cahill, Ralf Babinsky, Hans J. Markowitsch, and James L. McGaugh, "The Amygdala and Emotional Memory," *Nature* 377 (September 1995): 295-96.

116. See Elizabeth A. Phelps, Kevin S. LaBar, Adam K. Anderson, Kevin J. O'Connor, Robert K. Fulbright, and Dennis D. Spencer, "Specifying the Contributions of the Human Amygdala to Emotional Memory: A Case Study," *Neurocase* 4, no. 6 (1998): 527-40.

117. Yonelinas and Ritchey, "The Slow Forgetting," 259-67; see especially p. 261.

118. Yonelinas and Ritchey, "The Slow Forgetting," 259-67; see especially p. 261.

119. See, for instance, Alexander Todorov and Ingrid R. Olson, "Robust Learning of Affective Trait Associations with Faces When the Hippocampus is Damaged, But Not When the Amygdala and Temporal Pole are Damaged," *Social Cognitive and Affective Neuroscience* 3, no. 3 (September 2008): 195-203.

120. See, for instance, Sharot et al., "How Emotion Strengthens," e1068.

121. See, for instance, Elizabeth A. Kensinger and Suzanne Corkin, "Two Routes to Emotional Memory: Distinct Neural Processes for Valence and Arousal," *Proceedings of the National Academy of Sciences of the United States of America* 101, no. 9 (March 2004): 3310-15.

122. See, for instance, Florin Dolcos, Kevin S. LaBar, and Roberto Cabeza, "Interaction Between the Amygdala and the Medial Temporal Lobe Memory System Predicts Better Memory for Emotional Events," *Neuron* 42 (June 2004): 855-63.

123. See, for instance, Florin Dolcos, Kevin S. LaBar, and Roberto Cabeza, "Remembering One Year Later: Role of the Amygdala and the Medial Temporal Lobe Memory System in Retrieving Emotional Memories," *Proceedings of the National Academy of Sciences of the United States of America* 102, no. 7 (February 2005): 2626-31.

124. See, for instance, Duclos et al., "Interaction Between the Amygdala," 855-63.

125. See, for instance, Maureen Ritchey, Florin Dolcos, and Roberto Cabeza, "Role of Amygdala Connectivity in the Persistence of Emotional Memories Over Time: An Event-Related fMRI Investigation," *Cerebral Cortex* 18 (November 2008): 2494-2504.

126. See, for instance, Lisa Kilpatrick and Larry Cahill, "Amygdala Modulation of Parahippocampal and Frontal Regions During Emotionally Influenced Memory Storage," *NeuroImage* 20, no. 4 (December 2003): 2091-99.

127. Yonelinas and Ritchey, "The Slow Forgetting," 259-67; see especially pp. 261-62.

128. Joyce Graciela Martínez-Galindo and Selene Cansino, "Positive and Negative Emotional Contexts Unevenly Predict Episodic Memory," *Behavioural Brain Research* 291 (2015): 89-102; see especially p. 89.

129. See, for instance, Alexandre Schaefer, Claire L. Pottage, and Adam J. Rickart, "Electrophysiological Correlates of Remembering Emotional Pictures," *NeuroImage* 54, no. 1 (January 2011): 714-24.

130. See, for instance, Manuel G. Calvo and David Beltrán, "Recognition Advantage of Happy Faces: Tracing the Neurocognitive Processes," *Neuropsychologia* 51, no. 11 (September 2013): 2051-61.

131. See, for instance, Jan W. Van Strien, Sandra J. E. Langeslag, Nadja J. Strekalova, Liselotte Gootjes, and Ingmar H. A. Franken, "Valence Interacts with the Early ERP Old/New Effect and Arousal with the Sustained ERP Old/New Effect for Affective Pictures," *Brain Research* 1251 (2009): 223-35.

132. See, for instance, Elizabeth J. Maratos and Michael D. Rugg, "Electrophysiological Correlates of the Retrieval of Emotional and Non-Emotional Context," *Journal of Cognitive Neuroscience* 13, no. 7 (October 2001): 877-91.

133. See Susanne Erk, Sonja Martin, and Henrik Walter, "Emotional Context During Encoding of Neutral Items Modulates Brain Activation Not Only During Encoding But Also During Recognition," *NeuroImage* 26, no. 3 (July 2005): 829-38.

134. Martínez-Galindo and Cansino, "Positive and Negative Emotional Contexts," 89-102; see especially pp. 89-90.

135. See, for instance, Susanne Erk, Markus Kiefer, Jo Grothe, Arthur P. Wunderlich, Manfred Spitzer, and Henrik Walter, "Emotional Context Modulates Subsequent Memory Effect," *NeuroImage* 18, no. 2 (February 2003): 439-47.

136. Martínez-Galindo and Cansino, "Positive and Negative Emotional Contexts," 89-102; see especially p. 90.

137. See Erk et al., "Emotional Context Modulates," 439-47.

138. See Markus Kiefer, Stefanie Schuch, Wolfram Schenck, and Klaus Fiedler, "Emotion and Memory: Event-Related Potential Indices Predictive for Subsequent Successful Memory Depend on the Emotional Mood State," *Advances in Cognitive Psychology* 3, no. 3 (2007): 363-73.

139. Martínez-Galindo and Cansino, "Positive and Negative Emotional Contexts," 89-102; see especially p. 90.

140. Jaclyn Hennessey Ford, Donna Rose Addis, and Kelly S. Giovanello, "Differential Effects of Arousal in Positive and Negative Autobiographical Memories," *Memory* 20, no. 7 (2012): 771-78l; see especially p. 772.

141. See, for instance, Gerald L. Clore and Justin Storbeck, "Affect as Information about Liking, Efficacy, and Importance," in *Hearts and Minds: Affective Influences on Social Cognition and Behavior*, ed. Joseph Forgas (New York: Psychology Press, 2006), 123-42.

142. See, for instance, Shelley E. Taylor and Jonathon D. Brown, "Illusion and Well-Being: A Social Psychological Perspective on Mental Health," *Psychological Bulletin* 103, no. 2 (1988): 193-210.

143. See Anne S. Rasmussen and Dorthe Bersten, "Emotional Valence and the Functions of Autobiographical Memories: Positive and Negative Memories Serve Different Functions," *Memory & Cognition* 37, no. 4 (June 2009): 477-92.

144. See, for instance, Jennifer M. Talarico, Dorthe Berntsen, and David C. Rubin, "Positive Emotions Enhance Recall of Peripheral Details," *Cognition and Emotion* 23, no. 2 (February 2009): 380-98.

145. See Susan Bluck and Karen Z. H. Li, "Predicting Memory Completeness and Accuracy: Emotion and Exposure in Repeated Autobiographical Recall," *Applied Cognitive Psychology* 15, no. 2 (March/April 2001): 145-58.

146. See Elizabeth A. Kensinger and Daniel L. Schacter, "When the Red Sox Shocked the Yankees: Comparing Negative and Positive Memories," *Psychonomic Bulletin & Review* 13, no. 5 (2006): 757-63.

147. Ford et al., "Differential Effects of Arousal," 771-78l; see especially p. 772.

148. See Clore and Storbeck, "Affect as Information," 123-42.

149. See Alisha C. Holland and Elizabeth A. Kensinger, "Emotion and Autobiographical Memory," *Physics of Life Reviews* 7, no. 1 (March 2010): 88-131.

150. See, for instance, Shannon Tromp, Mary P. Koss, Aurelio Jose Figueredo, and Melinda Tharan, "Are Rape Memories Different? A Comparison of Rape, Other Unpleasant, and Pleasant Memories Among Employed Women," *Journal of Traumatic Stress* 8, no. 4 (October 1995): 607-27.

151. See, for instance, J. Mark G. Williams, Thorsten Barnhofer, Catherine Crane, Dirk Hermans, Filip Raes, Ed Watkins, and Tim Dalgleish, "Autobiographical Memory Specificity and Emotional Disorder," *Psychological Bulletin* 133, no. 1 (January 2007): 122-48.

152. See Cecilia Au Yeung, Tim Dalgleish, Ann-Marie Golden, and Patricia Schartau, "Reduced Specificity of Autobiographical Memories Following a Negative Mood Induction," *Behaviour Research and Therapy* 44, no. 10 (November 2006): 1481-90.

153. Ford et al., "Differential Effects of Arousal," 771-78l; see especially p. 772-73.

154. Franca Tani, Carole Peterson, and Andrea Smorti, "Empathy and Autobiographical Memory: Are They Linked?" *The Journal of Genetic Psychology* 175, no. 3 (2014): 252-69; see especially p. 252.

155. See, for instance, Katherine Nelson and Robyn Fivush, "The Emergence of Autobiographical Memory: A Social Cultural Developmental Theory," *Psychological Review* 111, no. 2 (2004): 486-511.

156. See, for instance, Robyn Fivush, Catherine A. Haden, and Elaine Reese, "Elaborating on Elaborations: Role of Maternal Reminiscing Style in Cognitive and Socioemotional Development," *Child Development* 77, no. 6 (November/December 2006): 1568-88.

157. See, for instance, Allyssa McCabe, Carole Peterson, and Dianne M. Connors, "Attachment Security and Narrative Elaboration," *International Journal of Behavioral Development* 30, no. 5 (2006): 8-19.

158. See, for instance, Deborah J. Laible, Gustavo Carlo, and Scott C. Roesch, "Pathways to Self-Esteem in Late Adolescence: The Role of Parent and Peer Attachment, Empathy, and Social Behaviors," *Journal of Adolescence* 27, no. 6 (December 2004): 703-16.

159. See, for instance, Maayan Davidov and Joan E. Grusec, "Untangling the Links of Parental Responsiveness to Distress and Warmth to Child Outcomes," *Child Development* 77, no. 1 (January/February 2006): 44-58.

160. See, for instance, Laible et al., "Pathways to Self-Esteem," 703-16.

161. See, for instance, Qing Zhou, Nancy Eisenberg, Sandra H. Losoya, Richard A. Fabes, Mark Reiser, Ivanna K. Guthrie, Bridget C. Murphy, Amanda J. Cumberland and Stephanie A. Shepard, "The Relations of Parental Warmth and Positive Expressiveness to Children's Empathy-Related Responding and Social

Functioning: A Longitudinal Study," *Child Development* 73, no. 3 (May/June 2002): 893-915.

162. See Richard Jackson Harris, Steven J. Hoekstra, Christina L. Scott, Fred W. Sanborn, Laura A. Dodds, and Jason Dean Brandenburg, "Autobiographical Memories for Seeing Romantic Movies on Date: Romance Is Not Just for Women," *Media Psychology* 6, no. 3 (August 2004): 257-84.

163. See Susan Bluck, Jacqueline M. Baron, Sarah A. Ainsworth, Amanda N. Gesselman, and Kim L. Gold, "Eliciting Empathy for Adults in Chronic Pain through Autobiographical Memory Sharing," *Applied Cognitive Psychology* 27, no. 1 (January/February 2013): 81-90.

164. See Bluck et al., "Eliciting Empathy for Adults," 81-90.

165. Tani et al., "Empathy and Autobiographical Memory," 252-69; see especially pp. 252-53.

166. Tani et al., "Empathy and Autobiographical Memory," 252-69; see especially p. 253.

167. See Mark H. Davis, *Empathy: A Social Psychological Approach* (Madison, WI: Westview Press, 1996).

168. Davis, *Empathy*; see especially pp. 1-22.

169. Brendan Gaesser, "Constructing Memory, Imagination, and Empathy: A Cognitive Neuroscience Perspective," *Frontiers in Psychology* 3 (January 2013): 1-6; see especially p. 3.

170. See, for instance, R. Shayna Rosenbaum, Donald T. Stuss, Brian Levine, and Endel Tulving, "Theory of Mind is Independent of Episodic Memory," *Science* 23 (November 2007): 1257.

171. See, for instance, Michael V. Lombardo, Jennifer L. Barnes, Sally J. Wheelwright, and Simon Baron-Cohen, "Self-Referential Cognition and Empathy in Autism," *PLoS ONE* 2, no. 9 (September 2007): e883.

172. See Lombardo et al., "Self-Referential Cognition," e883.

173. See, for instance, Brendan Gaesser, Daniel C. Sacchetti, Donna Rose Addis, and Daniel L. Schacter, "Characterizing Age-Related Changes in Remembering the Past and Imagining the Future," *Psychology and Aging* 26, no. 1 (March 2011): 80-84.

174. See Daniel Grühn, Kristine Rebucal, Manfred Diehl, Mark Lumley, and Gisela Labouvie-Vief, "Empathy Across the Adult Lifespan: Longitudinal and Experience-Sampling Findings," *Emotion* 8, no. 6 (December 2008): 753-65.

175. See, for instance, Grit Hein and Robert T. Knight, "Superior Temporal Sulcus—It's My Area: Or Is It?" *Journal of Cognitive Neuroscience* 20, no. 12 (December 2008): 2125-36.

176. Gaesser, "Constructing Memory, Imagination, and Empathy," 1-6; see especially p. 3.

177. See Carrie L. Masten, Sylvia A. Morelli, and Naomi I. Eisenberger, "An fMRI Investigation of Empathy for 'Social Pain' and Subsequent Prosocial Behavior," *NeuroImage* 22 (2011): 381-88.

178. Masten et al., "An fMRI Investigation of Empathy," 381-88.

179. See Vani A. Mathur, Tokiko Harada, Trixie Lipke, and Joan Y. Chiao, "Neural Basis of Extraordinary Empathy and Altruistic Motivation," *NeuroImage* 51 (2010): 1468-75.

180. See David M. Amodio and Chris D. Frith, "Meeting of Minds: The Medial Frontal Cortex and Social Cognition," *Nature Reviews Neuroscience* 7 (April 2006): 268-77.

181. See, for instance, Robert B. Cialdini, Stephanie L. Brown, Brian P. Lewis, Carol Luce, and Steven L. Neuberg, "Reinterpreting the Empathy-Altruism Relationship: When One Into One Equals Oneness," *Journal of Personality and Social Psychology* 73, no. 3 (1997): 481-94.

182. See, for instance, Bryan T. Denny, Hedy Kober, Tor D. Wager, and Kevin N. Ochsner, "A Meta-analysis of Functional Neuroimaging Studies of Self- and Other Judgments Reveals a Spatial Gradient for Mentalizing in Medial Prefrontal Cortex," *Journal of Cognitive Neuroscience* 24, no. 8 (2012): 1742-52.

183. Masten et al., "An fMRI Investigation of Empathy," 381-88.

184. See Daniel L. Schacter, Donna Rose Addis, and Randy L. Buckner, "Episodic Simulation of Future Events: Concepts, Data, and Applications," *Annals of the New York Academy of Sciences* 1124 (2008): 39-60.

185. Gaesser, "Constructing Memory, Imagination, and Empathy," 1-6; see especially pp. 3-4.

186. Jefferson A. Singer, Pavel Blagov, Meridith Berry, and Kathryn M. Oost, "Self-Defining Memories, Scripts, and the Life Story: Narrative Identity in Personality and Psychotherapy," *Journal of Personality* 81, no. 6 (December 2013): 569-82; see especially p. 574.

187. Singer et al., "Self-Defining Memories," 569-82; see especially pp. 574-75.

188. Singer et al., "Self-Defining Memories," 569-82; see especially p. 575.

189. See Tali Boritz, Lynne Angus, Georges Monette and Laurie Hollis-Walker, "An Empirical Analysis of Autobiographical Memory Specificity Subtypes in Brief Emotion-Focused and Client-Centered Treatments of Depression," *Psychotherapy Research* 18, no. 5 (September 2008): 584-93.

190. See Williams et al., "Autobiographical Memory Specificity," 122-48.

191. See Leslie S. Greenberg and Lynne E. Angus, "The Contributions of Emotional Processes to Narrative Change in Psychotherapy: A Dialectical Constructivist Approach," in *The Handbook of Psychotherapy*, ed. Lynne E. Angus and John McLeod (Thousand Oaks, CA: SAGE Publications, 2004), 330-50.

192. Tali Zweig Boritz, Lynne Angus, Georges Monette, Laurie Hollis-Walker, and Serine Warwar, "Narrative and Emotion Integration in Psychotherapy: Investigating the Relationship Between Autobiographical Memory Specificity and Expressed Emotional Arousal in Brief Emotion-Focused and Client-Centred Treatments of Depression," *Psychotherapy Research* 21, no. 1 (January 2011): 16-26; see especially p. 17.

193. David C. Rubin, Andriel Boals, and Rick H. Hoyle, "Narrative Centrality and Negative Affectivity: Independent and Interactive Contributors to Stress Reactions," *Journal of Experimental Psychology: General* 143, no. 3 (June 2014): 1159-70; see especially p. 1160.

194. See, for instance, Richard A. Bryant, "Psychological Interventions for Trauma Exposure and PTSD," in *Post-Traumatic Stress Disorder*, ed. Dan J. Stein,

Matthew J. Friedman, and Carlos Blanco (Hoboken, NJ: Wiley-Blackwell, 2011), 171-202.

195. Rubin et al., "Narrative Centrality and Negative Affectivity," 1159-70; see especially p. 1160.

196. Rubin et al., "Narrative Centrality and Negative Affectivity," 1159-70; see especially p. 1160.

197. See Jule Specht, Boris Egloff, and Stefan C. Schmukle, "Stability and Change of Personality Across the Life Course: The Impact of Age and Major Life Events on Mean-Level and Rank-Order Stability of the Big Five," *Journal of Personality and Social Psychology* 101, no. 4 (2011): 862-82.

198. See Christin M. Ogle, David C. Rubin, and Ilene C. Siegler, "Changes in Neuroticism Following Trauma Exposure," *Journal of Personality* 82, no. 2 (April 2014): 93-102.

199. Rubin et al., "Narrative Centrality and Negative Affectivity," 1159-70.

200. See Adriel Boals and Darnell Schuettler, "A Double-Edged Sword: Event Centrality, PTSD and Posttraumatic Growth," *Applied Cognitive Psychology* 25, no. 5 (September/October 2011): 817-22.

201. Rubin et al., "Narrative Centrality and Negative Affectivity," 1159-70; see especially pp. 1166-67.

202. It is worth noting that memory judgments can be scored for accuracy while remember-know judgments cannot. See Hintzman, "Memory Judgments", 165-77; see especially p. 175.

203. Hintzman, "Memory Judgments," 165-77.

204. Roland G. Benoit, Karl K. Szpunar, and Daniel L Schacter, "Ventromedial Prefrontal Cortex Supports Affective Future Simulation by Integrating Distributed Knowledge," *Proceedings of the National Academy of Sciences of the United States of America* 111, no. 46 (November 2014): 16550-55; see especially p. 16550.

205. See Benoit et al., "Ventromedial Prefrontal Cortex," 16550-55.

206. Benoit et al., "Ventromedial Prefrontal Cortex," 16550-55; see especially p. 16550.

207. See, for instance, Anita Tusche, Jonathan Smallwood, Boris C. Bernhardt, and Tania Singer, "Classifying the Wandering Mind: Revealing the Affective Content of Thoughts During Task-Free Rest Periods," *NeuroImage* 97 (2014): 107-16.

208. Benoit et al., "Ventromedial Prefrontal Cortex," 16550-55; see especially pp. 16550-51.

209. Benoit et al., "Ventromedial Prefrontal Cortex," 16550-55; see especially pp. 16551-53.

210. Benoit et al., "Ventromedial Prefrontal Cortex," 16550-55; see especially p. 16551.

211. Sheri J. Y. Mizumori, "Context Prediction Analysis and Episodic Memory," *Frontiers in Behavioral Neuroscience* 7 (October 2013): 1-10; see especially p. 5.

212. Mizumori, "Context Prediction Analysis," 1-10; see especially p. 5.

213. See Eve Marder and Jean-Marc Goaillard, "Variability, Compensation and Homeostasis in Neuron and Network Function," *Nature Reviews Neuroscience* 7 (July 2006): 563-74.

214. Mizumori, "Context Prediction Analysis," 1-10; see especially pp. 5-6.

215. Mizumori, "Context Prediction Analysis," 1-10; see especially p. 6.

216. Mizumori, "Context Prediction Analysis," 1-10; see especially p. 6.

217. Antoine Bechara, Hanna Damasio, and Antonio R. Damasio, "Role of the Amygdala in Decision-Making," *Annals of the New York Academy of Sciences* 985 (2003): 356-69; see especially p. 357.

218. See Antonio Damasio, *Descartes' Error: Emotion, Reason, and the Human Brain* (New York: Grosset/Putnam Group, 1994); see especially chs. 8-9.

219. See Antoine Bechara, Antonio R. Damasio, Hanna Damasio, and Steven W. Anderson, "Insensitivity to Future Consequences Following Damage to Human Prefrontal Cortex," *Cognition* 50 (1994): 7-15.

220. Bechara et al., "Insensitivity to Future Consequences," 7-15.

221. Bechara et al., "Role of the Amygdala," 356-69; see especially pp. 357-58.

222. Bechara et al., "Role of the Amygdala," 356-69; see especially p. 360.

223. See Bechara et al., "Role of the Amygdala," 356-69; see especially p. 361.

224. See Bechara et al., "Role of the Amygdala," 356-69; see especially p. 361.

225. Bechara et al., "Role of the Amygdala," 356-69; see especially p. 361.

226. Diana Robertson, John Snarey, Opal Ousley, Keith Harenski, F. DuBois Bowman, Rick Gilkey, and Clinton Kilts, "The Neural Processing of Moral Sensitivity to Issues of Justice and Care," *Neuropsychologia* 45 (2007): 755-66; see especially p. 755.

227. Robertson et al., "The Neural Processing," 755-66; see especially pp. 755-56.

228. Robertson et al., "The Neural Processing," 755-66; see especially p. 756.

229. See, for instance, Jorge Moll, Jorge, Roland Zahn, Ricardo de Oliveira-Souza, Frank Krueger, and Jordan Grafman, "The Neural Basis of Human Moral Cognition," *Nature Reviews Neuroscience* 6 (October 2005): 799-809.

230. Robertson et al., "The Neural Processing," 755-66; see especially pp. 759-62.

231. See Gregory V. Jones and Maryanne Martin, "Primacy of Memory Linkage in Choice Among Valued Objects," *Memory & Cognition* 34, no. 8 (2006): 1587-97.

232. Jones and Martin, "Primacy of Memory Linkage," 1587-97.

233. Jones and Martin, "Primacy of Memory Linkage," 1587-97 see especially p. 1595.

234. Jones and Martin, "Primacy of Memory Linkage," 1587-97 see especially p. 1595.

235. Sanghoon Han, Akira R. O'Connor, Andrea N. Eslick, and Ian G. Dobbins, "The Role of Left Ventrolateral Prefrontal Cortex During Episodic Decisions: Semantic Elaboration or Resolution of Episodic Interference," *Journal of Cognitive Neuroscience* 24, no. 1 (January 2012): 223-34; see especially p. 223.

236. See Han et al., "The Role of Left Ventrolateral," 223-34.

237. Han et al., "The Role of Left Ventrolateral," 223-34; see especially p. 223.

238. Han et al., "The Role of Left Ventrolateral," 223-34; see especially pp. 228-31.

239. Daniel L. Schacter, Roland G. Benoit, Felipe De Brigard and Karl K. Szpunar, "Episodic Future Thinking and Episodic Counterfactual Thinking: Intersections Between Memory and Decisions," *Neurobiology of Learning and Memory* 117 (2015): 14-21; see especially p. 14.

240. See, for instance, Kathy D. Gerlach, David W. Dornblaser, and Daniel L. Schacter, "Adaptive Constructive Processes and Memory Accuracy: Consequences of Counterfactual Simulations in Young and Older Adults," *Memory* 22, no. 1 (January 2014): 1-26.

241. Schacter et al., "Episodic Future Thinking," 14-21; see especially pp. 14-17.

242. See, for instance, Stanley B. Klein, "The Complex Act of Projecting Oneself into the Future," *Wiley Interdisciplinary Reviews: Cognitive Science* 4, no. 1 (January/February 2013): 63-79.

243. See, for instance, Arnaud D'Argembeau and Martial Van der Linden, "Phenomenal Characteristics Associated with Projecting Oneself Back into the Past and Forward into the Future: Influence of Valence and Temporal Distance," *Consciousness and Cognition* 13, no. 4 (2004): 844-58.

244. See, for instance, Donna Rose Addis, Alana T. Wong, and Daniel L. Schacter, "Age-Related Changes in the Episodic Simulation of Future Events," *Psychological Science* 19, no. 1 (2008): 33-41.

245. See Felipe De Brigard and Kelly S. Giovanello, "Influence of Outcome Valence in the Subjective Experience of Episodic Past, Future, and Counterfactual Thinking," *Consciousness and Cognition* 21 (2012): 1085-96.

246. De Brigard and Giovanello, "Influence of Outcome Valence," 1085-96.

247. Schacter et al., "Episodic Future Thinking," 14-21; see especially pp. 18-19.

248. Schacter et al., "Episodic Future Thinking," 14-21; see especially p. 19.

249. Schacter et al., "Episodic Future Thinking," 14-21; see especially p. 19.

250. See Nathalie Camille, Giorgio Coricelli, Jerome Sallet, Pascale Pradat-Diehl, Jean-René Duhamel, and Angela Sirigu, "The Involvement of the Orbitofrontal Cortex in the Experience of Regret," *Science* 304 (May 2004): 1167-70.

251. See Giorgio Coricelli, Hugo D. Critchley, Mateus Joffily, John P. O'Doherty, Angela Sirigu, and Raymond J. Dolan, "Regret and Its Avoidance: A Neuroimaging Study of Choice Behavior," *Nature Neuroscience* 8 (2005): 1255-62.

252. Schacter et al., "Episodic Future Thinking," 14-21; see especially pp. 19-20.

# 4.
# THE CASE AGAINST AFFECTIVE DEGENERATION: THE MORAL SIGNIFICANCE OF EMOTIONAL RATIONALITY FOR ETHICAL DECISION MAKING

In light of the foregoing chapter's conclusions – namely, that MM disintegrates the episodic memories whereby individuals derive the capacity for moral judgments, rational reflection, and emotional intelligibility – this chapter aims to make the case against affective degeneration by underscoring the moral significance of emotional rationality for ethical decision making. To that end, it endeavors, first, to identify and unpack the interrelationships between emotion, reason, and choice. Second, it discusses the requisite dependence of subjective emotion on objective rationality. Third, it explores the correlation of emotional rationality and morality. The chapter concludes by highlighting the necessity of emotional rationality for ethical decision making.

## 4.1. Emotion, Reason, and Choice

The relationship between emotion and reason has been a major topic in Western philosophy since its genesis.[1] However, the relationship between emotional choice and rational choice is a more recent concern.[2] Both reason and rationality are primarily normative inasmuch as they inform agents of the options that ideally should be pursued in the effort to secure sought-after ends. Their explanatory use arises when the agent takes the normative suggestion and tests it by confronting the prescribed behavior with its observed counterpart. In economics, for instance, rational-choice explanations, based on the assumption that agents will maximize utility, was the standard understanding of behavior until 1980, when it became subject to criticism from a number of scholars who subsequently developed new models of behavior. Broadly speaking, these alternative, neurocognitively-geared models constitute what has loosely become known as the field of behavioral economics.[3]

### 4.1.1. Emotion and Reason in Ancient Greek Philosophy

The ancient Greeks had no word equivalent to the Latinate "emotion." The term commonly used in its place, "pathos," generally indicates "that which happens to a person or thing."[4] The writings of Plato and Aristotle develop conceptions of relevant cognitions, and exemplify the need to be reflective about the fundamental nature and role of cognition within emotion. Insofar as the experiences of pleasures and opinions depend on a living body (animated by the soul), Plato draws a corollary when he divides not soul from body, but a rational soul – competent to determine what is best for the entire soul – from a non-rational soul, which he further divides into appetite and spirit. Appetite concerns a certain class of desires (epithumiai). Objects are considered desirous (epithumētikon) by virtue of the intensity of desire accompanying them. Though not emotions themselves, intense desires are accompanied by emotion: in respect of its appetite, the soul hungers, thirsts, loves, and feels the excitement (eptoētai) of other desires. Appetite is cognitive inasmuch as it is a companion to certain repletions and pleasures. To desire something is, for Plato, to desire it as pleasant, and such desiring involves thinking it to be pleasant. Through the example of Socrates' anger, Plato later chronicles the spirit as a distinct part of the soul. When anger is directed at oneself, it gives rise to the emotion of shame. Provoked by wrongs to be resisted, it speaks in terms of "oughts," and becomes by nature the ally of reason: it receives the values of others and translates them into its own terms. On this basis, it is tempting to conceive of Plato's spirit as the emotive part of the soul. In Plato, then, what emotions lack in rationality, they make up for in a phenomenology born of imaginative recall that combines pleasure and pain.[5]

Aristotle connects the passivity of emotions with their physicality.[6] For Aristotle, all emotions of the soul involve the body – anger, good temper, fear, pity, confidence, joy, love, and hatred – and upon the experience of emotion the body is affected in a certain way. Aristotelian deliberation is a process of calculation and reflection that leads the agent from contingent goals (suggested by one's character in the perceived circumstances) to a means that acceptably achieves it. The ultimate end is acting well, which demands a receptivity to the emergence of pros and cons. The practical wisdom (phronēsis) exercised within deliberation tests the goal (in the circumstances) by discovering whether to permit it to be achieved acceptably. Yet the initial selection of a goal emerges out of an agent's perception of the situation. Hence, it is not simply a matter of identifying the neutral facts of the case, but of being attracted and repelled by the possibilities that painful or pleasant perceptions present. Put simply, the practical eye is the eye of the heart, not merely of the head. For Aristotle, phronēsis is rooted in educated affective responses. Education takes place through experience that conditions the agent in choosing and acting well through correct training in response to pleasure and pain. Achieving the mean in relation to an affection is a matter of being disposed to it, not indiscriminately, but as is best from one situation to another. Hence, ethical education in Aristotle is emotional education. Discarding too simple a Socratic focus on the contents of thought, Plato and Aristotle embrace the interconnections within the emotions of body and soul, perception and imagination, and feeling and thinking.[7] In the end, emotions are not simply things felt; they are also the very framework of thinking.

For Plato, rational knowledge is the highest good.[8] As highest good, it serves as the measure of worth for everything else. Thus, until one possesses rational knowledge, the acquisition thereof must be the singular goal that trumps all others, for all things are made good solely thereby. Inquiry – which arises from the awareness that one does not know something in particular – serves as the provisional good until one possesses the highest good: rational knowledge. The ability to reason about and secure definitive knowledge of something, Plato suggests, is directly connected to the capacity to overcome the distinction between subject and object. In brief, to know an object is for that object to be present for the subject in the same way that the object is present in itself. Yet the possibility of reason's self-awareness depends on the separation between subject and object. The thinker cannot fully take on the object while simultaneously maintaining the hypothetical character of the investigation. Plato contends, therefore, that reason is the grasp of an object as a collection of parts, and that in order for this type of work to take place, there must be a higher faculty at work as well – one responsible for the "oneness" of the subject's grasp. This higher faculty, "nous," is responsible for grasping the oneness of an object, and must overcome the separation of subject and object to do so. For Plato, the moment of insight, which marks the overcoming of differences between the distinct parts being considered, is the operation of nous, and in that act both the opposition between the parts of the object and the subject and object is overcome.[9]

The nucleus of Aristotle's concept of rationality lies in his theory of first principles, which in turn comprises his theory of epistēmē.[10] The Greek word epistēmē has a broad range of meanings, ranging from (rational) "knowledge" at the weakest to "science" or "understanding" at the strongest. For Aristotle, epistēmē is an organized body of knowledge arranged in a hierarchical and explanatory structure, with lower-level items being explained by higher-level ones, and those in turn being explained by still higher-level ones, until at the apex first principles (nous) are reached, which are themselves indemonstrable. However, first principles are not cognitive starting points; rather, individuals must start from appearances (phainomena) that are not themselves (at least initially) objects of epistēmē, and from them work up dialectically to a grasp of first principles. The phainomena from which individuals begin include not only the empirical data of sense-perception, but also endoxa: "reputable beliefs."[11] Of immediate import to the present context is Aristotle's concept of practical rationality (phronēsis): the intellectual virtue or excellence that enables individuals to plan or deliberate well about what is good or useful for living well and therefore being happy.[12] Confined to identifying the proportionate means to ends, practical rationality is concerned with right means, whereas moral virtue is concerned with right ends. Aristotle compares the process of rational deliberation to the process of scientific discovery. Just as problem-solving terminates in the recognition of something ultimate, which forms the first step in the construction of a figure, so practical deliberation terminates in the recognition of something ultimate, which marks the "first cause" in action. Through detailed and rigorous discussions of the relationship between the actions taken by moral agents and the practical reasoning leading up to it, Aristotle concludes that the employment of practical rationality in the planning process can only be completed at the time of action, and that it includes, as its terminus, a practical syllogism.[13]

### 4.1.2. The Emotional and Rational Tenets of Choice

In the past decade, multiple studies have suggested that activity in a small number of brain areas encode reward quantities during tasks related to choice.[14] Areas such as the parietal cortex appear to encode how many milliliters of water a choice (between clearly enough and clearly too much) will yield to a thirsty person, and areas such as the ventral striatum and the medial PFC appear to encode the amount of money an option will yield. Indeed, there is now broad consensus in the neuroscience of choice community that reward magnitude is clearly represented in a small number of well-identified areas. In their 2012 study, Dino Levy and Paul Glimcher conduct a meta-analysis using data from recent fMRI studies that suggest that one of these reward encoding areas – the VMPFC/OFC – can be understood as representative of the value of almost all reward types on a common scale that predicts behaviorally-observed comparison and choice.[15] The idea of common currency representation at a strictly theoretical level is not new. In his 1947 economic study, Paul Samuelson proved that decision makers who are internally consistent in their choices behave precisely as if they were applying a single common scale for the representation of value.[16] To be sure, the assumption that the chooser is rational is not a necessary condition for a common currency representation. Since Samuelson's proof, almost all theories of choice – from expected utility theory through prospect theory, including modern reinforcement learning algorithms – have shared the belief that in order to choose, the different elements of each option must eventually be converged, however uniquely and insufficiently, into a single value for the actual process of comparison. Nevertheless, there is no evidence to suggest that neural currency of values arises solely in the subregion of the VMPFC/OFC. Any common currency observed in the brain must reflect the activation of multiple cognitive processes.[17]

A long tradition of research on judgment and decision making (JDM), originating from choice or preference theory in microeconomics and decision theory in philosophy, supposes that uncertain decisions are rooted in cognitive processes involving means-end reasoning, logical inference, mental effort, and precision computation according to cost-benefit calculi.[18] In the 1990s, however, JDM models increasingly integrated emotional processes, influenced by a reconsideration of emotion in neuroscience. As these models evolved, a steadfast emphasis on emotional contributions to JDM was an approximate, heuristic process that delivered rapid evaluations without mental effort. Additionally, JDM researchers increasingly identified conflict in decision making as the product of divergence between rational and emotional evaluations, and pathological decision making as the result of affect heuristics. Notwithstanding the popularity and straightforward appeal of distinguishing between rational and emotional contributions to JDM, several fundamental issues remain unresolved. Contemporary theories can be characterized in terms of the representations they posit, and it remains unclear in what ways rational and emotional contributions to JDM differ along these dimensions.[19] Perhaps the most surprising discovery to date is that core emotional structures, including the midbrain dopamine system and insula, decompose under certain choice contexts along the statistical dimensions that comprise the cornerstone of financial decision theory. Previous accounts of emotion as value-stimulus associations tended to focus on the OFC as the cardinal structure involved in representing reward expectation, and viewed midbrain dopamine areas as involved in reward-related learning rather than in the representation of reward expectation as such. Nevertheless, recent findings suggest that value encoding in the midbrain dopamine areas potentially underlie an early implicit encoding that is communicated to the OFC, where it guides both emotional and rational choice.[20]

Emotional choice arises from belief insofar as emotions have cognitive causal antecedents.[21] Although this is the typical and most important case in human affairs, it is not the only one. As neurological studies of fear and panic attest, mere perception without any propositional content can also trigger emotion. A sign on the wall indicating the dangers of smoking, for instance, causes greater emotional arousal when accompanied by a color photo of a cancerous lung. Insofar as beliefs can serve as either the effect or cause of emotion, the relation between beliefs and the emotions they generate are remarkably fine-grained. In most cases, emotional choice will be minimally rational: the action performed will be the best means to realize the agent's desire, given the belief with which it corresponds. However, on Humean premises, the fact that desire is induced by emotion need not undermine the full rationality of the action.[22] In many psychological studies of emotion, valence – or position on the pleasure-pain dimension – is used as the independent variable. In this model, positive emotions tend to make individuals more risk-averse, whereas negative emotions tend to cause risk-seeking behavior. Simultaneously, positive and negative emotions respectively generate optimistic and pessimistic cognitive biases. Hence, when subjects are not explicitly told the odds of winning and losing, but have to assess them from the evidence provided, cognitive bias and risk attitudes work in opposite directions. Compared with the fine-grained classifications of emotions based on cognitive antecedents or action tendencies, valence is a coarse-grained category. In studies employing cognitive antecedents as the independent variable, the impact of emotion on risk attitudes appears in a different light. Whereas fearful individuals typically expressed pessimistic attitudes and risk-averse choices, angry individuals expressed optimistic risk calculi and risk-seeking choices. In this case, emotional-induced risk dispositions and emotion-induced cognitive bias work in the same direction.[23]

The standard model of rational choice is defined in terms of the relation obtaining among action, beliefs, desires, and information.[24] The beliefs and desires of rational agents cause them to choose a course of action that can be rationalized in light of the information at hand. Rational agents choose the action that best realizes their desires, given their beliefs about what their options are and about the consequences of pursuing them. The beliefs are themselves inferred from the available data by the procedures that are most likely to yield true beliefs. Prior to belief formation, rational agents gather more evidence in an amount that is optimal in light of their desire and the expected costs and benefits of gathering new information. Thus understood, rationality is a matter of process and outcome. Whereas a direct influence on beliefs is inconsistent with rational belief formation, an indirect influence mediated by information gathering is not necessarily irrational. An individual who cares little about the environment, for instance, is rationally unmotivated to invest resources to find the best ways to recycle waste. It will prove useful, here, to distinguish between two kinds of preferences: substantive and formal. Substantive preferences relate to specific pairs of options, such as one political candidate over another. Instances of formal preferences involve the rate at which the future is discounted, including considerations of risk aversion, loss aversion, and the like. Although formal preferences are somewhat domain-specific – e.g., an individual may use different discount rates for future health and future income – they have a wider scope than substantive preferences. Rational choice, then, is typically defined by the interaction between substantive and formal preferences. This suggests that rational agents do as well as they can in light of the cognitive resources and propensities with which they find themselves endowed.[25]

## 4.2. Emotion and Rationality

Neurocognitive studies within the last several years suggest that the primary moral-psychological deficiency in contemporary society lies not in knowing, but in feeling, what is moral.[26] This feeling, associated with the prefrontal cortex and amygdala and frequently referred to as "moral emotion,"[27] is the springboard from which cognitive recognition that a particular act is immoral is translated into specific behavioral inhibition.[28] The complexity of emotion has rendered it neuroethically problematic for several reasons, many of which continue to serve as the impetus for its rejection in the realm of rationality. Hence, emotion is widely conceived as relatively useless – or at least not particularly constructive – to ethical decision making. Emotion has been critically described as overly perceptive, personal,[29] unstable, intense, partial, and fleeting.[30] What remains scantily investigated, however, is the more positive – and often overlooked – objectively-beneficial features of emotion, namely, its cognitive complimentarily, evaluative faculties, motivating power, strength, and tendency to capture critical, otherwise inexplicit elements of reality.[31] These uncharted features hint at the idea that emotion may inherently possess a particular, if peculiar, rationality that, in turn, renders it practically useful.

### 4.2.1. The Functional Ontology of Rational Emotion

The functional ontology of emotion has haunted philosophers and psychologists for centuries,[32] and typical replies reduce emotions to one of their components, such as a capacity, feeling, or cognitive state.[33] Emotions are arguably the most sophisticated cognitive phenomena insofar as they involve various types of entities and states that belong to diverse ontological levels. Emotions typically occur when individuals perceive positive or negative significant changes in their personal situations or in the situations of those most immediately related to them. Major positive or negative concerns significantly improve or interrupt a stable situation relevant to one's concerns. Similar to a triggered alarm, emotions signal that something needs attention, and the corresponding level of attention is directly proportionate to one's personal concerns. These concerns are short- or long-term dispositions toward preferences for particular states of the world and the self. Inasmuch as emotions elicit an event's significance, emotional meaning is primarily comparative. One's emotional environment contains not only what is and what will be experienced, but also all that could be or all that one wishes will be experienced. For the entire emotional system, then, all such possibilities are at once available and compared with each other. The import of comparative concern in the ontology of emotion is simultaneously connected with the central role of changes in generating rational emotions. That is, an event can be perceived as a significant change only insofar as it is compared against an empirically valid, evaluative framework.[34]

Four general ontological characteristics of emotion are (i) ephemerality, (ii) intensity, (iii) partiality, and (iv) brevity.[35] Regarding the first, in light of the critical role that change plays in generating emotions, the ephemerality of cognitive (and physiological) systems is a basic characteristic of rational emotion. Emotions indicate a transition in which the preceding context has changed but no new context has yet been stabilized. Like a looming storm, emotions are mercurial states that signify a rational agitation of value. This intense, occasional, and limited experience is, paradoxically, precisely what gives rise to emotion's rational endurance, evenness, and permanence. Regarding the second, emotions are appropriately characterized as intense reactions. Within emotional experience, the cognitive system has not yet adapted to the given change and, due to its significance, the change requires the mobilization of many resources. It is no wonder, therefore, that emotions are associated with urgency. For the emotional individual, there is no such thing as a minor concern; if the concern is minor, it is not emotional. Hence, a typical characteristic of rational emotion is its magnifying nature: everything looms larger in light of temporarily disordered values. Regarding the third, emotions are partial in two fundamental senses: they are focused on a narrow target, such as one personal or a specific group thereof, and they express personal and interested perspective. Emotions direct and color one's attention by rationally selecting what attracts and holds it. In this sense, emotions address objective (practical) concerns from a subjective perspective, and thereby express values and preferences indiscriminately. Finally, regarding the fourth, typical emotions are essentially states with relative brevity. The mobilization of emotional resources to focus on one event cannot last forever. Even the most rational system cannot be unstable for a long period and still function normally; it may implode due to continuous increase in emotional intensity. If emotions were to endure for a lengthy period regardless of what was occurring in the correlative environment, they would not possess adaptive rational value. Nevertheless, the fact that emotions are transient – lasting from seconds to hours to days – does not imply that their impact is merely transient. On the contrary, a brief emotional state can have profound and long-lasting behavioral implications in the realm of rational analysis.[36]

In addition to the four general ontological characteristics of emotion mentioned above, there are at least four functional components of emotion that shed light on its overarching rational ontology, namely, (i) cognition, (ii) evaluation, (iii) motivation, and (iv) feeling.[37] The cognitive component of emotion supplies the required information about a given situation. No affective attitude toward something can emerge without some preexisting information concerning it. While the cognitive component of emotion is occasionally distorted by its inherent partiality, closeness, and intensity, the unique nature of emotion has numerous vital cognitive advantages. While being close, for instance, one may better understand critical aspects of the situation at hand. Moreover, emotional arousal significantly increases attentional capacity. Accordingly, emotions typically increase memory. The evaluative component of emotion is perhaps its most significant rational feature. Every emotional expression entails a certain rational evaluation. In a state devoid of an evaluative component, or one in which its weight is marginal, existential indifference pervades. The evaluative component appraises the "cold" information presented by the cognitive component (vis-à-vis its implications for subjective well-being). The presence of an evaluative component is what distinguishes the emotion of hope, for instance, from that of expectation: individuals do not hope for something unless they first evaluative it as somehow favorable, whereas the expectation of something entails no comparable evaluation. Nevertheless, the evaluative component intrinsic to a particular emotional state should not be conflated with the rational moral evaluation of the entire state.[38]

The motivational component of emotion refers to the desire or readiness to maintain or change present, past, or future circumstances.[39] In some emotions, such as anger or fear, the desire typically manifests in overt behavior; in other emotions, such as envy or hope, the behavioral element is less evident and appears merely as a desire. Emotions are not theoretical states; rather, they involve practical concerns associated with a readiness to act. Since emotions are rationally evaluative attitudes, involving positive or negative stances toward particular objects, they also entail either taking action or being disposed to act in a manner compatible with the corresponding evaluation. In this way, emotions typically express one's most profound values and attitudes and, as such, express deep commitments. Lastly, the feeling dimension of emotion is a primitive mode of consciousness associated with one's own state. It is the lowest level of consciousness; unlike higher levels of awareness, such as those found in perception, memory, and thinking, the feeling dimension has no significant cognitive content. It expresses a subjective state, but is not itself directed toward any state (or object) in particular. Since this dimension is a mode of consciousness, one cannot be unconscious of it; there are no "unfelt" feelings. One indication for the absence of the intentional dimension in feelings is that individuals cannot speak of reasons for experiencing feelings as they can for emotions. That is, individuals cannot be reasoned out of their toothache as they can their hatred. Unlike rational emotions, then, feelings are not intrinsically subject to normative appraisal. Nevertheless, two different emotions, such as grief and pride, can occasionally be distinguished by virtue of the feeling component with which they are bound up. In other instances, the feeling component may be inadequate in this respect. The same emotion, such as joy, may share a variety of feelings, and the same feeling may be shared by different emotions, such as shame, remorse, regret, and guilt. Although many emotions embrace a variety of feelings within their scope, their range is generally restricted to a particular set of characteristic feelings. Hence, as several empirical studies demonstrate,[40] feeling cannot serve as the rational basis for distinguishing the precise nature of an emotional object.[41]

### 4.2.2. The Epistemology of Rational Emotion

As the critical filter through which human beings make sense of the world, rational emotion possesses an inherent epistemological axiom: without information gleaned from the emotional sense that imbues interaction and institutions with meaning, life would be reduced to hollow shells and randomly-acting forms.[42] Without epistemic emotions, for instance, one would see a lost child and not understand her cries, or hear the quiver in the voice of another without imagining the cause or foreseeing its impact on his immediate behavior. In brief, without recourse to one's vast emotional database, one would tumble helplessly into the gaps left open by language. Hence, individuals have instrumentally epistemic emotions: they are curious about the answers to questions of practical importance to them. When curiosity strikes, only truthful explanation will satisfy inquiry, not the possibility of belief formation. One obvious corollary is that knowledge slakes curiosity. While there are many accounts of knowledge in contemporary epistemology, the relevant alternatives theory provides a particularly helpful frame on which several current ideas about knowledge can be presented. According to this account, when individuals form beliefs, there are various possibilities that must be excluded if a specific belief is to count as knowledge. Occasionally, the fact that an alternative ought to be excluded is a mechanical matter of good epistemic training. At other times, it requires individuals to be subtly in tune with the demands of the situation. This inevitably entails abandoning some possibilities in favor of others, and one is likely to worry about some of these abandonments, perhaps allowing room for the haunting thought that some subtle interaction has gone unnoticed. Individuals naturally feel responsible for the accuracy of the results, concerned about abnormalities for which there is no good explanation, fascinated by preliminarily results, attracted to lines of investigation that might settle questions, wary of the potential of other results to prove distracting, satisfied that some possibilities have been ruled out, and unsatisfied with respect to the ability to exhaustively investigate all possible results. Worry (i.e., haunting, obsession), concern (i.e., responsibility), interest (i.e., fascination, attraction, wariness), and (dis)satisfaction are epistemic emotions that possess a common rational theme.[43]

Knowledge requires the exploration of a maze of possibilities, some consistent with the facts known, and some incompatible with them. Much of this exploration is impossible without the initial prompting, pushing, or goading of epistemic emotions such as worry, fascination, and curiosity.[44] Hence, the relationship between rational emotion and moral virtue finds its most vivid exemplification in the epistemic domain. The ordinary human operation of numerous virtues involves the activation of rational emotions that move agents to required patterns of action. For instance, kindness usually involves the sentiment of affection; fair-mindedness often involves the sense of injustice; courage typically involves the sense of outrage. In none of these circumstances is emotion required for the operation of the virtue, but most individuals would find it difficult to sustain the virtue without being simultaneously capable of the emotion. The same is true with knowledge. There are countless circumstances in which unmotivated epistemic virtues are driven by epistemic emotions, but wherein a subtle or sophisticated network of alternative possibilities needs to be explored. Moreover, the responsibility for an object of concern cannot easily be separated from the responsibility for an aspect of the enquiry. For example, the moral emotion of social (in)justice, with a particular population and situation as its objects, generates the rational emotion of epistemic responsibility for gathering and ordering the value-laden information relevant thereto. Like other epistemic emotions, the sense of responsibility can generate open-ended projects with unspecified aims. The rational emotion of moral responsibility is associated with epistemic worry. If one feels responsible for knowing something, one worries about possible leads that have not been explored and possibilities that have not been excluded. These possibilities typically nag at oneself, leading to scrupulous investigation, checking and rechecking, and imagination. The desired result is for the moral mind to be finally at rest, having achieved a desirable result, perhaps only insofar as the possibilities have been pursued in an epistemically responsible way. Regardless of whether moral responsibility is a function of epistemic virtue, it is above all an epistemically rational emotion.[45]

Accepting that rational emotion is indispensable to reliable knowledge does not mean, of course, that uncritical affection can be substituted for dispassionate investigation.[46] That is, although emotions are epistemologically indispensable, they are not epistemologically indisputable. Like all human faculties, they may be misleading, and their data, like all data, are subject to reinterpretation and revision. Insofar as emotions are not presocial, physiological responses to unequivocal situations, they are open to challenge on various grounds. They may be dishonest or self-deceptive, incorporate inaccurate or partial perceptions, or may be constituted by oppressive values. Accepting the indispensability of rational emotion to knowledge means no more or less than that discordant emotions must be attended to seriously and respectfully rather than condemned, ignored, discounted, or suppressed. Just as rational emotion may contribute to the development of knowledge, so the growth of knowledge may contribute to the development of rational emotion. For instance, the powerful insights of medical sociology often stimulate new emotional responses to past and present situations. Inevitably, one's emotions are affected by the knowledge that the sickest members of the global community are also the most poor. The new emotions invoked by this insight are likely, in turn, to stimulate further observations and insights, and these may generate new directions in both moral theory and political practice. Hence, there is a continuous feedback loop between one's emotional constitution and one's theorizing such that each continually shapes and sharpens the other and is, in principle, inseparable from it.[47]

The ease and speed with which human beings can reeducate their emotions is, unfortunately, hardly ideal: emotions are only partially in the control of the individuals who possess them.[48] Although affected by new information, they are habitual responses not quickly unlearned. Even when individuals come to the conscious belief that their fear or shame or revulsion is rationally unwarranted, they may still continue to experience emotions inconsistent with their conscious moral behavior. On the one hand, the persistence of these recalcitrant emotions likely demonstrates how fundamentally human beings have been constituted by their dominant worldview. On the other hand, it may indicate superficiality or other inadequacy in one's emerging theories and politics. Individuals can only start from where they are – an often cruelly disproportionate society that has shaped their bodies, minds, perceptions, values, emotions, language, and systems of knowledge. Hence, adequate epistemological models of rational emotion must (i) display the continuous interaction between how individuals understand the world and who they understand themselves to be as people; (ii) show how emotional responses to the world change as individuals conceptualize it differently, and how their changing emotional responses subsequently stimulate new insights; and (iii) demonstrate the need for moral theory to be self-reflexive – to focus not only on the outer world but also on individuals and their relation to that world. As a result, adequate epistemological models of rational emotion invariably prove how the reconstruction of knowledge is inseparable from the reconstruction of self. Therefore, time spent in analyzing rational emotion and uncovering its sources is properly viewed neither as irrelevant to theoretical investigation nor as a prerequisite for it. Instead, efforts to reinterpret and refine rational emotion are necessary to theoretical investigation, just as efforts to reeducate emotions are necessary for meaningful moral activity. Critical reflection on the epistemology of rational emotion is not a self-indulgent substitute for analysis and action, but a practice indispensable for robust moral theory and social transformation.[49]

### 4.2.3. The Phenomenology of Rational Affective Awareness

In his early writings, Robert Solomon insists on a clear distinction between emotions and feelings: emotions have a judgmental quality that feelings do not.[50] For Solomon, although feelings often accompany emotions, the relationship is one of association rather than constitution. Feelings are mere bodily reactions, whereas emotions are conceptually sophisticated intentional states that have objects outside of the body. By claiming that emotions are judgments, Solomon does not intend to suggest that they are attitudes one adopts on the basis of experiences of the world. Rather, for Solomon, emotions are responsible for the sense that things matter in various different ways. Hence, emotions do not give life meaning relative a standpoint external to experience, but are experienced as the significance of things in the world. As such, they serve as the "meaning in life."[51] Solomon acknowledges Heidegger's distinction between mood and emotion as a source of inspiration for his account of emotion and world-meaning.[52] For Heidegger, moods are phenomenologically deeper than emotions insofar as emotions are only intelligible in the context of mood.[53] This suggests that the emotional states individuals refer to themselves as being "in" generally go deeper than those states they "have." When individuals have an emotion, they are already in a situation. The sense of being in that situation, Heidegger appreciates, depends upon mood. Heidegger's conception of mood (Stimmung) is premised on the acknowledgement that individuals do not experience the world as disinterested spectators; they find themselves in it. They are not in the world the way in which an object is in a container: experiencing oneself as part of the world is not finally a matter of registering one's spatiotemporal location in relation to other entities. Rather, one is situated in the world inasmuch as one is purposively entangled with it.[54] Hence, human beings experience people, objects, events, and situations in the world in terms of different kinds of significant possibility – different ways of mattering.[55]

   Heidegger's primary assertion is that intentionally directed emotions are responsible for the diverse ways in which human beings are able to experience the world as mattering.[56] Employing the example of fear, Heidegger notes that the experience of being afraid (of something) presupposes an apprehension of distinctive possibilities: "different possibilities of Being emerge in fearing."[57] That is, in order to be afraid, one must already locate oneself in a world in which being in danger or under threat are possibilities. Some being – perhaps oneself – must matter in a particular way for fear to be possible. The point applies more generally still: different emotions presuppose a range of unique ways in which things can matter, such as having practical significance, posing a threat, or being intriguing. According to Heidegger, then, this rational mood determines the space of possible manifestations of concern. It is uncontroversial to claim that to have an emotion with a specific content requires having a particular set of concerns. However, Heidegger's claim does not relate simply to the rational contents of emotions, but also to the kinds of rational emotions individuals are capable of experiencing. Jean-Paul Sartre's "nausea" is a rational mood along such lines, where a background sense of purpose and function is altogether removed, resulting in the appearance of everything as alien and contingent. Things are experienced as bereft of familiarity, appearing instead as "soft, monstrous masses, in disorder—naked, with a frightening, obscene nakedness."[58] In Sartrean nausea, the world is not merely a place without purpose or function, but a state in which purpose and function are inconceivable. This fleeting mood – wherein excitement, delight, or disappointment are impossible due to the temporary inability to experience events through concerns rooted in emotional presuppositions – marks a shift in the range of possible rational emotions. Rather than being "an inner condition which then reaches forth in an enigmatic way and puts its marks on Things and persons,"[59] moods are, for Heidegger, a fundamental sense of belonging to a meaningful world and a condition for having intentional states: "The mood has already disclosed, in every case, Being-in-the-world as a whole, and makes it possible first of all to direct oneself toward something."[60] Thus, rational emotion constitutes a phenomenological context in which intentionally directed experience is possible.[61]

In a Heideggerian model, rational affective awareness occurs when individuals attend to some aspect of their emotional state, experience, or behavior and are thereby able to explicitly report it.[62] The capacity to attend to and report one's emotion in this way involves the employment of a theory or conception of emotion. The diverse abilities of individuals to report their emotional states will thus vary according to different attentional styles and various theories and concepts of emotion. For one's emotional state to play a role in deliberative rationality, one must be aware of it. This is due, in part, to the fact that identifying one's state as an emotion enables one to detach from the "compellingness" of any action urge that is part of the affective state.[63] This permits one to inhibit the action urge – considering it, instead, as an option – rather than simply acting it out. This rational affective awareness can be distinguished at three levels. According to John Lambie, level 0 is the state of not having an emotion; level 1 is being in an emotional state, but not being aware of it as such; and level 2 is being in an emotional state and being aware of it as such.[64] Agents in level 2 are able (in principle) to be fully and deliberatively rational with regard to selecting actions on the basis of affective states. They can choose whether to act on the basis of emotion, and can account for any subjective emotional biases when executing a decision. Agents at level 1 are only partially rational with regard to the role of emotion in decision making. Their emotional states act as determinants in action selection, but agents are unable to account for their emotions as such in any practical deliberations, and thus are hindered in their ability to inhibit or override actions prompted from their emotions. Infants, for instance, operate at level 1 insofar as they cannot report their emotional states. Moreover, individuals who experience emotions (i.e., have conscious emotional feelings) but are not aware of them as such can only operate at level 1. Agents at level 0 may similarly be decisionally impaired if they are situated in an environment wherein rational affective awareness would serve as a critical heuristic.[65]

Rational affective awareness improves decision making insofar as awareness enables deliberation.[66] Although there are certain circumstances in which deliberation may not be rational (e.g., when the costs of deliberation outweigh the potential benefits), deliberation predominantly improves rationality, and the capacity to deliberate is necessary for coordinating short- and long-term goals in a rational life plan (i.e., when short- and long-term goals are in conflict). Deliberation creates the need for explicit reasons for actions; that is, it forces one to choose the basis on which to act. This, in turn, invokes the normative question of what one ought to do. Thus, it is only through deliberation that there is explicit rational engagement on the part of the moral agent. Rational affective awareness serves as the foundational structure that creates the possibility of deliberation, allowing one to choose whether to act on a desire or believe a perception. This structure is the distancing created by the focal attention necessary for rational affective awareness. In a state of detached affective awareness, the structure of one's phenomenal field is twofold: there exists an observer and an observed. If what one is observing are one's own states (e.g., one's thoughts, desires, actions, or feelings), then such awareness is considered "reflective consciousness."[67] By contrast, in an immersed mode of attending, the distance in the phenomenal field between the observer and the observed disintegrates. As a result, there is an absorption in the object of attention, the self is perceptually recessive, and one's thoughts and impulses are not seen as such: the former are "believed in" and the latter are "acted out." The distancing of detached self-awareness brings a rational, affective mental state (e.g., desire) into view. In sum, rational affective awareness allows (i) potentially goal-incongruent action tendencies to be identified and, hence, to be inhibited; (ii) knowledge of one's own biases to be employed in adjusting action plans; (iii) theories of one's self and one's behavior to be revised in accord with self-observational data; and (iv) actions to be consciously chosen on the basis of concrete moral reasons.[68]

### 4.2.4. Emotional Rationality as Necessary Condition for Evaluative Judgment

For Blaise Pascal, the emotional life is not chaotic or disorderly, but possesses the capacity to know the truth, corresponding to the rational faculty.[69] It is on this basis that Pascal could emphatically state both that "the heart has its reasons, which reason does not know,"[70] and "we know the truth, not only by reason, but also by the heart."[71] Following Pascal, Franz Brentano, largely regarded as the forerunner of the phenomenological movement, was the first to explicitly acknowledge the role of emotional rationality in moral knowledge. For Brentano, all cognitive phenomena are characterized by their intentional nature. Among these intentional acts are those pertaining to loving and hating, which are capable of evoking favorable and unfavorable feelings toward things or acts in their value-aspects. In this evaluative system, to say that an action is good or a thing is beautiful is to acknowledge that it is "correctly" loved. Here, the direct object of love is not the value itself but the goods or actions in which value arises through the act of loving. This Brentonian conception gave rise to the view of both Edmund Husserl and Max Scheler in this regard. For Husserl, "the heart and the will must have their analogous but specific forms of rationality," and the valuable properties of things are disclosed by emotions and feelings.[72] Scheler expresses solidarity with Husserl in this vein, and fortifies his personal insight by appealing to the analogous views of Pascal. In his major ethical work, *Formalism in Ethics and the Non-Formal Ethics of Values*,[73] Scheler distinguishes the sort of feelings that enable individuals to apprehend values from mere psychological states, the latter of which is non-cognitive in nature. Following Scheler, Nicolai Hartmann incorporates this view in his value-Platonism theory, positing that values are beheld in the acts of feeling.[74] In order to illuminate emotional rationality as a necessary condition of evaluative judgment, it is critical to understand what Scheler and Hartmann intend to communicate through their respective uses of "value."[75]

Both Scheler and Hartmann agree that values are not "furnitures" of the world: they neither reside nor are found in the world, and arise neither from desire nor interest.[76] Values are not, in other words, the products of subjectivism and psychologism; they are distinct from their carriers. Scheler and Hartmann disagree, however, concerning the positive characterization of values. Scheler distinguishes values from goods, noting that the former are "value-things" and the latter are "thing-value." Drawing an analogy with colors, he comments that values "exist" the way a pure color of spectrum exists,[77] and that values are "genuinely objective objects,"[78] similar to the way color species is an object (given through an intuitive act) for Husserl.[79] For Scheler, "value-qualities ... are ideal objects" as are qualities of color and sound,[80] and subsistent or autonomous beings: "all norms, imperatives, demands, etc. ... have their foundation in an autonomous *being*, the being of values."[81] On the contrary, Hartmann steadfastly maintains a status for values like that of Platonic ideas.[82] This implies, first, that Hartmann attributes to values the status of ideal being similar to the way Plato conceives Ideas as Ideal Being. The ideality of values, for Hartmann, consists in their self-existent character. This means that values are independent of everything, including the subject who passes judgment, the judgment itself, the subject's actual conduct, and valuable things themselves. This timeless independence of the world is the meaning of Hartmann's "ideal being." Second, he attaches a notion of "ground" to values like Plato's ideas insofar as values are "that 'through which' everything that participates in them is exactly as it is, namely, valuable."[83] Finally, like Plato's ideas, values are a priori for Hartmann: "[values] must be an a priori condition."[84] Despite the respective differences between the non-Platonic and Platonic views of Scheler and Hartmann, both agree that values are not subject to change or mutation and primarily apprehended though non-conceptual means – that is, though "emotional sensing" or feeling-acts.[85]

Building on Scheler and Hartmann, this text argues that emotions motivate commitments to import by virtue of "being" – that is, on the basis of their evaluative content.[86] Evaluative judgments, in turn, motivate individuals to act both directly and indirectly insofar as they succeed in committing individuals to something of import. Comprehending this requires an examination of the rational interconnections between emotions, desires, and evaluative judgments and so into the role evaluative judgments play in constituting import. Of particular significance is the way the linguistic concepts of evaluative judgment come to inform emotions, thereby making intelligible how emotions play a much more substantial role in practical reason than the "auxiliary" role (as adaptive mechanisms) that neo-Jamesian accounts permit. Rational emotions are commitments to the import of their foci – and thereby to their targets – having the evaluative property expressed by their formal objects. Such evaluations made within the matrix of rational emotions define the moral agent's "evaluative perspective."[87] In making evaluative judgments, individuals also articulate an evaluative perspective. The claim being made is that evaluative judgments, emotions, and desires together define a unified evaluative perspective: it is possible to make judgments with the same intentional content as one's emotions and desires (and vice versa) such that each is a commitment to the same import. This truth is revealed by the manifold rational interconnections obtaining among them. Here, an example from Solomon is helpful.[88] If Mr. X believes that Ms. Y has stolen his car, he is liable to be angry at her, thereby evaluating her, the target, as having offended him. If Mr. X subsequently discovers that Ms. Y did not steal it, his anger ought to disappear: this revised understanding of his circumstances should alter his emotional response, which is based on that same understanding. Consequently, if Mr. X continues to be angry at Ms. Y after making these new judgments in light of further information about the actual perpetrator, then his anger is irrational, other things being equal.[89]

In light of Solomon's example, at least two implications are noteworthy for the rational interconnections between emotions and evaluative judgments. First, if something has import by virtue of a projectable, rational pattern of emotions that share a common focus, then one ought to judge – other things being equal in the appropriate circumstances – that it has import.[90] One's worldview is typically unified rather than bifurcated, and so to misjudge this is to fail in one's genuine commitment to sustaining this evaluative perspective. Second, and conversely, to commit oneself in judgment to the import of something is to commit to sustaining the relevant evaluative perspective not only by making subsequent evaluative judgments when appropriate, but also by having the relevant emotions. This is true not simply when the judgment corresponds to the focus of a rational emotion, but also when it corresponds to the emotion's formal object. Indeed, without a general resonance between evaluative judgments and rational emotions, the commitment to import one undertakes in judgment or experiences in emotion would be defective. The conclusion here is that emotions and evaluative judgments are rationally connected not merely insofar as they can conflict (requiring the subject to propose a rational resolution by modifying one or the other) but, more fundamentally, inasmuch as they together define a unified worldview – a single evaluative perspective – which can diverge only irrationally. Hence, as human beings with the capacity for evaluative judgment, import is constituted by projectable, rational patterns of emotions. This conclusion refutes neo-Jamesian accounts, which understand the interconnections between emotions and judgments fundamentally in causal or dispositional rather than rational terms. As a condition of the possibility of being committed to import, individuals are motivated to act in the appropriate circumstances. By making emotionally rational evaluative judgments and thereby committing oneself to import, one has the capacity to act freely and willfully.[91]

### 4.3. Emotional Rationality and Morality

Emotions play many roles in human life, and none more important than forming the enduring individual and social bonds necessary to secure adequate moral judgment. Although the philosophical critique of emotional rationality is often mistakenly attributed to Kant,[92] his claim that consciousness of one's obligations depends on the capacity to feel them projects the immediate significance of emotions in the moral arena. Here, two points deserve notation. The first is that emotions play a major part in motivating moral behavior. Persons are motivated to provide assistance to others by virtue of their affection, affinity, or compassion.[93] Emotions also motivate individuals to pursue justified punishment or revenge, both of which are moralistic behaviors.[94] Second, emotions are, as suggested above, critical to moral epistemology. Moral evaluation is frequently associated with perception, and this capacity is considered to possess an emotional foundation. In other words, something is considered prima facie good or bad, right or wrong in light of the emotional response it elicits. Properly tempered by objective rationality, emotions serve as tools that perceive essential moral data and alert individuals to the presence of significant moral events.[95]

### 4.3.1. The Cognitive Dimensions of Emotionally Rational Moral Intuition

Moral stimuli are inherently emotionally charged, and a large corpus of neurological research has confirmed that rational emotions influence moral cognition.[96] Investigative paradigms have ranged from the passive viewing of pictures portraying moral violations to complex ethical decision-making tasks, including moral dilemmas and theory of mind exercises wherein affective reactions permeate and influence higher-level cognition. Previous neuroimaging studies have identified a specific cortical network recruited in moral processing, including the medial PFC, right TPJ, posterior cingulate, precuneus, and STS.[97] This research has also demonstrated that brain regions involving moral thought overlap with cognitive networks associated with reasoning and processing of social and emotional content, suggesting that emotional rationality plays a critical role in moral judgment. Such emotionally rational moral intuition has become an intriguing topic attracting significant attention within the domain of contemporary cognitive neuroscience. Jonathan Haidt has proposed that moral intuition refers to fast, automatic, and typically unconscious emotionally-laden processes.[98] In contrast to intuition, Haidt suggests that moral reasoning – a cognitive process – is a controlled and "cooler" (i.e., less affective) process involving conscious mental activity. For Haidt, then, moral reasoning is a post-hoc process in which one searches for evidence to support an intuitive moral reaction.[99] However, Joshua Greene and colleagues posit a converse theoretical perspective on the interactions between moral intuition (presumed by Haidt to be a strictly affective process in his social-intuitionist model) and moral reasoning (presumed by Haidt to be a strictly cognitive process).[100] Focusing on fMRI studies of moral dilemmas, the authors illustrate that increased cognitive processing in the DLPFC predicts utilitarian moral behavior. This suggests that cognitive processes can sharpen and even override emotional responses, favoring personal moral violations (utilitarian judgments) when the benefits sufficiently outweigh the costs. On this basis, Greene and colleagues propose that both emotional and rational processes are vital for moral judgments.[101] Thus understood, emotionally rational morality is a post hoc process that refines and edits initial moral intuition.[102]

Despite the considerable interest in emotionally rational moral intuition and the wealth of historical research on the intersections of emotion, reason, and morality, the specific neural correlates of emotionally rational moral intuition remain poorly understood.[103] Of particular contemporary interest is the time course of data processing in moral cognition, which can be elucidated through ERPs due to the high temporal resolution of ERP signals.[104] At least three ERP components are related to emotionally rational morality. The first component is N1, a negative wave peaking approximately 100 milliseconds after stimulus onset. Some studies have revealed that brain regions such as the posterior STS/interior parietal region are recruited during the early stages of moral-related data processing, as quickly as sixty-two milliseconds post-stimulus.[105] Recently, one study found that the N1 component could distinguish between a morally good and bad action in a picture-viewing task.[106] This research suggests that emotionally rational moral intuition is a fast and automatic process likely indexed by N1. The second component of relevance is N2, a negative wave peaking between 200 and 350 milliseconds after stimulus onset. N2 has been shown to index cognitive control, negative emotional processing, and the perception of others' pain. A contemporary study by Keith Yoder and Jean Decety reports that scenarios of morally good actions elicited higher N2 amplitudes than did morally bad actions.[107] Hence, N2 could potentially index emotionally rational responses triggered by moral stimuli. The third component of relevance is P2, a positive wave peaking at approximately 200 milliseconds after stimulus onset. In the study of Michela Sarlo and colleagues, a larger P2 was found in the frontopolar and frontal areas when subjects decided on instrumental rather than incidental dilemmas.[108] Taken together, the empirical evidence from ERP studies on N1, N2, and P2 further suggests that emotionally rational moral intuition is a fast, initial moral process.[109]

In the 2015 study of Dan-Yang Gui and colleagues, which investigates how processing moral content is influenced by emotional arousal, early frontal N1 peak amplitudes elicited by pictures with moral content were significantly lower than those elicited by pictures without moral content.[110] This suggests that emotionally rational moral thoughts occur at early stages of the ethical decision-making process. The fact that moral stimuli are processed at such an early stage is consistent with the findings of several previous studies. Jean Decety and Stephanie Cacioppo, for instance, found that intentionally and accidentally harmful actions can be distinguished as quickly as sixty-two milliseconds after participants have viewed morally-laden scenarios.[111] In light of these results, it may be postulated that emotionally rational moral intuition, as a fast and automatic process, may be indexed by the early ERP component N1. Until recently, emotionally rational moral intuition was merely a presumption from numerous behavioral studies. The empirical existence of emotionally rational moral intuition has not been shown via fMRI insofar as the intuitive processes occur at an early time state, which cannot be captured by low temporal resolution. The evidence presented by Gui and colleagues for the neural underpinnings of moral intuition are consistent with the model of Marc Hauser, which proposes that moral events trigger a causal intentional analysis prior to the involvement of emotional and rational reasoning.[112] Gui and colleagues further substantiate this claim by pointing to the standardized low resolution brain electromagnetic tomography analysis (sLORETA), which indicates that the most likely source of the N1 component was in the right inferior parietal lobule (BA 40), a brain region indispensable for causal intentional analysis underlying the process of moral judgment in the Hauser model.[113] Nevertheless, given the indubitable complexity and multidimensionality of rational affectivity, this hardly suggests that rational emotions play no role in a sophisticated moral intuition.

The PCC is a vital node in the default mode network involved in internally directed cognition (e.g., self-awareness and autobiographical memories), regulating the focus of attention and signaling environmental change.[114] In the very early stages of moral cognition, the PCC may serve as a bridge between internally mental representation and externally sensory information. The postcentral gyrus is located in the primary somatosensory cortex, and the involvement of the somatosensory cortex described in the study of Gui and colleagues is consistent with previous studies suggesting that abstract thoughts regarding moral concepts are grounded in sensory experiences.[115] Hence, emotionally rational moral intuition may also be represented in the early somatic perception of an external world inasmuch as physical and moral experiences are psychologically interwoven. Moreover, the frontal P2 and widely distributed N2 and late positive potential (LPP) were influenced by affective arousal, indicating that emotionally rational intuition has a high impact on moral thinking at these stages of neural processing. Lastly, the sLORETA analysis of the N2 and LPP demonstrated that the source of maximal difference between high and low emotional arousal conditions was located in the parahippocampal gyrus (BAs 28 and 34). The parahippocampal gyrus has dense interconnectivity with the amygdala, and plays a principal role in emotional picture perception. The emotionally rational intuitive processes indexed by the LPP are memory related and thus serve to evoke rich associations. This last finding suggests that emotionally rational moral intuition not only entails an arousal-related response in the amygdala, but also an engagement of emotional- and memory-associated hippocampus activities. Taken together, the foregoing results substantiate the conclusion that rational moral intuition is the intuitive detection of social norms entrenched in the minds of social beings and derived from social evolution.[116]

### 4.3.2. The Psychology of Emotionally Rational Moral Value

The role of emotional rationality in the psychological appraisal of moral value has been the focus of philosophical dispute for decades.[117] Critically, the question of whether emotional rationality influences moral judgment or whether it merely motivates morally relevant action has been raised. Notwithstanding the underwhelming evidence on either side of the issue, it has long been recognized that violating moral norms is extraordinarily emotionally taxing. Although emotional rationality does not always prevent individuals from contemplating morally contemptuous actions, the experience of guilt and shame typically compel individuals to stop short of immoral action. Indeed, psychopaths – who lack the capacity for empathy and guilt – often fail to inhibit their violent tendencies. These realities provide the neuropsychological backdrop against which the role of emotionally rational moral value can be understood. Neurobiological data have been offered in support of the claim that affective circuits are integral to the rational evaluation of affective moral claims. For example, value judgments concerning affective moral claims (e.g., "the elderly are burdensome on society") show greater activity in the frontal polar cortex (FPC) and medial frontal gyrus than do value judgments about nonmoral claims (e.g., "pencils cannot write well").[118] Moreover, affective moral stimuli evoke increased functional connectivity between the left FPC, OFC, ACC, and anterior temporal cortex, and limbic structures such as the thalamus, midbrain, and basal forebrain.[119] Finally, personal moral dilemmas that are acutely affective in nature (e.g., the trolley problem) more selectively recruit emotional circuits than do their impersonal counterparts (e.g., the bystander dilemma). As interesting as these neurobiological data are, however, they simply indicate that perceived deontological value violations are associated with strong affective responses – a perspective that few could doubt or deny.[120]

A stronger test of the hypothesis that emotional rationality is the source of moral value judgments is highlighted by studies of patients with adult-onset, bilateral damage to the VMPFC. These patients consistently exhibit (i) a dampening of social emotions, indicated by behavioral and psychological measures; (ii) an incapacity to reestablish emotional representations previously associated with punishment and reward; (iii) an incapacity to anticipate future outcomes, punishments, and rewards; and (iv) a lack of inhibitory control.[121] On this basis, Damasio argues, for instance, that emotional rationality is typically integral to moral cognition and, on this basis, hypothesizes that moral value judgments are likely to rely on the affective processes implemented by the VMPFC.[122] Consistent with this argument, frontotemporal dementia (FTD), resulting from the deterioration of the prefrontal and anterior temporal cortex, generates blunted emotion, disregard for others, and a willingness to engage in moral transgression. Moreover, FTD patients demonstrate a pronounced tendency to adopt utilitarian alternatives in personal moral dilemmas.[123] With these results in mind, Michael Koenigs and colleagues examined the value judgments of the aforementioned VMPFC patients for moral and nonmoral dilemmas.[124] Each dilemma was classified as involving personal or impersonal violations, with personal violations further subdivided into low conflict (distinguished by short reaction times and low variance in subject judgments) and high conflict cases. VMPFC patients were indistinguishable from controls, save for their escalated tendency to endorse utilitarian outcomes in high conflict dilemmas. Overall, these data indicate that notwithstanding the experience of flattened socio-affective profiles, VMPFC patients judge most moral dilemmas as do healthy controls. Moreover, given the limited range of deviation exhibited by VMPFC patients, it is plausible to conclude that these individuals fail to treat morally salient features of high conflict dilemmas as such. If this interpretation is correct, then moral cognition would yield deviant outputs as a result of deviant inputs, rather than as a result of deficient moral processing per se. However, the dampening of negative affect would yield more permissible judgments due to a failure to focus on the antecedently morally salient features of the scenarios, thereby proving impedimentary to a comprehensive and mature apprehension of emotionally rational moral value.[125]

Emotionally rational moral value judgments are associated with unique patterns of cognition.[126] Cognitive appraisals, or the way individuals interpret and make meaning of their environment, trigger and persist throughout the experience of emotionally rational moral reflection. In turn, these affective-rational appraisal tendencies define how specific emotions color subsequent value judgments by prioritizing specific concerns semantically related to the emotion's appraisals. Fear, for instance, is induced and partly characterized by value appraisals that events are uncertain and outside of one's control. As a result, when an individual experiences fear, subsequent value judgments reflect increased concern about uncertainty and reduced control, even when the object of value judgment is unrelated to the original cause of fear. In studies investigating this phenomenon, participants triggered to experience fear in one situation were later discovered to prefer risk-averse options related to the events that elicited the fear. Conversely, participants triggered to experience anger preferred risk-seeking options, consistent with anger's constituent value appraisals of high certainty and control.[127] Similar appraisal patterns have helped to clarify the distinct role of emotionally rational moral value, including causal attributions and endowment effects. This appraisal tendency perspective sets the stage for predictions about how emotional rationality prioritizes specific moral concerns and thus promotes different value judgments. These predictions derive from research highlighting that certain emotions arise from appraisals with diverse moral themes. Appraisal themes tie the emotion to the particular concerns known to underlie value judgments across cultures, such as justice, purity, and hierarchy.[128]

Emotional rationality is an embodied phenomenon, engaging ancient mammalian response systems that involve skeletal muscle movements and activity within the central and peripheral nervous systems.[129] The somatosensory components of emotionally rational moral value judgments can shape memory, attitudes, information processing, and decision making, and therefore serve as cardinal guides to social interaction. A significant portion of the mounting evidence for the embodiment of emotional rationality is rooted in the exploration of the role autonomic physiology plays in value judgments. Research on the somatic marker hypothesis evinces that physiological responses indicative of negative emotional arousal – viz., spikes in skin conductance – can signal the positive and negative value of events. For instance, Sid Carter and Martha Pasqualini observed that study participants who exhibited higher skin conductance just before making disadvantageous moves in a gambling exercise soon developed premonitions that such moves were detrimental and avoided them.[130] These findings reveal embodiment effects for global affective states. There are analogous effects for distinct emotionally rational moral values. When individuals demonstrate a somatic element of a distinct emotional state, even beneath the level of awareness, their value judgments shift as is they were subjectively experiencing the emotion. For example, one study measured whether unwittingly posing the prototypical facial expression of anger would lead to greater attributions of human agency for various events – a causal attribution pattern that emerges when individuals experience anger. As expected, participants covertly manipulated to pose anger rated future life events as driven more by human agency (and therefore less by situational causes) relative to participants posing sadness.[131] Finally, new data suggests that disgust-related physiological responses strengthen value judgments. In one study, disgust primed through repulsive films or offensive odors led to harsher critique of others' moral transgressions, particularly for individuals highly sensitive to their own somatic changes – preliminary evidence that bodily aspects of disgust underlie the tendency to strengthen moral judgment.[132] This literature has significant implications for the study of the effectual distinctiveness of emotional rationality on moral value judgments. The embodied elements of emotional rationality – including facial or skeletal muscle movements, and activation in peripheral physiology and the neuroendocrine system – prepare the individual to meet specific challenges and opportunities, most of which are inextricably laced with moral value.[133]

### 4.3.3. The Rationality of Affective Moral Objectivity

While little attention has been heretofore paid to John Rawls' account of the moral emotions,[134] a comprehensive analysis of the affective tone of his theory of justice gleans significant insight into the rationality of affective moral objectivity. While moral emotions do not play a central role in Rawlsian justice, he nevertheless attributes an important role to them – or, more precisely, to moral sentiments, conceived as the general dispositions they manifest – in moral life and rational choice.[135] His developmental account defends the stability of a "well-ordered society" – that is, one arranged in accordance with principles of justice. For Rawls, social stability amounts to the likelihood that a set of social arrangements give rise to sentiments that support (rather than undermine) justice, such as a desire to act in abidance with its principles and an inclination to experience guilt upon violating them. Rawls argues that his two principles of justice (i.e., the principle of equal liberty and the difference principle) effect enough stability to make a society founded on them viable, and interprets guilt as developing into multiple stages of (eventually widespread) emotionally rational moral value through the promotion of increasingly cultivated cognition within conditions of love and trust that subsequently aim to increase self-esteem. Rawls's account of the development of emotionally rational more value, and thus an appropriate sense of justice, begins from a general assumption of rational, and therefore proportionate, emotions.[136]

Beyond its exhortation to take seriously the principles of justice with which societies must requisitely comply in order to live well, it presupposes communities comprised of supportive families, peers, and other cooperative social groups who first agree to abide by these operative concepts of equality and goodness. However, because the breach of normative principles is inevitable even in well-ordered societies, Rawls contends that guilt is the emotion that ideally serves as the guiding force to rehabilitate behavior to its intended moral state. The role of emotion in securing morality is, then, for Rawls, visibly justificatory, and it provides essential support to the operative principles of normative morality. In this way, Rawls' provides vital insight into the emotional framework of normative morality with the assignment of objective status to affective intuitions, which consequently hints at their rational evaluative capacities. While Rawls' social principles permit various moral views as reasonable alternatives, his personal view favors a Kantian respect for persons. On this basis, Rawls assigns two primary structural features to emotions in defense of a normative view. First, emotions are evidently justificatory, albeit secondary to his argument for the two principles of justice. Insofar as Rawls' developmental account is needed to establish the stability of a society ordered in accordance with the principles, emotions provide the essential support without which principles would not be worth instituting. Second, Rawls sees the role he assigns to emotion in support of normative principles as harmonious with the assignment of objective status to them. The claim of rational affective objectivity emerges explicitly in his earlier writings on Kantian constructivism in metaethics,[137] but as interpreted in a sense distinct from factual truth, with Rawls' principles of justice seen as limited to a particular cultural setting, namely, a democratic pluralist society. The contractual basis of Rawls' theory – in the rational choice of principles in an original position characterized by the ignorance of their distinguishing features – is understood as a reasonable procedure under these circumstances for constructing principles of justice. The principles therefore count as objective, without any claim of correspondence to an independent order of moral facts. It is their basis in a rational decision-making process that confers objectivity.[138]

Against this Rawlsian backdrop, affective development may be viewed as a way of building social norms into objective touchstones of emotionally rational moral reflection.[139] In the process of prudential reasoning, emotionally rational moral objectivity brings the future to bear on the present circumstances, reflecting envisioned consequences of actions in more or less immediate affective comfort or discomfort. In relation to ethics, emotionally rational moral objectivity reflects a social perspective within the individual standpoint, concretizing for the agent the consequences of action for others. Through this process, it imbues moral reasons with additional practical reasons in the agent's interests – that is, reasons for sustaining or minimizing affect. That one might feel guilty after choosing wrongful action, for instance, is a reason for one to avoid performing it, given that the experience of guilt is an unpleasant emotion that one has reasons to avoid experiencing. Thus understood, affective objectivity does not function simply as a blind psychological force acting upon human beings, but rather plays a secondary role within practical moral reasoning. To be sure, explicit appeal to one's feelings would keep one from acting exclusively for moral reasons in the way favored by the Kantian approach expressed above. Nevertheless, what is typically in question is the availability of emotions as reinforcement for moral reasons that individuals might be otherwise inclined to discount. Hence, an objective moral code that is not learned in conjunction with rational emotions would be at a serious disadvantage with respect to viability. "Wrong" and similar terms and concepts would carry no more affective developmental weight than would "out of bounds" in a trivial game in which any particular agent might be disinclined to play.[140]

Emotions constrain the content of moral codes insofar as norms based on considerations too remote from ordinary concerns or too complex to be learned would not be viable.[141] However, the significant point concerning the rationality of affective moral objectivity is that coexisting emotions do not themselves supply the context of robust ethical analysis. Instead, certain emotions come to incorporate moral reasons that are capable of independent formulation. The moral reason for experiencing guilt – for instance, that the act one is pondering would cause someone disproportionate harm, or that it would therefore be wrong – supplies the objective content of the emotion. Despite having concurrently developed with the element of rational affect – that is, with the motivational force dependent on a particular history – the emotion in question can be fine-tuned separately at a more advanced stage. In effect, emotions involve two layers of rationality, corresponding to both affective and evaluative objectivity. At the advanced stage, individuals can separate the element of affect from moral reasons originating initially from the fundamental precepts of caretakers, but finally capable, as Rawls suggests, of reflecting principles too complex to be accessible to ordinary moral consciousness. Whatever objectivity applies to affective moral reasons from these sources – interpreted, perhaps, in realist rather than Rawlsian constructivist terms – will not be undermined by their motivational basis in emotion. Indeed, moral action relies on emotions as a cardinal heuristic that, in the first instance, is epistemic: emotional rationality aids one in the objective formation of practical moral judgments.[142]

### 4.3.4. Emotional Rationality as Canon of Moral Motivation

Emotional rationality has a long history of being understood as the canon of moral evaluation and, as a corollary, moral motivation. If individuals rejoice over an event, they typically see it as good, and they usually wish to pursue similar events in the future; if they grieve over an event, they typically see it as bad, and they usually wish to avoid similar events in the future.[143] Plato held that the human capacity to love could be cultivated, through right rearing and philosophical education, to become accurate in its evaluations, and to thereby motivate individuals to respond to the genuinely good with emotional pleasure to the genuinely bad with displeasure.[144] His presupposition was that goodness and badness were objective qualities, and that emotional rationality could, therefore, serve epistemically or, at a minimum, as an aspect of epistemic power. Plato noted that the same event – for instance, the victory of a particular team in an athletic event – may occasion rejoicing in some and grieving in others, and that this is due to a lack of perspective on the part of the two groups of perceivers – that is, to the fact that such individuals are not epistemically and morally mature, or to the fact that they are not expressing the best part of themselves at the moment. Some influential philosophers in the eighteenth century were of opposite mind to Plato, the most notable of which was David Hume. Among other things, Hume wanted to humanize the phenomena of value and the way in which value perception contributed to subsequent moral motivation. He observed that even the sages of the world sometimes contradict one another in their moral sentiments and motivations. In "Of the Standard of Taste," Hume points out that the Greek hero Ulysses seems to delight in lies and fictions, often employing them without necessity or advantage; that the Qur'an's prophet Mohammed bestows praise on instances of treachery, inhumanity, cruelty, revenge, and bigotry as compatible with civilized society; and that the Greek stoic Epictetus scarcely mentioned the sentiment of compassion without simultaneously warning against it.[145] On similar grounds, it is presumable that Hume would be equally repelled by the moral approvals or disgusts of Friedrich Nietzsche's sage Zarathustra, or Fyodor Dostoyevsky's warm appreciation of the more coenobitic traits of Alyosha Karamazov. The anti-relativist tone of Hume's remarks indicates that the above characters are in serious error about values.[146] If the distinctions on which moral judgments turn are created by properly-ordered emotions, then they must be created according to a rational set of morally qualified motivations.

While competing ethical theories frequently differ on the question of which actions and concerns should be considered morally relevant, there is agreement that any action vying to be considered moral should have certain minimal characteristics, at least two of which are principal.[147] First, any moral action should be intentional. Moral actions cannot be produced accidentally or result from causes outside of the agent's consciousness. In other words, moral actions must be the result of reasons. Second, the reasons that generate moral actions must themselves be moral; that is, they must be related to what is morally good or bad in the agent's understanding. If either of these conditions is not met, moral praise, blame, or responsibility cannot be spoken of. Thus understood, motivation rooted in emotional rationality may be considered moral insofar as it leads to actions in their formal moral quality – as morally good or bad. That is, an action must derive intentionally from moral motivation. Moreover, the agent's reason for action must simultaneously serve as its motivation, and this motivation-reason must be seen by the agent as morally relevant. In considering these assumptions, the question arises over whether rational emotions provide moral motivation, giving origin to moral action and ultimately a sustained moral life. Against the backdrop of traditional psychology, the answer would be negative. Historically, emotion has been seen as arising unintentionally, as generating its action tendencies automatically, and as being regulated by processes that are considered to be mostly unconscious. Additionally, the meaning of emotion is essentially given by those concerns from which the emotion derives. Therefore, an emotion would have moral meaning on the condition that it reflected moral concerns and moral motivation. In sum, it appears that rather than explaining moral motivation, rational emotion prerequires moral motivation in order to explained in moral terms.[148]

The foregoing explanation may seem too glib vis-à-vis the question of why rational emotion could not be accepted as a form of moral motivation if, having been elicited from spontaneous concerns, it generates positive actions with prosocial consequences. In addressing that question, it is important to note, first, that consequences do not determine whether an action is moral or immoral when viewed objectively and externally.[149] For instance, the objective observer may act concurrently as moral agent, realizing only after the fact that the actions performed were good. What would be required is that the agent understood, before the action, that the anticipated consequences were morally good and intended them by performing the action. In this case, the consequences are seen from, and form a part of, the agent's moral concerns out of which rational emotions arose. This analysis suggests that the moral quality of rational emotions cannot be grounded in their natural existence or natural production of certain effects. A naturally elicited empathic response is not moral per se, even if it generates prosocial behavior. Emotional rationality becomes moral when it is inserted into the special perspective of moral goodness that transforms its naturalness and spontaneity into cultivated intention. This is not to say that spontaneous emotions could not be the precursors of moral concerns. A natural tendency to respond empathetically may predict a subsequent development of altruistic concerns, just as early natural desires for revenge may be related to a future concern for justice. It is an empirical question as to whether these relations exist or not. To become moral, however, "raw" emotion must be shaped, refined, and directed according to an objective framework that is not naturally given. This work of reorientation demonstrates the moral limits of spontaneity, and is particularly relevant to the psychological work on the capacities of children younger than three or four.[150]

There is still another way of understanding emotional rationality as the canon of moral motivation and, finally, moral action. On this view, spontaneously elicited emotions naturally lack the intentionality and the particular moral meaning required to produce moral motivation, but the agentic regulation of spontaneous emotions would emphasize and become the focus of explanation.[151] Here, there are two ways for emotional rationality to have moral meaning and therefore contribute to moral motivation. In one way, emotional rationality would be shaped and directed by agentic regulatory processes. Insofar as this form of regulation is consciously motivated, it can be guided by concerns that are specifically moral. For instance, one may actively work at creating or amplifying feelings of guilt or compassion, sensing that in so doing one would be more likely to pursue a course of action considered moral or refrain from one considered immoral. In some cases, the psychological characteristics of the affective response may be impacted by regulatory processes. In other cases, the spontaneous emotion may remain fundamentally unchanged, albeit consciously accepted, owned, and integrated into one's moral concerns. According to these accounts, the moral significance of emotional rationality and its capacity to serve as the canon of moral motivation depends on the prior existence of moral concerns, even when affective responses consequently reinforce these concerns and their effectiveness in appropriate action formation. Thus understood, it is clear to see how emotional rationality is inherent to the process of practical reasoning, and insists on personal responsibility as the basis for rational and moral goals. However, this can only be done within a motivational context in which concerns for rationality and morality are already present. Nevertheless, it would be a grave mistake to reduce emotional rationality to its instrumental function. Emotional rationality can be an intrinsic component of moral concerns: the experience of guilt, for example, for acting against one's moral values seems to be an essential part of caring about morality. Hence, emotional rationality can have purely expressive value, regardless of whether any action is called for or follows. Within the functionalism of traditional psychology, it is all too tempting to underestimate and neglect the motivational significance that symbolic expressions possess, not only in social expression, but also in the intimacy of self-reflection and self-appraisal.[152]

## 4.4. Emotional Rationality and Ethical Decision Making

Insofar as the VMPFC is essential for decision making, and because both rational and emotional systems are active within the VMPFC, ethical decision making can be understood as a complex process in which emotional rationality plays a decisive part.[153] Indeed, this is what leads Damasio to conclude that patients who suffer damage to and dysfunction in the VMPFC are rendered impaired in both rational and emotional processes.[154] Hence, emotional impairments are intimately linked with irrational moral choices. Given the scientific evidence that the thalamic-amygdala – a primitively hard-wired neurological system – may potentially function independently of cognition, philosophical arguments such as that posited by David Hume become more understandable.[155] For Hume, reason is but the slave of emotion; therefore, moral decision making is driven primarily by affection.[156] Against this idea, Peter Singer concludes that the only way to avoid moral skepticism in decision making is to detach moral judgments that are owed to cultural history from those that possess a rational basis.[157]

### 4.4.1. The Cognitive Neuroscience of Emotionally Rational Ethical Decision Making

The somatic marker hypothesis is based on the assumptions that (i) human reasoning and decision making depend on multiple levels of conscious and unconscious neural operation; (ii) cognitive operations, irrespective of their content, depend on support processes such as attention, working memory, and emotion; and (iii) reasoning and decision making depend on the availability of knowledge about situations, actors, options for action, and outcomes.[158] Throughout higher-order cortices and subcortical nuclei, this knowledge is stored in dispositional form and made explicit in varied and complex motor responses. The results of motor responses, including those generated unconsciously, can be represented through images, and the visual product can be classified in one of four ways: (i) as innate and acquired knowledge concerning bioregulatory processes and biophysical states and actions, including those that manifest as emotions; (ii) as knowledge about entities, facts (including relations and rules), actions, and action complexes (including narratives), which typically become known as images; (iii) as knowledge about the interconnections between (i) and (ii), as reflected in individual experience; and (iv) as knowledge resulting from the categorizations of items in (i), (ii), and (iii).[159]

The VMPFC is a repository of dispositionally recorded linkages between factual knowledge and bioregulatory states.[160] Structures in the VMPFC provide the emotionally rational substrate for learning the association between certain classes of complex moral situations, on the one hand, and the type of bioregulatory state typically associated with that class of circumstance in past subjective experience, on the other hand. The ventromedial sector houses links between facts that compose particular moral scenarios as well as emotionally rational sentiments previously paired with it in an individual's contingent experience. The linkages are dispositional insofar as they do not contain the representation of facts or sentiments of emotional rationality explicitly, but rather contain the potential to reactivate emotionally rational sentiments by acting on the appropriate cortical or subcortical structures. The experience acquired vis-à-vis a complex moral scenario and its components – including a configuration of actors and actions requiring a substantive ethical response, and a set of immediate and long-term outcomes for each response option – is processed in sensory, imagetic, and motor terms and subsequently recorded in dispositional and categorized form. However, inasmuch as the experience of some of these components has been associated with emotionally rational moral responses triggered from cortical and subcortical sites that are dispositionally prepared to respond, it is proposed that the VMPFC establishes a link between the disposition for certain aspects of a situation – for instance, the long-term outcome for a type of response option – and the disposition for the type of emotional rationality that in past experience has been associated with similar moral situations.[161]

When individuals face morally complex situations for which some factual aspects have been previously categorized, the pertinent dispositions are activated in higher-order cortices.[162] This leads to the recall of pertinently associated facts, which are experienced in imagetic form. Simultaneously, the related ventromedial prefrontal linkages are activated, and the emotionally rational disposition apparatus is competently triggered as well. The result of these combined actions is the reconstruction of a previously learned factual-emotional set. The reactivation of emotionally rational sentiments by cortical or subcortical structures mentioned above can be carried out via the "body loop," in which the soma actually changes in response to the activation and the ensuing changes are relayed to somatosensory cortices, or via the "pseudo body loop," in which the soma is bypassed and reactivation signals are conveyed to the somatosensory structures that then adopt the appropriate pattern. From both evolutionary and ontogenetic perspectives, the body loop is the original mechanism but has been superseded by the pseudo body loop and is possibly used less frequently. The result of either the body loop or the pseudo body loop may become overt (i.e., conscious) or remain covert (i.e., non-conscious). The establishment of a somatosensory pattern appropriate to the situation is co-demonstrated and qualified by factual evocations pertinent to the moral dilemma at hand. This constrains the process of reasoning over multiple potential ethical decisions and future outcomes. For instance, when the somatosensory image that defines a certain emotionally rational response is juxtaposed to the images that describe a related scenario of future outcome and that activated the emotionally rational response via the ventromedial linkage, the somatosensory pattern marks the scenario as morally good or morally bad.[163]

When the process of reasoning about multiple ethical decisions is covert, the somatic state operates as an alarm or incentive signal. In that instance, the somatic state is alerting the individual to the moral goodness or badness of a certain ethical option-outcome decisional pair. The device produces its result at the openly cognitive level. When the process is covert, the somatic state constitutes a biasing signal. Using a non-conscious influence (e.g., through a non-specific neurotransmitter system), the device influences the cognitive processing of emotional rationality. Certain ethical option-outcome decisional pairs can be rejected or endorsed, and pertinent facts can be more effectively processed. This suggests that somatic markers typically help constrain the ethical decision-making space by making that space manageable for logic-based, emotionally rational analyses. In situations having remarkable moral uncertainty about the future and in which the decision should be influenced by previous subjective experience, such constraints permit the moral agent to decide with ethical efficiency within short time intervals. Conversely, lesions of the VMPFC interfere with the normal processing of somatic or emotional signals but leave other cognitive functions minimally affected. This damage leads to pathological impairments in the emotionally rational ethical decision making process that acutely compromises the efficiency of daily life. The somatic marker hypothesis is therefore consistent with the views of others who invoke a primary role or mood, affect, and emotion in moral reflection.[164] Through the development of the pseudo loop body operation, then, the somatic marker hypothesis provides the principal guide for emotionally rational ethical decision making.[165]

### 4.4.2. Emotional Rationality as Ethical Insight and Practical Reason

Novel ethical insight occurs when an emotionally rational interpretation of a situation or solution to a moral problem unexpectedly springs into conscious awareness, seems obviously correct, and is accompanied by the subjective experience frequently referred to as the "aha" phenomenon.[166] Although the necessity of some components of ethical insight have been disputed, most scholars agree on the three fundamental characteristics mentioned above. Contemporary theorists add to this basic understanding the requirement that the moral agent must experience a restructuring of mind and heart with regard to some aspect of the moral problem or the solution in order to achieve the apex of ethical insight. Grounded in emotional rationality, this form of insight is often contrasted with a more formally analytic problem solving in which the problem solver consciously and deliberately manipulates elements in pursuit of a logical solution. However, the possibility that insightful ethical problem solving processes share the same neural mechanisms must also be considered. Theorists belonging to the conventional analytic camp have argued that the cognitive programs by which moral problems are solved via ethical insight are identical to those used in search solutions, and that it is only the affective experience that distinguishes the processes. Considering the undeniably large neurobiological overlap, it makes sense than an affective experience (similar to that occurring with an "aha"), and possibly other cognitive processes typically associated with ethical insight, could operate when achieving solutions with conventional analysis. Historically, researchers have cited the distinct qualities of emotional rationality within the ethical insight experience and the inexplicability of its process as evidence that the cognitive mechanisms involved in achieving ethical insight solutions must be distinct from non-insight solving. Theorists adhering to this view belong to the unconventional affective camp, and evidence to support their perspective has steadily accumulated. Recently, demonstrations that the right cerebral hemisphere makes a unique contribution to ethical insight not evident in analytic processes have led most researchers to accept that ethical insight is a markedly emotionally rational process.[167]

When Gestalt psychologists first began to discuss insight almost a century ago, they distinguished between insight and analysis based primarily on the diverse affective experience that accompanies insight but also because insight often develops incidentally.[168] Their view was supported not only by shared phenomenological experience, but also by famous anecdotes of scientific insights that yielded creative solutions to vexing problems, including Archimedes' eureka moment in the bathtub, Newton's falling apple, and Poincaré's bus ride. One commonality among these experiences is that these individuals were not consciously considering the problem at the time they experienced insight. In contrast, insofar as analytic thought is deliberate by definition, solution by incidental analysis proves to be oxymoronic. Problem solving via emotionally rational ethical insight is also different from analytic problem solving to the extent that ethical insight involves processing that renders it inaccessible to metacognition. When Janet Metcalfe and David Wiebe asked research participants to periodically judge how close they were to achieving a solution during the course of a problem solving exercise, they found that prior to analytic solutions, subjects reported gradually increasing closeness to solution. In contrast, prior to insight solutions, subjects reported little or no progress toward solution right up until the point at which solution was imminent.[169] On this basis, Metcalfe and Wiebe concluded that insight was a fundamentally different process that involves critical mechanisms of emotional rationality that occur outside conscious awareness. Employing a somewhat different paradigm in which participants were interrupted during solution and asked to verbally describe their solution efforts, Jonathan Schooler and Joseph Melcher found a difference in conscious access to solution data between insight and analytic problem solving.[170] In an earlier study, Schooler and colleagues found that verbalization during solving interfered with the solution of insight problems but not analytic problems.[171] Based on these findings, Schooler and colleagues propose a verbal overshadowing effect on insight due to the inaccessibility of critical insight processes to verbal description. A more recent investigation suggests that verbal overshadowing might be specific to visual problems, regardless of insight or analytic solution strategy.[172] Nevertheless, much of the research on metacognition and verbal overshadowing of insight suggests that steps in the ethical insight process are beyond conscientious analysis. Furthermore, it is possible that the inaccessibility of emotionally rational ethical insight processes to linguistic analysis may be due to the involvement of the less-verbal right brain hemisphere.[173]

Beyond serving as the cardinal mechanism for ethical insight, emotional rationality also functions as the linchpin of practical reason during ethical decision making. Moral agents employ practical reasoning to shape or guide future actions. Insofar as practical reason bears on actions yet to be carried out, it cannot bear on action tokens: there are no relevant, individuable act tokens at the time practical reasoning occurs. Hence, practical moral reasoning must bear on act types, including the attitudes with which they correspond. By these means, emotional rationality provides reasons for thinking that certain types of action or attitude are required or forbidden, recommended or inadvisable.[174] The bearing of emotional rationality on ethical decision making can be captured in simple form by the first or major premise of an Aristotelian syllogism – Aristotle's three line schema of practical reasoning. In contemporary terms, the major premise expresses a pro or con attitude toward something that the second or minor premise then informs on how to attain or avoid. In other words, it evaluates something as good or bad, desirable or worth avoiding – or simply as an object of a present desire or aversion. Fear, for instance, may be said to represent its object (i.e., what one is fearful of or about: a certain event, or the likely cause thereof) as a threat and thus a possible harm. Anger represents its object (e.g., another person, tied with the performance of some action) as already a cause of harm or offense that now calls for retaliation. Despite differences between the propositional content of these two emotions – one describes a rational possibility while the other describes an actual fact – both include a reason for avoiding some future contingency: the presence or occurrence of what is feared or a failure to retaliate for the cause of anger.[175]

Emotional rationality also involves a corresponding element, derived from pleasure or pain in the Aristotelian account of emotions, which amounts to a good or bad state of the agent and hence supplies a reinforcing reason for moral action.[176] In decision-theoretical terms, emotional rationality modifies the "payoff structure" of the situation: the array of potential costs and benefits of alternate responses open to the moral agent. The discomfort experienced in fear or anger, on the one hand, provides the moral agent with the further (pro tanto) reason for acting to change the situation that provokes discomfort. On the other hand, the experience of joy and pride, as positive evaluations of some state of affairs or of oneself, do not provide a reason to change anything, but their effect yields a further reason to sustain the conditions that make the ethical evaluation appropriate. However, the focus of the moral agent's attention in each of these circumstances is the evaluative content of emotional rationality, not the experience of one's own state of feeling. Apart from its role as a source of reinforcing reasons, emotional rationality is useful to practical reasoning insofar as it serves to cognitively maintain the evaluative content of a particular emotion without explicit ethical reflection. While driving down the Autobahn, for instance, one does not need to deliberate at length about the possible negative consequences of swerving and its relevance to the task of steering straight. To the extent that steering straight is not just automatic, and one wanders out of one's lane, an anxious awareness of the possibility of accident brings one quickly back to the task at hand. This way of anticipating practical eventualities in daily life corresponds roughly to Damasio's understanding of emotions as somatic markers.[177] Damasio's view is put forth to explain why cases of affective impairment due to brain lesions (such as the famous nineteenth-century example of Phineas Gage) involve a loss of practical reasoning ability. On this account, emotional rationality serves to mark practically significant thoughts with bodily indicators of past experience. Thus understood, characteristic thoughts have come to be the contents of emotional rationality and part of what identifies them as the types of moral emotion they are. It is only in a qualified sense that an irrational evaluative stance could be said to justify a moral action. Hume's famous denial that the passions can be rationally assessed at all was grounded in his interpretation of them as non-representational. However, contemporary neuroscientific data makes clear that emotional rationality represents the evaluative propositions that make substantive practical ethical decision making possible.[178]

### 4.4.3. Emotionally Rational Moral Reflection as Ethical Problem Solving

Since Haidt's influential critique of rationalist moral psychology, the roles of reflection and reasoning in moral judgment and, hence, ethical problem solving, have remained unclear.[179] While many researchers, including Haidt himself, believe that reflection and reasoning play significant roles in moral judgment, the evidence for the claim remains surprisingly limited. Thus, the present aim is to document and characterize emotionally rational moral reflection as the necessary condition for ethical problem solving. Viewed through the lens of Haidt's social intuitionist model (SIM), emotionally rational moral reflection is predominantly intuitive, driven primarily by automatic affective responses that are effortless and produced by unconscious processes. Numerous studies now attest to the influence of automatic affective responses on moral reflection.[180] In accord with the SIM, emotionally rational moral reflection typically serves to rationalize moral judgments that were previously made intuitively. The SIM also posits that one may occasionally influence one's own judgments directly through intentional reflection or indirectly through one's intuitional reasoning. Finally, the SIM allows that one's expressed reasoning can influence others' judgment by influencing their reasoning. Other theories assign to emotionally rational moral reflection a more prominent position, at least rhetorically. For example, Greene and colleagues' dual process theory is consistent with the ubiquity of utilitarian moral reasoning.[181] Moreover, the hypothesis of David Pizarro and Paul Bloom suggest that emotionally rational moral reflection plays a prominent role in shaping moral intuitions,[182] and the emphasis by Shaun Nichols and Ron Mallon on the roots of moral rules in emotions suggest their rational application via sophisticated ethical problem solving techniques.[183]

Joseph Paxton and Joshua Greene recently reviewed evidence for the influence of moral reflection on ethical reasoning and concluded that the data is suggestive but limited.[184] For instance, Greene and colleagues have implicated controlled cognitive processes in moral judgment using fMRI and reaction time data,[185] while Adam Moore and colleagues have produced consistent results by examining individual differences.[186] Controlled regulatory processes have also been implicated in morally questionable behavior, such as rationalizing and moral hypocrisy.[187] However, the nature of these controlled processes remain unclear. More specifically, it may be argued that these controlled processes simply manage competing intuitions and do not involve genuine moral reflection. Individuals modify their subsequent judgments when implicitly instructed to make "rational, objective" judgments,[188] reject judgments that are inconsistent with competing judgments (when the inconsistency is highlighted by investigators),[189] and appear to override certain negative implicit attitudes.[190] While there is evidence that attitudes can, in general, be modified by reasoned arguments,[191] moral attitudes seem to be particularly stubborn in this regard. Hence, as the necessary condition of ethical problem solving, emotionally rational moral reflection requisitely transcends the task of managing competing intuitions. In the study of Paxton and colleagues, for instance, there was no evidence to suggest that the cognitive reflection test (CRT) manipulation influenced moral judgment by favoring one intuition over another.[192] In fact, the CRT manipulation was designed to induce a general distrust of intuition. Thus, its efficacy implies that the induced utilitarian judgments are, to some extent, counterintuitive and not merely driven by competing intuitions. Moreover, in demonstrating the persuasive power of an abstract argument based on the evolutionary theory of incest taboo,[193] Paxton and colleagues illustrate that it is possible to persuade individuals by appealing to both the rational and affective dimensions of moral reflection in ethical problem solving exercises.

Given the highly subjective nature of what constitutes a moral problem, researchers who study ethical problem solving have often presented individuals with novel problems (capable of being solved) and attempt to find regularities in the resulting problem-solving behavior.[194] Notwithstanding the variety of possible problem scenarios, scholars have identified important regularities in the emotionally rational moral reflection processes by which individuals (i) represent, or understand, problem situations, and (ii) search for possible ways to attain their goal. Originating with Gestalt psychologists who pioneered the extension of organizational principles of visual perception to the domain of problem solving, ethical problem representation is a model constructed by the moral agent that summarizes the agent's understanding of the problem components – that is, the initial state, the goal state, and the set of possible operators the agent may apply to get from the initial state to the goal state. Ethical problem components differ in the extent to which they are well defined: some components leave little room for interpretation, whereas other components may be ambiguous and therefore require the moral agent to provide a concrete definition. Through the process of emotionally rational moral reflection, the agent's representation of the problem guides the search for a possible solution. In turn, this search may change the representation of the problem and lead to a new search. Such a recursive process of emotionally rational moral reflection, representation, and judgment continues until the ethical problem is solved or until the moral agent decides to abort the goal.[195]

In 1945, Karl Duncker documented the interplay between representation and search based on his careful analysis of one individual's ethical solution to the "radiation problem."[196] This problem requires using radiation to destroy a patient's life-threatening stomach tumor without mortally harming the patient. At sufficiently high intensity, the radiation will successfully destroy the tumor. However, at that same intensity, it will also destroy the healthy tissue surrounding the tumor that is required for postoperative survival. At lower intensity, the radiation will not harm the healthy tissue, but it will also not destroy the tumor. Duncker's analysis revealed that the agent's solution attempts were guided by three distinct ethical problem representations. He depicted these solution attempts as an inverted "search tree" in which the three main branches correspond to the three general ethical problem representations. The desired solution appears on the rightmost branch of Duncker's tree, within the general problem representation in which the agent aims to lower the intensity of the radiation on its way through healthy tissue. The actual solution is to project multiple bursts of low-intensity radiation at the tumor from several points around the patient by use of lens technology. The low-intensity radiation will converge on the tumor, where their individual intensities will sum to a level sufficient to destroy the tumor. The relevance of perception and background knowledge to emotionally rational moral reflection and ethical problem solving illuminates the fact that, when individuals attempt to devise ways to reach their moral goals, they draw on, and engage in, a variety of affective and rational cognitive activities. If Duncker is correct, such goal-related activities invariably include (i) placing moral objects into categories and making inferences based on category membership; (ii) making inductive references from multiple phenomenological instances; (iii) reasoning by affective analogy; (iv) identifying the objective cause of events and their subjective consequences; (v) deducing the logical implications of given moral information; (vi) making emotionally rational moral judgments with concern for their social repercussions; and (vii) diagnosing deficient ethical problem solving techniques in light of personal history and moral norms.[197]

### 4.4.4. The Normative Logic of Emotionally Rational Ethical Decision Making

The normative logic of emotionally rational ethical decision making is concerned primarily with how emotionally rational moral reasoning should be conducted and executed in pursuit of the good. It grounds itself in the objective conclusions of logic, probability, and rational choice.[198] As such, this comprehensive system of decision making is indispensable to a robust neuroethical understanding of how human beings reason in at least three ways. First, normative ethical decision making (NEDM) theories attempt to provide a standard against which moral behavior can be addressed. That is, they offer an analysis of what a system is supposed to do: it determines what counts as successful performance and what counts as error. The juxtaposition of intentional behavior with various NEDM theories is also crucial for framing key questions about thinking and reasoning. For example, to ask whether, or to what extent, individuals reason deductively requires an analysis of what deduction is, and deduction is a logical notion. Similarly, ethical decision making is organized around comparisons with the theory of "correct" decision making, a concept that derives from emotionally rational choice. Second, NEDM theories provide a starting point for descriptive theories of moral reasoning, not merely a standard of comparison. Thus, just as a natural starting point for a theory of the operation of a calculator is that it follows, perhaps to some approximation, the laws of arithmetic, it is natural to assume that an emotionally rational moral agent follows, to some approximation, the normative logic of ethical decision making. Third, unless individuals' words and behavior obey the constraints of emotional rationality, their words and actions are rendered, quite literally, meaningless. This is why an individual randomly generating sentences or moving arbitrarily cannot usefully be attributed moral intentions, beliefs, or goals, at least according to most philosophical accounts. To sharpen the point: to see a human body as an individual with a mind, rather than a random collection of physiological responses, requires that the individual can be attributed some measure of emotional rationality, and NEDM theories attempt to make headway in explaining the nature of such constraints. Similarly, returning to the calculator analogy, without some degree of conformity with the laws of arithmetic, it would be impossible to interpret the keys and displays of the calculator as pertaining to arithmetic operations.[199]

Intuitions concerning the emotional rationality of any moral judgment are crucially influenced by the normative background against which those judgments make sense.[200] This type of global relationship, between a single judgment and the morass of background knowledge, is inevitably difficult to analyze. The logic of NEDM theory focuses on the local consistency (and consequence) relationships between judgments, which depend on those judgments alone, so that background knowledge becomes superfluous and irrelevant. Such local relationships can only depend on the structure of normative judgments; they cannot depend on the subject of the judgments inasmuch as understanding the subject of a judgment requires references to background knowledge (among other things). For instance, there is something wrong if a moral agent believes the following: (i) all human beings are dangerous; (ii) newborn children are human beings; (iii) newborn children are not dangerous. Roughly, the first two beliefs appear to imply that newborn children are dangerous, and yet the third belief is that they are not. To be sure, a minimal consistency condition for any agent is that it should not be possible to believe both P and not P insofar as this marks a contradiction. This local inconsistency holds completely independently of any facts or beliefs about danger, human beings, or newborn children. Normative logic transforms emotionally rational moral intuitions by translating sentences of natural language (expressing judgment) into formulae of precisely defined formal language and specifying inferential rules over those formulae. The notion that emotionally rational ethical decision making works in this way is embodied in the mental logic approaches in the psychology of normative reasoning. From the view of cognitive neuroscience, normative logic vis-à-vis emotionally rational ethical decision making can be viewed as providing a formal mechanism for detecting inconsistencies by determining the sets of judgments from which a contradiction can arise. Nevertheless, normative logic cannot be viewed as an independent theory of moral reasoning – that is, as a theory of how judgments should be changed in light of moral reflection. In other words, the normative logic of emotionally rational ethical decision making indicates inconsistency has struck, but it does not simultaneously demonstrate how consistency should be restored.[201]

The classical philosophical treatment of decision making, known as the "rational theory of choice" or the "standard economic model," posits that individuals have orderly preferences that obey simple and intuitive axioms.[202] In this model, when faced with a choice, individuals are assumed to gauge the subjective utility of each choice and to choose the alternative with the highest amount thereof. In the face of uncertainty about whether particular outcomes will obtain, they are thought to calculate the subjective expected utility of an outcome, which is the sum of its subjective utilities over the possible outcomes weighted by the estimated probability of occurrence of the outcomes. Deciding, then, is simply a matter of choosing the option with the greatest expected utility. Choice is therefore thought to reveal an individual's subjective utility functions and, hence, the individual's underlying preferences. However, while highly compelling in principle, this text contends that the classic philosophical understanding of decision theory is deeply flawed, particularly as it related to ethical decision making. Throughout history, persistent neuroeconomic critiques have been forwarded against the conventional view delineated above, most of which address the inadequacy with which it describes how decisions are made. As early as the mid 1950s, for instance, it was suggested to replace the rational model with a framework that accounted for a variety of human resource constraints, such as bounded attention and memory capacity, as well as limited time.[203] According to the bounded rationality view, it is unreasonable to expect decision makers to exhaustively compute the expected utility of others. Other critiques have focused on systematic violations of even the most fundamental requirements of rational choice theory. According to the theory, for example, preferences should remain unaffected by logically inconsequential factors, such as the specific method used to elicit preferences or the precise manner in that options are described. However, a recent series of persuasive demonstrations revealed that choices failed to obey simple consistency requirements and were instead affected by nuances of the decision context that were not subsumed by the normative accounts.[204] This suggests that emotionally rational moral preferences are constructed, not merely revealed, in making ethical decisions that may lead to significant and systematic departures from normative predictions.[205]

Beyond these distinctions, emotional rationality also influences the associations or normative moral images that come to mind in ethical decision making. Insofar as images can be consulted quickly and effortlessly, an affective heuristic has been proposed wherein the affective assessment of options and outcomes guides decisions.[206] Moreover, anticipatory emotions (e.g., emotional reactions to being in risky situations) can influence the cognitive appraisal of ethical decision situations and can affect moral choice, just as drives and motivations can influence reasoning more generally. Research on the normative logic of emotionally rational ethical decision making is active and growing. Among interesting current developments is a thriving stream of investigations into the systematic associations and disassociations between experienced utility (i.e., the hedonic experience an option actually brings) and decision utility (i.e., the utility implied by the decision). As research on emotionally rational forecasting has repeatedly demonstrated, however, misprediction of utility is common. Perhaps the most consistent finding concerns the belief that future ethical decisions will have a greater and longer-lasting impact than they actually do. Individuals also mispredict that emotionally rational ethical decisions will make them happier, thus pursuing experiences that are less satisfying than they could be. A productive avenue for future research on emotionally rational ethical decision making would involve the investigation of these affective forecasting errors with an eye toward helping individuals make moral judgments that will maximize both personal and social well-being. Among other current and future directions, researchers are using insights about attention, perception, memory, goal pursuit, and learning to procure a more sophisticated understanding of how moral judgments and ethical decisions are formed and why they sometimes prove inconsistent. Along related lines, an active direction for decision making research lies in the area of neuroeconomics, which gains insights into emotionally rational moral judgments and ethical decisions by exploring their neural underpinnings. The product of this research is, and will continue to be, quite distant from the elegant normative logic of emotionally rational ethical decision making. At the same time, acknowledged departures from normative logic need not weaken the normative force of emotionally rational ethical decision making, as normative theories are themselves empirical projects, capturing what, upon careful reflection, individuals consider ideal. As human beings improve their understanding of how emotionally rational ethical decisions are made, they may be able to formulate prescriptive procedures to guide moral agents, in light of their limitations, to better capture their normative wishes.[207]

## 4.5. Conclusion

As the second link in the book's logic chain, this chapter endeavored to underscore the moral significance of emotional rationality for ethical decision-making and to thereby make the case against direct and intentional affective degeneration via MM. It suggested, first, that insofar as belief serves as either the impetus for or consequence of any given emotion, the relation between beliefs and the choices they generate are remarkably fine-grained. On the surface, emotional choice may manifest as minimally rational: the action performed, stripped down, seems simply to be the best means to realize the agent's desire, given the belief with which it corresponds. However, the fact that desire is induced by emotion does not undermine the rational choice of the action as such. Indeed, agents are able to choose rationally only inasmuch as they rely on the emotional resources and propensities with which they find themselves endowed. Second, the chapter addressed the mutual dependence of emotion and rationality. Like a looming storm, emotions signify rational agitations of value. These intense experiences are precisely what gives rise to emotion's rational endurance, evenness, and permanence. Moreover, since emotions are rationally evaluative attitudes, they entail either taking action or being disposed to act in a manner compatible with the corresponding evaluation. In this way, emotions express one's most profound values and rational attitudes.

Third, the chapter addressed the intimate correlation between emotional rationality and morality. A morality that is emotionally rational is associated with unique patterns of cognitive evaluation that define how particular emotions color subsequent value judgments by prioritizing specific concerns semantically related to the emotion's appraisals. These appraisal patterns, which tie subjective emotions to objective concerns, have helped to clarify the distinct role of emotional rationality in moral discernment and set the stage for predictions about how emotional rationality prioritizes certain moral valuations and promotes normative judgments across cultures. Finally, the chapter addressed the interplay between emotional rationality and ethical decision making. Beyond serving as the cardinal mechanism for ethical insight, emotional rationality also functions as the linchpin of practical reason during ethical decision making. Moral agents employ practical reasoning to shape or guide future actions, and insofar as practical reason concerns actions yet to be carried out, it necessarily bears on action typologies, including the attitudes with which they correspond. As such, emotional rationality provides reasons for thinking that certain types of action or attitude are required or forbidden, recommended or inadvisable.

Against this backdrop, at least three conclusions can be drawn. First, emotional rationality serves as the necessary condition for evaluative judgments. Yet the practice of MM effectively eliminates – and, worse, proactively intends to eliminate – the very means by which rational moral reflection is possible. Second, the affective reasons that generate moral actions must themselves be moral; that is, they must be related to what is morally good or bad in the agent's understanding. Hence, when emotional rationality is acutely blunted through MM, so too is one's capacity to seek the good. Finally, emotional rationality has a function beyond serving as a canon of moral motivation; it also influences the associations or normative moral images that come to mind in ethical decision making. Insofar as these images depend on the structure of normative judgments and not on the subject of the judgments (inasmuch as understanding the subject of a judgment requires references to background knowledge), the degeneration of emotional rationality through MM also precludes the ability to do distinguish between normative and relative structures of morality. Continuing in this vein of thought, the book now turns to critically examine the immediate consequence of affective degeneration: the decaying of narrative identity.

### 4.6. Notes

1. A. W. Price, "Emotions in Plato and Aristotle," in *The Oxford Handbook of Philosophy of Emotion*, ed. Peter Goldie (New York: Oxford University Press, 2010), 121-42; see especially pp. 121-22.

2. This text will not endeavor to explore the distinction between reason and rationality except to clarify that while the idea of reason is normative in purpose, rationality primarily serves to explain behavior. See Jon Elster, "Emotional Choice and Rational Choice," in *The Oxford Handbook of Philosophy of Emotion*, ed. Peter Goldie (New York: Oxford University Press, 2010), 263-81; see especially pp. 263-64.

3. Elster, "Emotional Choice and Rational Choice," 263-81; see especially pp. 263-64.

4. Price, "Emotions in Plato and Aristotle," 121-42; see especially p. 121.

5. Price, "Emotions in Plato and Aristotle," 121-42; see especially pp. 122-30.

6. Price, "Emotions in Plato and Aristotle," 121-42; see especially p. 131.

7. Price, "Emotions in Plato and Aristotle," 121-42; see especially pp. 131-41.

8. Daniel Bloom, "The Self-Awareness of Reason in Plato," *Journal of Cognition and Neuroethics* 2, no. 1 (2014): 95-103; see especially p. 100.

9. Bloom, "The Self-Awareness of Reason," 95-103; see especially pp. 100-03.

10. Roderick T. Long, *Reason and Value: Aristotle versus Rand* (Washington, DC: The Atlas Society, 2000), 17-29; see especially p. 17.

11. Long, *Reason and Value*, 17-29; see especially p. 17.

12. Fred D. Miller, "Rationality and Freedom in Aristotle and Hayek," *Reason Papers* 9 (Winter 1983): 29-36; see especially p. 30.

13. Miller, "Rationality and Freedom," 29-36; see especially pp. 30-31.

14. Dino J. Levy and Paul W. Glimcher, "The Root of All Value: A Neural Common Currency for Choice," *Current Opinion in Neurobiology* 22 (2012): 1027-38; see especially p. 1027.

15. See Levy and Glimcher, "The Root of All Value," 1027-38.

16. See Paul A. Samuelson, *Foundations of Economic Choice* (Cambridge, MA: Harvard University Press, 1947).

17. Levy and Glimcher, "The Root of All Value," 1027-38; see especially pp. 1028-35.

18. Steven R. Quartz, "Reason, Emotion, and Decision Making: Risk and Reward Computation with Feeling," *Trends in Cognitive Sciences* 13, no. 5 (April 2009): 209-15; see especially p. 209.

19. Quartz, "Reason, Emotion, and Decision Making," 209-15; see especially p. 209.

20. Quartz, "Reason, Emotion, and Decision Making," 209-15; see especially p. 214.

21. Elster, "Emotional Choice and Rational Choice," 263-81; see especially p. 268.

22. Elster, "Emotional Choice and Rational Choice," 263-81; see especially pp. 268-69.

23. Elster, "Emotional Choice and Rational Choice," 263-81; see especially pp. 274-75.

24. Elster, "Emotional Choice and Rational Choice," 263-81; see especially p. 264.

25. Elster, "Emotional Choice and Rational Choice," 263-81; see especially pp. 265-67.

26. Adrian Raine and Yaling Yang, "Neural Foundations to Moral Reasoning and Antisocial Behavior," *Social Cognitive and Affective Neuroscience* 1, no. 3 (October 2006): 203-13; see especially pp. 209-10.

27. For a superb singular analysis of "moral emotion," see Jesse J. Prinz, "The Moral Emotions," in *The Oxford Handbook of Philosophy of Emotion,* ed. Peter Goldie (New York: Oxford University Press, 2010), 519-38.

28. Raine and Yang, "Neural Foundations to Moral Reasoning," 203-13; see especially p. 209.

29. Interestingly, the perceptivity and personal nature of emotion has been used to critique it on grounds that it is, ipso facto, overly subjective. Yet it seems quite contrarily the case that perceptivity and reasonable idiosyncrasy would hinder something more than prove a redeeming virtue – here, integral authenticity.

30. Ben-Ze'ev, "The Thing Called Emotion," 41-62.

31. Ben-Ze'ev, "The Thing Called Emotion," 41-62; see especially p. 61.

32. For a fine overview of the psychology of emotional rationality, see Gerald L. Clore, "Psychology and the Rationality of Emotion," *Modern Theology* 27, no. 2 (April 2011): 325-38.

33. Ben-Ze'ev, "The Thing Called Emotion," 41-62; see especially p. 41.

34. Ben-Ze'ev, "The Thing Called Emotion," 41-62; see especially pp. 41-44.

35. Ben-Ze'ev, "The Thing Called Emotion," 41-62; see especially p. 45.

36. Ben-Ze'ev, "The Thing Called Emotion," 41-62; see especially pp. 45-46.

37. Ben-Ze'ev, "The Thing Called Emotion," 41-62; see especially p. 47.

38. Ben-Ze'ev, "The Thing Called Emotion," 41-62; see especially pp. 47-48.

39. Ben-Ze'ev, "The Thing Called Emotion," 41-62; see especially p. 48.

40. See, for instance, Rainer Reisenzein, "Pleasure-Arousal Theory and the Intensity of Emotions," *Journal of Personality and Social Psychology* 67, no. 3 (1994): 525-39.

41. Ben-Ze'ev, "The Thing Called Emotion," 41-62; see especially pp. 48-50.

42. Erin Ryan, "The Discourse Beneath: Emotional Epistemology in Legal Deliberation and Negotiation," *Harvard Negotiation Law Review* 10, no. 231 (Spring 2005): 231-85; see especially p. 232.

43. Adam Morton, "Epistemic Emotions," in *The Oxford Handbook of Philosophy of Emotion*, ed. Peter Goldie (New York: Oxford University Press, 2010), 385-99; see especially pp. 385-93.

44. Morton, "Epistemic Emotions," 385-99; see especially p. 394.

45. Morton, "Epistemic Emotions," 385-99; see especially pp. 394-96.

46. Alison M. Jaggar, "Love and Knowledge: Emotion in Feminist Epistemology," *Inquiry* 32, no. 2 (1989): 151-76; see especially p. 169.

47. Jaggar, "Love and Knowledge," 151-76; see especially pp. 169-70.

48. Jaggar, "Love and Knowledge," 151-76; see especially p. 170.

49. Jaggar, "Love and Knowledge," 151-76; see especially pp. 170-71.

50. See Robert C. Solomon, *The Passions: Emotions and the Meaning of Life* (Cambridge, UK: Hackett Publishing Company, 1993).

51. Solomon, *The Passions*, 1-26; see especially p. 7.

52. Solomon, *The Passions*, 49-66; see especially p. 50.

53. Martin Heidegger, *Being and Time*, trans. John Macquarrie and Edward Robinson (Cambridge, MA: Blackwell Publishers Ltd., 1962).

54. This is the case, for example, with items of equipment, which knit together in holistic teleological frameworks that reflect potential activities. See Heidegger, *Being and Time*, 91-148; see especially pp. 95-102.

55. Matthew Ratcliffe, "The Phenomenology of Mood and the Meaning of Life," in *The Oxford Handbook of Philosophy of Emotion*, ed. Peter Goldie (New York: Oxford University Press, 2010), 349-71; see especially pp. 350-54.

56. Ratcliffe, "The Phenomenology of Mood," 349-71; see especially p. 355.

57. Heidegger, *Being and Time*, 169-224; see especially p. 181.

58. Jean-Paul Sartre, *Nausea*, trans. R. Baldick (London: Penguin Books, 1963); see especially p. 183.

59. Heidegger, *Being and Time*, 169-224; see especially p. 176.

60. Heidegger, *Being and Time*, 169-224; see especially p. 176.

61. Ratcliffe, "The Phenomenology of Mood," 349-71; see especially pp. 355-57.

62. John A. Lambie, "On the Irrationality of Emotion and the Rationality of Awareness," *Consciousness and Cognition* 17 (2007): 946-71; see especially p. 959.

63. See Lambie, "On the Irrationality of Emotion," 946-71; see especially p. 959.

64. See Lambie, "On the Irrationality of Emotion," 946-71; see especially p. 959.

65. Lambie, "On the Irrationality of Emotion," 946-71; see especially p. 959.

66. Lambie, "On the Irrationality of Emotion," 946-71; see especially p. 967.

67. Lambie, "On the Irrationality of Emotion," 946-71; see especially p. 967.

68. Lambie, "On the Irrationality of Emotion," 946-71; see especially pp. 967-68.

69. Benulal Dhar, "The Phenomenology of Value-Experience: Some Reflections on Scheler and Hartmann," *Indian Philosophical Quarterly* 26, no. 2 (April 1999): 183-97; see especially p. 184.

70. Blaise Pascal, *Pensees,* ed. Robert Maynard Hutchins, vol. 33 of *Great Books of the Western World* (Chicago: Encyclopedia Britannica, 1952), 222; see especially sec. IV, para. 277.

71. Pascal, *Pensees,* 223; see especially para. 282.

72. See John J. Drummond, "Moral Objectivity: Husserl's Sentiments of the Understanding," *Husserl Studies* 12, no. 2 (1995): 165-83; see especially p. 170.

73. See Max Scheler, *Formalism in Ethics and the Non-Formal Ethics of Values: A New Attempt toward the Foundation of an Ethical Personalism,* trans. Manfred S. Frings and Roger L. Funk (Evanston, IL: Northwestern University Press, 1973).

74. See Nicolai Hartmann, *Ethics,* trans. Stanton Colt (London: George Allen & Unwin Ltd., 1958).

75. Dhar, "The Phenomenology of Value-Experience," 183-97; see especially pp. 184-85.

76. Dhar, "The Phenomenology of Value-Experience," 183-97; see especially p. 185.

77. See Scheler, *Formalism in Ethics,* 12-22; see especially p. 12.

78. Scheler, *Formalism in Ethics,* 253-64; see especially p. 255.

79. See Imtiaz Moosa, "A Critical Examination of Scheler's Justification of the Existence of Values," *The Journal of Value Inquiry* 25, no. 1 (1991): 23-41.

80. See Scheler, *Formalism in Ethics,* 12-22; see especially p. 21.

81. Scheler, *Formalism in Ethics,* 163-202; see especially pp. 186-87.

82. Hartmann, *Ethics;* see especially p. 184.

83. Hartmann, *Ethics;* see especially p. 185.

84. Hartmann, *Ethics;* see especially p. 193.

85. Dhar, "The Phenomenology of Value-Experience," 183-97; see especially pp. 185-87.

86. Bennett W. Helm, "Emotions and Motivation: Reconsidering Neo-Jamesian Accounts," in *The Oxford Handbook of Philosophy of Emotion,* ed. Peter Goldie (New York: Oxford University Press, 2010), 303-23; see especially p. 314.

87. See Helm, "Emotions and Motivation," 303-23; see especially p. 315.

88. Solomon, *The Passions,* 180-92; see especially p. 185.

89. See Helm, "Emotions and Motivation," 303-23; see especially p. 315.

90. Helm, "Emotions and Motivation," 303-23; see especially pp. 316-17.

91. Helm, "Emotions and Motivation," 303-23; see especially pp. 317-18.

92. Immanuel Kant, *The Metaphysics of Morals,* trans. M. J. Gregor (Cambridge: Cambridge University Press, 1996); see especially p. 160.

93. A common contextual critique is that emotions are egotistic and not genuinely altruistic; therefore, they cannot be considered "rational." The point here is not to prove or disprove this claim, but to contend that, in either case, emotions are important. See Prinz, "The Moral Emotions," 519-38; see especially p. 520.

94. As Prinz clarifies further, the motivation to punish poor behavior, as an essential element of global systems of criminal justice, is often retributive in nature, and, as such, emotional. See "The Moral Emotions," 519-38; see especially p. 520.

95. Prinz, "The Moral Emotions," 519-38; see especially pp. 520-21.

96. Dan-Yang Gui, Tian Gan, and Chao Liu, "Neural Evidence for Moral Intuition and the Temporal Dynamics of Interactions Between Emotional Processes and Moral Cognition," *Social Neuroscience* [Epub ahead of print] (August 2015): 1-15; see especially p. 1.

97. See, for instance, Jorge Moll, Ricardo de Oliveira-Souza, and Roland Zahn, "The Neural Basis of Moral Cognition: Sentiments, Concepts, and Values," *Annals of the New York Academy of Sciences* 1124 (March 2008): 161-80.

98. See Jonathan Haidt, "The New Synthesis in Moral Psychology," *Science* 316 (May 2007): 998-1002.

99. See Haidt, "The New Synthesis in Moral Psychology," 998-1002.

100. See Joshua D. Greene, Leigh E. Nystrom, Andrew D. Engell, John M. Darley, and Jonathan D. Cohen, "The Neural Bases of Cognitive Conflict and Control in Moral Judgment," *Neuron* 44 (October 2004): 389-400.

101. See Greene et al., "The Neural Bases," 389-400.

102. Gui et al., "Neural Evidence for Moral Intuition," 1-15; see especially pp. 1-2.

103. Gui et al., "Neural Evidence for Moral Intuition," 1-15; see especially p. 2.

104. It is worthy of note that while experimental conditions can be distinguished by ERPs at time T, this does not necessarily indicate that neural processing has been completed.

105. See, for instance, Jean Decety and Stephanie Cacioppo, "The Speed of Morality: A High-Density Electrical Neuroimaging Study," *Journal of Neurophysiology* 108 (September 2012): 3068-72.

106. See Keith J. Yoder and Jean Decety, "Spatiotemporal Neural Dynamics of Moral Judgment: A High-Density ERP Study," *Neuropsychologia* 60 (July 2014): 39-45.

107. See Yoder and Decety, "Spatiotemporal Neural Dynamics," 39-45.

108. See Michela Sarlo, Lorella Lotto, Andrea Manfrinati, Rino Rumiati, Germano Gallicchio, and Daniela Palomba, "Temporal Dynamics of Cognitive-Emotional Interplay in Moral Decision-Making," *Journal of Cognitive Neuroscience* 24, no. 4 (March 2012): 1018-29.

109. Gui et al., "Neural Evidence for Moral Intuition," 1-15; see especially pp. 2-3.

110. Gui et al., "Neural Evidence for Moral Intuition," 1-15; see especially p. 9.

111. Decety and Cacioppo, "The Speed of Morality," 3068-72.

112. See Marc D. Hauser, "The Liver and the Moral Organ," *Social Cognitive and Affective Neuroscience* 1, no. 3 (December 2006): 214-20.

113. Gui et al., "Neural Evidence for Moral Intuition," 1-15; see especially p. 9.

114. Gui et al., "Neural Evidence for Moral Intuition," 1-15; see especially pp. 9-10.

115. See, for instance, Claudia Denke, Michael Rotte, Hans-Jochen Heinze, and Michael Schaefer, "Belief in a Just World is Associated with Activity in Insula and Somatosensory Cortices as a Response to the Perception of Norm Violations," *Social Neuroscience* 9, no. 5 (2014): 514-21.

116. Gui et al., "Neural Evidence for Moral Intuition," 1-15; see especially p. 10.

117. Bryce Huebner, Susan Dwyer, and Marc Hauser, "The Role of Emotion in Moral Psychology," *Trends in Cognitive Sciences* 13, no. 1 (January 2009): 1-6; see especially p. 1.

118. See Jorge Moll, Ricardo de Oliveira-Souza, Paul J. Eslinger, Ivanei E. Bramati, Janaina Mourão-Miranda, Pedro Angelo Andreiuolo, and Luiz Pessoa, "The Neural Correlates of Moral Sensitivity: A Functional Magnetic Resonance Imaging Investigation of Basic and Moral Emotions," *The Journal of Neuroscience* 22, no. 7 (April 2002): 2730-36.

119. See Jorge Moll, Ricardo de Oliveira-Souza, and Paul J. Eslinger, "Morals and the Human Brain: A Working Model," *NeuroReport* 14, no. 3 (March 2003): 299-305.

120. Huebner et al., "The Role of Emotion in Moral Psychology," 1-6; see especially pp. 1-3.

121. See, for instance, Michael Koenigs, Liane Young, Ralph Adolphs, Daniel Tranel, Fiery Cushman, Marc Hauser, and Antonio Damasio, "Damage to the Prefrontal Cortex Increases Utilitarian Moral Judgements [sic]," *Nature* 446, no. 7138 (April 2007): 908-11.

122. See Damasio, *Descartes' Error;* see especially chs. 7-9.

123. See, for instance, Mario Mendez, Eric Anderson, and Jill S. Shapira, "An Investigation of Moral Judgement [sic] in Frontotemporal Dementia," *Cognitive and Behavioral Neurology* 18, no. 4 (December 2005): 193-97.

124. See Koenigs et al., "Damage to the Prefrontal Cortex," 908-11.

125. Huebner et al., "The Role of Emotion in Moral Psychology," 1-6; see especially p. 4.

126. Elizabeth J. Horberg, Christopher Oveis, and Dacher Keltner, "Emotions as Moral Amplifiers: An Appraisal Tendency Approach to the Influences of Distinct Emotions upon Moral Judgment," *Emotion Review* 3, no. 3 (July 2011): 237-44; see especially p. 238.

127. See, for instance, Jennifer S. Lerner and Dacher Keltner, "Fear, Anger, and Risk," *Journal of Personality and Social Psychology* 81, no. 1 (July 2001): 146-59.

128. Horberg et al., "Emotions as Moral Amplifiers," 237-44; see especially p. 238.

129. Horberg et al., "Emotions as Moral Amplifiers," 237-44; see especially p. 240.

130. Sid Carter and Marcia Smith Pasqualini, "Stronger Autonomic Response Accompanies Better Learning: A Test of Damasio's Somatic Marker Hypothesis," *Cognition and Emotion* 18, no. 7 (November 2004): 901-911.

131. See Dacher Keltner, Phoebe C. Ellsworth, and Kari Edwards, "Beyond Simple Pessimism: Effects of Sadness and Anger on Social Perception," *Journal of Personality and Social Psychology* 64, no. 5 (1993): 740-52.

132. See Simone Schnall, Jennifer Benton, and Sophie Harvey, "With a Clean Conscience: Cleanliness Reduces the Severity of Moral Judgments," *Psychological Science* 19, no. 12 (December 2008): 1219-22.

133. Horberg et al., "Emotions as Moral Amplifiers," 237-44; see especially pp. 240-41.

134. Patricia Greenspan, "Learning Emotions and Ethics," in *The Oxford Handbook of Philosophy of Emotion,* ed. Peter Goldie (New York: Oxford University Press, 2010), 539-59; see especially p. 550.

135. John Rawls, *A Theory of Justice,* rev. ed. (Cambridge, MA: Harvard University Press, 1998); see especially pp. 397-441.

136. Greenspan, "Learning Emotions and Ethics," 539-60; see especially pp. 549-52.

137. See John Rawls, "Kantian Constructivism in Moral Theory," *The Journal of Philosophy* 77, no. 9 (September 1980): 515-72.

138. Greenspan, "Learning Emotions and Ethics," 539-60; see especially pp. 550-52.

139. Greenspan, "Learning Emotions and Ethics," 539-60; see especially p. 554.

140. Greenspan, "Learning Emotions and Ethics," 539-60; see especially p. 554.

141. Greenspan, "Learning Emotions and Ethics," 539-60; see especially p. 555.

142. Greenspan, "Learning Emotions and Ethics," 539-60; see especially pp. 555-56.

143. Robert C. Roberts, "Emotions and the Canons of Evaluation," in *The Oxford Handbook of Philosophy of Emotion,* ed. Peter Goldie (New York: Oxford University Press, 2010), 561-83; see especially p. 561.

144. See, for instance, Plato's *Symposium,* and Books III and VII of his *Republic.*

145. See David Hume, "Of the Standard of Taste," in *Essays: Moral, Political, and Literary,* ed. Eugene F. Miller (Indianapolis: Liberty Classics, 1985), 226-49; see especially pp. 228-29.

146. Roberts, "Emotions and the Canons of Evaluation," 561-83; see especially pp. 561-63.

147. Augusto Blasi, "Emotions and Moral Motivation," *Journal for the Theory of Social Behaviour* 29, no. 1 (March 1999): 1-19; see especially p. 12.

148. Blasi, "Emotions and Moral Motivation," 1-19; see especially pp. 12-13.

149. Blasi, "Emotions and Moral Motivation," 1-19; see especially p. 13.

150. Blasi, "Emotions and Moral Motivation," 1-19; see especially pp. 13-14.

151. Blasi, "Emotions and Moral Motivation," 1-19; see especially p. 15.

152. Blasi, "Emotions and Moral Motivation," 1-19; see especially pp. 15-16.

153. Walter Glannon, *Brain, Body, and Mind: Neuroethics with a Human Face* (New York: Oxford University Press, 2011), 93-114; see especially pp. 99-100.

154. Antonio Damasio, "Neuroscience and Ethics—Intersections," *American Journal of Bioethics—AJOB Neuroscience* 7, no. 1 (2007): 3-7; see especially p. 4.

155. See David Hume, *A Treatise of Human Nature*, ed. P. H. Nidditch (Oxford: Clarendon Press, 1978); see especially at 2.3.3.

156. Glannon, *Body, Brain, and Mind,* 93-114; see especially pp. 102-104.

157. Peter Singer, "Ethics and Intuitions," *Journal of Ethics* 9 (2005): 331-52; see especially p. 351.

158. Antoine Bechara, Hanna Damasio, and Antonio R. Damasio, "Emotion, Decision Making and the Orbitofrontal Cortex," *Cerebral Cortex* 10 (March 2000): 295-307; see especially p. 295.

159. Bechara et al., "Emotion," 295-307; see especially pp. 295-96.

160. Bechara et al., "Emotion," 295-307; see especially p. 296.

161. Bechara et al., "Emotion," 295-307; see especially pp. 296-97.

162. Bechara et al., "Emotion," 295-307; see especially p. 297.

163. Bechara et al., "Emotion," 295-307; see especially p. 297.

164. See, for instance, Joseph LeDoux, *The Emotional Brain: The Mysterious Underpinnings of the Emotional Life* (New York: Touchstone, 1998).

165. Bechara et al., "Emotion," 295-307; see especially pp. 297-306.

166. J. Jason van Steenburgh, Jessica I. Fleck, Mark Beeman, and John Kounios, "Insight," in *The Oxford Handbook of Thinking and Reasoning*, ed. Keith J. Holyoak and Robert G. Morrison (New York: Oxford University Press, 2012), 475-91; see especially p. 475.

167. van Steenburgh et al., "Insight," 475-91; see especially pp. 475-76.

168. van Steenburgh et al., "Insight," 475-91; see especially p. 476.

169. See Janet Metcalfe and David Wiebe, "Intuition in Insight and Noninsight Problem Solving," *Memory and Cognition* 15, no. 3 (1987): 238-46.

170. See Jonathan W. Schooler and Joseph Melcher, "The Ineffability of Insight," in *The Creative Cognition Approach*, ed. Steven M. Smith, Thomas B. Ward, and Ronald A. Finke (Cambridge, MA: The MIT Press, 1995), 97-133.

171. See Jonathan W. Schooler, Stellan Ohlsson, and Kevin Brooks, "Thoughts Beyond Words: When Language Overshadows Insight," *Journal of Experimental Psychology: General* 122, no. 2 (1993): 166-83.

172. See Kenneth J. Gilhooly and P. Murphy, "Differentiating Insight from Noninsight Problems," *Thinking and Reasoning* 11, no. 3 (2005): 279-302.

173. van Steenburgh et al., "Insight," 475-91; see especially pp. 478-79.

174. Onora O'Neill, "Kant: Rationality as Practical Reason," in *The Oxford Handbook of Rationality*, ed. Alfred R. Mele and Piers Rawling (New York: Oxford University Press, 2004), 93-109; see especially p. 94.

175. Patricia Greenspan, "Practical Reasoning and Emotion," in *The Oxford Handbook of Rationality*, ed. Alfred R. Mele and Piers Rawling (New York: Oxford University Press, 2004), 206-221; see especially p. 207.

176. Greenspan, "Practical Reasoning and Emotion," 206-21; see especially p. 207.

177. See Damasio, *Descartes' Error;* see especially ch. 8.

178. Greenspan, "Practical Reasoning and Emotion," 206-21; see especially pp. 207-08.

179. Joseph M. Paxton, Leo Ungar, and Joshua D. Greene, "Reflection and Reasoning in Moral Judgment," *Cognitive Science* 36 (2012): 163-77; see especially p. 163.

180. See, for instance, Schnall et al., "With a Clean Conscience," 1219-22.

181. See Joshua D. Greene, Sylvia A. Morelli, Kelly Lowenberg, Leigh E. Nystrom, and Jonathan D. Cohen, "Cognitive Load Selectively Interferes with Utilitarian Moral Judgment," *Cognition* 107, no. 3 (June 2008): 1144-54.

182. See David A. Pizarro and Paul Bloom, "The Intelligence of the Moral Intuitions: Comment on Haidt (2001)," *Psychological Review* 110, no. 1 (2003): 193-96.

183. Shaun Nichols and Ron Mallon, "Moral Dilemmas and Moral Rules," *Cognition* 100, no. 3 (2006): 530-42.

184. See Joseph M. Paxton and Joshua D. Greene, "Moral Reasoning: Hints and Allegations," *Topics in Cognitive Science* 2, no. 3 (July 2010): 511-27.

185. See Greene et al., "Cognitive Load," 1144-54.

186. See Adam B. Moore, Brian A. Clark, and Michael J. Kane, "Who Shalt Not Kill? Individual Differences in Working Memory Capacity, Executive Control, and Moral Judgment," *Psychological Science* 19, no. 6 (2008): 549-57.

187. See Piercarlo Valdesolo and David DeSteno, "Moral Hypocrisy: Social Groups and the Flexibility of Virtue," *Psychological Science* 18, no. 8 (2007): 689-90.

188. See David A. Pizarro, Eric Uhlmann, and Paul Bloom, "Causal Deviance and the Attribution of Moral Responsibility," *Journal of Experimental Social Psychology* 39, no. 6 (2003): 653-60.

189. See Neeru Paharia, Karim S. Kassam, Joshua D. Greene, and Max H. Bazerman, "Dirty Work, Clean Hands: The Moral Psychology of Indirect Agency," *Organizational Behavior and Human Decision Processes* 109, no. 2 (2009): 134-41.

190. See Yoel Inbar, David A. Pizarro, Joshua Knobe, and Paul Bloom, "Disgust Sensitivity Predicts Intuitive Disapproval of Gays," *Emotion* 9, no. 3 (2009): 435-39.

191. See, for instance, Richard E. Petty and John T. Cacioppo, "The Elaboration Likelihood Model of Persuasion," *Advances in Experimental Social Psychology* 19 (1986): 123-205.

192. See Paxton et al., "Reflection and Reasoning in Moral Judgment," 163-77.

193. See Paxton et al., "Reflection and Reasoning in Moral Judgment," 163-77.

194. Miriam Bassok and Laura R. Novick, "Problem Solving," in *The Oxford Handbook of Thinking and Reasoning*, ed. Keith J. Holyoak and Robert G. Morrison (New York: Oxford University Press, 2012), 413-32; see especially p. 414.

195. Bassok and Novick, "Problem Solving," 413-32; see especially p. 414.

196. Bassok and Novick, "Problem Solving," 413-32; see especially p. 414.

197. Bassok and Novick, "Problem Solving," 413-32; see especially pp. 414-28.

198. Nick Chater and Mike Oaksford, "Normative Systems: Logic, Probability, and Rational Choice," in *The Oxford Handbook of Thinking and Reasoning*, ed. Keith J. Holyoak and Robert G. Morrison (New York: Oxford University Press, 2012), 11-21; see especially p. 11.

199. Chater and Oaksford, "Normative Systems," 11-21; see especially pp. 11-12.

200. Chater and Oaksford, "Normative Systems," 11-21; see especially p. 12.

201. Chater and Oaksford, "Normative Systems," 11-21; see especially pp. 12-15.

202. Robyn A. LeBoeuf and Eldar Shafir, "Decision Making," in *The Oxford Handbook of Thinking and Reasoning*, ed. Keith J. Holyoak and Robert G. Morrison (New York: Oxford University Press, 2012), 301-21; see especially p. 301.

203. See Herbert A. Simon, "A Behavioral Model of Rational Choice," *The Quarterly Journal of Economics* 69, no. 1 (February 1955): 99-118.

204. See Chater and Oaksford, "Normative Systems," 11-21.

205. LeBoeuf and Shafir, "Decision Making," 301-21; see especially pp. 301-02.

206. See Paul Slovic, Melissa Finucane, Ellen Peters, and Donald G. MacGregor, "The Affect Heuristic," in *Heuristics and Biases: The Psychology of Intuitive Judgment,* ed. Thomas Gilovich, Dale Griffin, and Daniel Kahneman (New York: Cambridge University Press, 2002), 397-420.

207. LeBoeuf and Shafir, "Decision Making," 301-21; see especially pp. 314-16.

# 5.
# THE CASE AGAINST NARRATIVE DECAYING: THE MORAL SIGNIFICANCE OF NARRATIVE IDENTITY FOR ETHICAL DECISION MAKING

In light of the foregoing chapter's conclusions – namely, that MM degenerates the emotional rationality by which individuals are capable of acting in a manner compatible with evaluative moral judgments – this chapter aims to make the case against narrative decaying by underscoring the moral significance of narrative identity for ethical decision making. To that end, it endeavors, first, to identify and unpack the interrelationships between life narratives, personal identity, and moral development. Second, it discusses narrative identity as the ripe fruit of autobiographical memory and emotional rationality. Third, it explores the correlation of narrative identity and experiential unpredictability. The chapter concludes by highlighting the imminent threat of MM to coherent narratives, subjective authenticity, and objective ethical decision making.

### 5.1. Narrative, Identity, and Development

The formal concept of narrative identity was first postulated in the twentieth century: Sigmund Freud wrote about dream narratives, Carl Jung explored universal life myths, Alfred Adler examined narrative accounts of earliest memory, and Henry Murray identified recurrent autobiographical themes in the Thematic Apperception Test.[1] Still, none of the traditional theories of personality in the first half of the twentieth century imagined human being as storytellers and human experience as a story to be told. The inaugural theories of narrative personality were developed in the late 1970s and early 1980s. Silvan Tompkins proposed a "script theory" of personality that conceived of the individual as a metaphorical playwright who organizes the emotional experiences of life in terms of salient "scenes" and recurrent "scripts."[2] In a somewhat similar line of thought, Dan McAdams formulated a "life-story" model of identity, suggesting that people living in modern society begin, in late adolescence and young adulthood, to understand their lives as ever-evolving stories that integrate the reconstructed past and the projected future in order to imbue life with degrees of unity and purpose.[3]

### 5.1.1. The Evolution of Narrative in History

It has been suggested that in the process of replacing supernatural with natural theories of disease causation, Hippocrates laid the groundwork for a practice of medicine in which clinicians rarely speak to patients.[4] There is certainly a strong history in Western medicine that regards medical practice as, first and foremost, the palpitation of the abdomen and the analysis of laboratory results, so that conversation of any sort is viewed peripherally. However, this tradition fails to account for what becomes readily evident on a perfunctory reading of Hippocrates and subsequent classics of medicine: notwithstanding the status of conversation with patients, conversation among clinicians is central to medical practice, and this conversation frequently takes the form of telling stories. From Hippocrates until fairly recent times, narrative case history has dominated medical thinking and has been the nucleus of much medical literature. As Kathryn Montgomery Hunter notes, there is an essential narrative core to the practice of medicine.[5] Montgomery Hunter discerned a parallelism between the ethical method of casuistry and the reasoning of clinicians, which explains in part why casuistry is aptly suited to serve as the methodology par excellence for clinical bioethics. Casuistry applies to both general maxims and large sets of cases that illustrate the application of moral rules with more or less precision. Moral wisdom lies not in the recollection of maxims, but arises from witnessing the complex interplay between maxims and cases. In similar fashion, medical practice lies not in the understanding of medical science, but involves the application of medical science to particular, individual cases. Without a repository of case exemplars to draw on, medicine could neither be taught nor practiced. For Montgomery Hunter, then, learning to become a clinician requires, first, learning to tell the story of a patient's illness in standardized format, and then learning the skill of comparing and contrasting that case presentation to a collection of parallel cases within the appropriate scientific category to discern the correct diagnosis and treatment.[6] The "right" answer in clinical reasoning requires both a thorough understanding of the basic science and also an appreciation of the unique features of the case at hand that might – and likely will – require that the standard treatment be modified.[7]

As mentioned above, prior to the mid 1980s, there was scant evidence in the medical literature to suggest any interest in narrative and story. Unsurprisingly, the first area of medicine to express a systematic interest in narrative was psychiatry, particularly psychoanalysis. James Hillman, for instance, was comfortable denying that psychoanalysis is empirical or scientific in any sense, asserting instead that it was a special form of "poiesis," or "making by imagination into words."[8] To do psychoanalysis in the 1970s was to create certain kinds of stories that have powerful impacts on the people about and to whom the stories are told. For Hillman, the stories appear superficially to be empirical and scientific because it is this appearance that gives them their power. Further, the stories are particularly powerful from the viewpoint of patients insofar as these individuals have, by virtue of their thoughts or behavior, frightened away others who might have offered help in understanding, or remained silent out of fear of driving others away. In stark contrast, Eugene Brody and Judith Tormey were unwilling to dismiss the empirical basis of psychoanalysis so readily. They maintained that "reality" constructed by psychoanalysis in the 1980s was intersubjective and dependent on both psychoanalytic theory and the continued reciprocity of the analyst-patient relationship. In this way, the analyst's interpretations and interventions serve to make behavior intelligible.[9] Narrative intelligibility is a function of deep-seated constructs shared by both analyst and patient alike. Thus understood, psychoanalytic theory is properly interpreted as a theoretical construction that serves as a principle of selection and organization in the formulation of a patient's life narrative. Brody and Tormey would later observe that insofar as psychoanalysis is fundamentally therapeutic (and not investigative), the way in which a patient's suffering responds to the newly created narrative forms an unavoidable criterion for the success of the endeavor.[10]

As research advances of the 1980s and early 1990s reestablished character traits as the dominant constructs in field of personality psychology, narrative approaches began to assume new roles in the medical literature.[11] Like character traits, life narratives speak to the organization and structure of life, but unlike character traits, narrative approaches to personality explicitly address issues of context. Strongly influenced by social constructivist perspectives on the self, leading theorists and researchers such as Hubert Hermans, Gary Gregg, Ruthellen Josselson, Michael Pratt, Bertram Cohler, and Avril Thorne developed narrative approaches to personality that placed the construction of life-stories more explicitly in the context of cultural discourse, emphasized the ways in which personal narratives make for numerous contextualized selves even as they serve to integrate lives in time, and highlighted the roles of gender, class, race, and social status in the construction and performance of life stories.[12] The nearly simultaneous publication of the first edition of Howard Brody's *Stories of Sickness* (in 1987) and Arthur Kleinman's *The Illness Narratives* (in 1988) was an early signal of a more general turn of interest toward narrative in medicine.[13] This "narrative turn" in healthcare mirrored a broadly based interest in narrative emerging concurrently in the fields of religion, psychology, anthropology, political science, linguistics, education, philosophy, and literature. This novel interest in narrative was a call to attend to stories in medicine and to think about medicine in an interdisciplinary manner. A flurry of scholarly studies on narrative appeared throughout the 1990s, including Montgomery Hunter's *Doctors' Stories* (1991),[14] Anne Hudson Jones' series of articles on literature and medicine in *The Lancet* (1996),[15] Hilde Lindemann Nelson's edited volume *Stories and Their Limits* (1997),[16] the British anthology *Narrative Based Medicine* (1998),[17] and Anne Hunsaker Hawkins' "pathography" accounts (1999).[18]

If narrative theories in the 1980s aimed to reveal the inner coherence of lives in response to the situationist critique, by the turn of the twenty-first century they had managed to appropriate and improve upon some of the main themes of the old situationist position, namely, the emphasis on local meanings, contingent performance, and the role of historical and cultural contexts in the exercise and evolution of personality.[19] The ways in which narrative theories and methods have helped to recontextualize personality psychology in recent years are evident in a variety of diverse studies and research programs. Researchers have demonstrated how character traits and personal needs are expressed through particular kinds of life stories, and how character traits pair with narratives to predict psychological well-being and other important life outcomes.[20] Moving beyond character traits, researchers have also examined how particular values and moral orientations are reflected in and shaped by life narratives, family stories, and broader community and societal myths. Narrative approaches have been similarly employed in the study of difficult life events and major life decisions, revealing how individuals make sense of adversity and change and the ways in which sense-making influences the maturation of personality. Moreover, while narrative approaches have enriched nomothetic research in personality psychology, the turn toward narrative has also revived a commitment to ideographic research. With its emphasis on exploring qualitative data about individual lives, narrative methods have provided researchers with new tools for examining the particularities of a single case. If personality psychology is to make significant headway in the future, it must rely on narrative approaches to reconcile its historical divide between nomothetic and ideographic ways of understanding human beings.[21]

### 5.1.2. The Ontology of Narrative Identity

As the study of narrative has swept across a wide range of academic disciplines, a variety of implicit and explicit views have developed about what narratives are and why they are important for human self-understanding.[22] There seems to be relative consensus that narrative does not merely delineate what happens, but draws out or creates meaningful connections between events and experiences, thereby rendering them intelligible. Nevertheless, there is no consensus on the precise nature of this organizing activity. In an effort to elucidate the varying approaches to this question, theorists have been divided into those who conceive of narrative as a cognitive instrument for imposing meaningful order onto human reality or experience, and those who consider it to be primarily an ontological category that characterizes the human way of being in the world – that is, something constitutive of human existence. These can be referred to, respectively, as the epistemological and ontological position on the significance of narrative for human existence. However, the attempt to draw a sharp distinction between ontological and epistemological approaches to the narrative dimensions of human existence is particularly problematic from a phenomenological-hermeneutic perspective. In this vein of thought, the basic structure of interpretation – the hermeneutic "understanding-something-as-something" – is seen to characterize all experience, even the most rudimentary sense perception. Hence, the process of interpreting experiences is not an additional procedure of knowing but constitutes the original structure of being in the world.[23] Charles Taylor summarizes this view by positing the claim that human beings are "self-interpreting animals" – beings constituted by their ways of interpreting themselves and the world.[24] Drawing on the thought of Martin Heidegger, Hans-Georg Gadamer, and Hannah Arendt, Paul Ricoeur's *Time and Narrative*[25] and *Oneself as Another*[26] develop a hermeneutic theory of narrative subjectivity (identity) that emphasizes not only the culturally and historically grounded character of (self-) interpretation, but also the way in which cultural webs of narratives take part in shaping the horizon of interpretation, mediating the relationship of individuals to both the world and themselves. If cultural narratives affect the ways in which individuals experience things, there can be no pure, raw, immediately-given experiences, the narrative interpretation of which would necessarily be a matter of retrospective distortion. Instead, human beings are always entangled in stories, wearing personal narratives in a dialogical relation to cultural narratives, both of which are objects of unremitting reinterpretation.[27] Hence, as Ricoeur asserts, "our existence cannot be separated from the stories that we tell of ourselves."[28]

From the viewpoint of narrative hermeneutics, the debate concerning whether human beings live or tell narratives is based on a questionable opposition: it is not true that life "in itself" would naturally follow the structure of narrative, but neither is it true that individuals first live and then sew lived experiences into a story.[29] Rather, living and telling are interwoven with one another in a complex movement of reciprocal determination. On this view, the narrative interpretation of experience is not a process of fabricating something true and real, but is instead constitutive of being; as Jerome Bruner puts it: "life is not 'how it was' but how it was interpreted and reinterpreted, told and retold."[30] While narration constitutes something – that is, creates meaning rather than merely reflecting or imitating something that exists independently of it – this does not mean that narrative interpretations falsify experience or are somehow external or secondary to it. Human beings are the subjects of life stories that are constantly being told and retold in the process of being lived. This understanding emphasizes that human experience (in se) involves a process of repeated interpretation and sense-making. The relationship between subjective experience and narrative identity can be clarified in relation to the fundamental phenomenological-hermeneutic notion of interpretation. If experience (always) has the structure of interpretation, then narratives can be viewed as having the structure of a "double hermeneutic" insofar as they are interpretations of experiences that are already interpretations: they knit experiences together by demonstrating how they are related and by creating meaningful connections between them.[31] Ricoeur uses the notion of "mimesis II" to characterize the way in which literary and historical narratives configure everyday interpretations of action ("mimesis I"). His notion of refiguration, or "mimesis III," refers to the process by which individuals interpret these literary and historical narratives from the horizon of their concrete life situations and thereby reinterpret their experience in light of these cultural narratives. This process can be characterized in terms of a "triple hermeneutic."[32] In this way, Ricoeur notes, narrative "carries us beyond the oppressive order of our existence to a more liberating and refined order."[33]

Following the central claim of Edmund Husserl's notion of the life-world, it can be argued that meaningful thought arises from contact with objects in a shared world so that one's neural network is shaped and realizes the learned skills of perception, cognition, and action.[34] Kant takes this thought over from Aristotle's discussion of second nature – i.e., socialized nature – shaped in human children so that they react to the world in ways that reflect cultural attunement.[35] The human nervous system is thereby organized to track events and objects around the subject and differentially weight the data patterns received so as to assimilate the modes of adaptation to the human life-world employed by human beings in a common social milieu. To read the narrative data patterns in that shared form of life, one relates them to what others say so that reproducible regularities of brain function are intricately interwoven with conversations or speech. The conscious narrative manifesting from one's discursively engaged life trajectory (emanating from and continually reshaping its neural substrate) is a selection producing a lived subjective identity. Cognitive neuroscience attests that consciousness is not determined solely by bottom-up but also top-down influences of meaning and narrative coherence. This top-down effect parallels the contextual effects on letter and word perception of whole language reading activities. The imaged brain is, therefore, as responsive to the whole context and identity of the individual being imaged as it is to stimuli perceived by the relevant receptor organs. While human beings demonstrate commonality in the way information is processed and cognitive tasks are approached, the task of constructing an individual life and subjectivity is done idiosyncratically and forges a neuroimaging profile of narrative that is as unique as a fingerprint. This furnishes an approach to neuroethics that takes account of philosophical and ethical aspects of narrative identity that can be overlooked in neuroscientific thinking.[36]

The subjective brain embodies the significance attached to things and the way it shapes itself in light of experience and inheritance.[37] An older individual, for instance, faced with the inability to cope at home, may see this in various ways: the loss of a distinctive place in the world, alienation from the familiar scaffolding environment that has been previously depended upon, release from cares and responsibilities that have become straining, or something else entirely. From this view, the world reflects a dance in which the significations surrounding a person interact with the events that obtrude and together give shape to a narrative conception of self. Such a life story makes individuals who they are and their brains take on the shape necessary to translate that story in actuality in the world in which they live, move, and have their being. These stories are shaped by sensibilities and standards of reason so that, as Taylor remarks, "making sense of my present action ... requires a narrative understanding of my life, a sense of what I have become which can only be given in a story."[38] That story concerns the issues of one's place relative to the good – a negotiation within a sociocultural context (with the brain serving as the vehicle for this activity) reflecting an inclusive and ongoing editorial transformation of life episodes so that its integrity is crucial to one's function as a person. When it becomes disorganized and unstable, the narrative identity of the individual begins to unravel. For an Aristotelian, being somebody-somewhere is the result of a brain shaped to realize a certain psyche partly as a process of self-creation. Human beings are creatures who present themselves to others in ways that reflect choices constrained by their neurobiological nature and historical contingencies. In all individuals there is a struggle between diverse narrative dimensions that is evident as they endeavor to become somebody-somewhere by structuring their neurocognitive architecture in a way that conduces to living a good life with good relationships.[39] This is what drives Carson to suggest that in order to make sense of themselves, human beings tell tales – tales of truth, tall tales, tales of wisdom and woe – and listen to the tales told by others. Stories, with their beginnings, middles, and ends, redeem life from contingency and make it something other than a meaningless succession of events.[40]

### 5.1.3. Narrative Identity and Moral Development

The use of narrative categories to understand the construal of lives through time is a productive heuristic in personality, self, and identity, and its appeal for understanding moral development is long standing.[41] Indeed, narrative is increasingly employed to account for moral personality, prosocial moral identity, and the characteristics of various moral exemplars. In the present context, the use of narrative is valuable for (at least) two reasons. First, narrative is deployed to account for a novel region of the moral life – harm-doing – and, second, for understanding a new construct – moral agency – that comes with it. It does so with rich possibilities of integration with other developmental achievements, systems, and literatures. Hardly an orphan construct, narrative moral agency is deeply informed by achievements in non-moral developmental processes. This much is significant for those who worry that moral psychology has become isolated from theoretical and empirical advances in other domains of psychology. Monisha Pasupathi and Cecilia Wainryb position morality within the landscape of research programs in moral development, and agency within research on closely allied constructs, such as efficacy control and self-determination.[42] For Pasupathi and Wainryb, moral agency is defined in terms of morally relevant actions that have implications for justice and care. The emphasis on justice and care is a traditional way to bind narrative identity with moral development. Causing harm intentionally puts one at odds with such moral principles, and the resulting tension invites narrative construction of moral agency. Failing in prosocial behavior does not require construction of moral agency insofar as prosocial acts are not obligatory (under the conception of morality forwarded by the cognitive developmental tradition, including domain theory).[43] However, the work of moral agency is not a simple matter of making domain distinctions. While moral agency is singular, its sources are plural. Hence, morally charged situations are as complex as the individuals caught up with them. While moral identity is concerned with individual differences in moral value centrality, research has not yet identified a convincing developmental pathway, and existing investigations are limited by their reference to prosocial but discretionary behavior. As Pasupathi and Wainryb observe, when confronted with moral failure, human beings tend to construct narratives that help them make sense of moral agency.[44] Narrative moral agency is, therefore, positioned in a way that stakes out new territory in the study of moral development.[45]

Insofar as moral agency without an agent is as inconceivable as an agent without personality and selfhood, the language of moral self-identity and moral personality is not easily dispensed with. Narratives of moral agency are constructed by persons of a certain kind – moral persons with steadfast commitments that serve as cardinal components of their self-understanding. Put simply, what defines moral agency is not the actions performed – actions do not speak for themselves – but the agents of the actions.[46] The key turn of significance for moral development is how early social-cognitive units are transformed from episodic into autobiographical memory. At some point, specific episodic memories must be integrated into narrative form that references a self whose story it is. Parents, for instance, assist children in organizing events into personally relevant biographical memories in accord with the frequency and sort of questions they ask about daily routines or recent experiences. Parental interrogatives – concerning, perhaps, what happened when one child pushed another, and what that child should do next – serve as a scaffold that helps children structure events in narrative fashion. In turn, this provides (as part of a self-narrative) action-guiding scripts – apologizing for causing disproportionate harm, for example – which becomes frequently practiced, overlearned, routine, habitual, and automatic. Parental interrogates might also include reference to norms, standards, and values so that the development of the moral self becomes part of the child's autobiographical narrative. The narrative self that emerges from early dialogic interactions with caregivers is, then, also a moral self. The source of narrative moral development lies in the shared, positive, affective relationship with one's caregivers. It manifests as an effect of what Kochanska calls the "mutually responsive orientation" that characterizes the interpersonal foundation of conscience.[47] This orientation is marked by shared, positive affect, mutually coordinated (enjoyable) routines, and a cooperative interpersonal set that delineates the shared willingness of the parent and child to initiate and reciprocate relational overtures. It is from within mutually responsive orientations, paired with their corresponding secure attachments, that children are eager to comply with parental expectations and standards. The committed adherence of children to the norms of values of caregivers consequently motivate internalized moral development and the work of conscience.[48]

In its broadest applications, narrative identity is a label attributed to the attempt to differentiate and integrate a sense of self along diverse social and personal dimensions.[49] Consequently, identities can be differentiated and claimed according to various sociocultural categories: gender, age, race, occupation, socio-economic status, ethnicity, religion, and the like. Any claim of narrative identity faces (at least) three dilemmas: (i) sameness of a sense of self across time in the face of constant change; (ii) uniqueness of the person vis-à-vis others (i.e., not being the same as everyone else); and (iii) the construction of agency as constituted by self (with a self-to-world direction) and world (with a world-to-self direction). For Michael Bamberg, narrative identity takes off from the "continuity/change dilemma," and from there ventures into issues of uniqueness (self-other differentiation) and agency. In contrast, notions of self and sense of self develop from the self/other and agency differentiation, and from there filter into the "diachronicity of continuity and change."[50] The engagement in activities that make claims regarding self-understanding require acts of self-identification by implementing and choosing from particular repertoires that identify and contextualize the narrative author along varying socio-cultural categories. It is helpful to consider these repertoires not as mental or linguistic schemata located in the mind, but rather as preconscious, unfixed, and open to change, depending on context and function. Narrating one's life, as an activity that makes claims about the sort of person one is, requires the ordering of characters in space and time. At the same time, self-narration unites two different ways of making sense: a scientific approach according to which events follow each other in a casual, non-teleological sense; and a hermeneutic, plot-governed approach from which events gain meaning retrospectively owing to the overarching contour by which they are configured. Additionally, personal narratives, whether fictional or factual, draw toward aspects of human life that mark something more than what is reportable or tellable, but, as Taylor remarks, life- and live-worthy.[51] Hence, self-narrating enables individuals to disassociate the speaking/writing self, thereby allowing them to take a reflective position vis-à-vis the self as character in past time-space, make those past events relevant for the act of telling (an activity in the present), and become oriented to a transcendent human good. Against this horizon, self-narration establishes itself as a privileged site for moral self-development and analysis.[52]

The ability to conceive of life as an integrated narrative forms the epicenter of what Erik Erikson calls "ego identity" (on the basis of which later moral identity may develop and run its course).[53] Erikson's underlying assumption is that at some point in socio- and ontogenesis, life begins to solidify into building blocks that, when placed in correct order, cohere: important moments tie into important events, events tie into episodes, and episodes tie into a life story. It is this analogy between life and story – or, better, the metaphoric process of seeing life as storied – which fueled the narrative turn of the 1970s. However, the strength of how scholars make use of this analogy varies. On the one hand, there is a relatively loose connection according to which human beings tell stories of their lives (and the lives of others) by using particular narrative formats. Lives can be told as following an epic format or as consisting of unconnected patches. On the other hand, lives are most often told by laying out characters and their (moral) development. Character requires an internal and external form of organization. The former is typically organized through a complex interiority, usually in the form of traits, which arranges actions and unfolding events as outcomes from motives that spring from this interiority. The latter, external offset of character development takes plot as the overarching principle that orders human action and answers the threat of discontinuous and meaningless life by a set of possible continuities. The interplay of human interiority and culturally available models of continuity (plot) gives narratives a powerful role in viewing life as a story. With narration thus defined, life transcends the animalistic and unruly body and gains the power to organize unorganized material into human temporality – the answer to non-human, a-temporal, discontinuous chaos. Another helpful application of the narrative metaphor for life emerges as the product of a narrative mode of thinking. In separate but similar ways, both Bruner[54] and Donald Polkinghorne[55] contend that there is a particular cognitive mode of making sense of the social world that is organized narratively. Theodore Sarbin affiliates the cognitive claim with a more strongly formulated ontological position and argues that human beings live in a story-shaped world and are guided by narratory principles.[56] Finally, Elliot Mishler propagates the use of autobiographic narrative interview data in a contextual approach that does not limit itself to the capture of human experience or look behind the author, but rather fosters inquiry into the realm of interaction and relationships.[57]

The foregoing conceptualizations reflect a particular cultural perspective. In contemporary American society, narratives about defying convention in order to follow intrinsic longings, or suffering through life's harshest tribulations only to emerge enhanced and integrated in the end, enjoy considerable cachet and admiration.[58]

Among the most popular of these narratives are stories of upward mobility, liberation, recovery, atonement, and self-actualization. In each of these forms, an intrinsically motivated individual overcomes intense suffering to experience an enhanced status or state – moving from rags to riches, slavery to freedom, sickness (or addiction) to health, sin to salvation, or immaturity to the full expression of the morally-developed self. In many of these stories, the agent experiences a negative perception of the heretofore narrative self and, as a result, works hard to redeem life and develop virtue in some way. In many others, however, the problem concerns suffering that comes to the agent through no individual fault – through unexpected illness, loss, poverty, or the like. Such redemptive narratives chart the individual's movement and identity over time from suffering to an improved disposition. In contemporary American society, then, redemptive life narratives seem to suggest that if the terrain of moral development is not steep, the narrative identity thereby produced cannot maximize its potential.[59]

## 5.2. Narrative Identity as Product of Autobiographical Memory and Emotional Rationality

According to Walter Fisher, human communication is tested against principles of coherence (i.e., narrative probability) and fidelity (i.e., truthfulness and reliability).[60] Regarding the former, coherence is the degree to which a story "hangs together" – that is, how probable or believable the story seems, both to oneself and to others, and whether the characters act in a consistent manner.[61] Just as individuals often arrive at firm conclusions through comparing the coherence of their stories with stories of similar detail, so too the coherence of narrative is tested when the beginnings, middles, and ends of a life story resonate with others that have trod similar ground. Regarding the latter, fidelity is the degree to which values expressed in a story ring true with what one regards as truthful and fair. In this sense, narrative strikes a responsive, emotional cord. An autobiographical narrative possesses fidelity when it offers good reasons to accept its underlying moral, which will ultimately serve to guide one's actions in the future.[62]

### 5.2.1. The Autobiographical and Emotionally Rational Ontology of Narrative Identity

For better or worse, the convergence of one's autobiographical memory and emotional rationality produces one's narrative identity

– one's conception of self. At core, the notion of the "narrative self" centers on the innate effort of human beings to understand and interpret the world through storytelling.[63] Building on Heidegger's claim that human beings are essentially "embodied conversations," and that the unity of conversation serves to support human existence,[64] Fisher remarks that individuals experience and comprehend life as a series of narratives that possess various beginnings, middles, and ends.[65] However, all conversations and the narratives to which they contribute are not equally valuable. That is, authentic narratives must be evaluated by applying the standards of "narrative rationality" to them.[66] Such rationality is the method by which narratives, autobiographical and emotional at core, are accorded their status as "true."[67] This "logic of good reasons" incorporates the means of analyzing and evaluating arguments with critical questions that can locate and weigh values. These questions concern fact, relevance, consequence, consistency, and transcendence. To be sure, narrative rationality does not deny that discourse (often) contains structures of autobiographical and emotional reason that can be identified as specific forms of argument and assessed as such. Rather, it honors this fact but goes beyond it to claim that reason occurs in human communication outside of traditional argumentative forms.[68]

At the end of the third volume of *Time and Narrative*, Ricoeur addresses the question of whether there exists a fundamental experience capable of integrating historical and fictional narratives.[69] He forms the hypothesis that the constitution of narrative identity, whether of an individual or a historical community, is the locus to search for the ontological fusion between history and fiction. Insofar as human beings possess an intuitive pre-understanding of this fusion, their lives become more readable (lisibles) when interpreted in function of the stories individuals tell about themselves. These "life stories" are rendered intelligible when they are applied to narrative models borrowed from history and fiction. The epistemological status of both autobiographical memory and emotional rationality seems to confirm Ricoeur's intuition. On this basis, it is plausible to affirm his theses that (i) knowledge of the self is an interpretation; (ii) the interpretation of the self, in turn, finds narrative (among other signs and symbols) to be a privileged mediation; and (iii) this mediation borrows from history as much as fiction. What is missing in this apprehension, however, is a clear understanding of what is at stake in the question of narrative identity as applied to both individuals and communities. Following the publication of the third volume of *Time and Narrative,* Ricoeur became acutely aware of the considerable difficulties attached to question of identity as such. He would become convinced that a strong and convincing plea could be made in favor of narrative identity if it could be demonstrated that this notion, along with the experience it designates, contributed to the resolution of difficulties pertaining to personal identity as discussed in the broader philosophical circumstances of the time. The conceptual framework under which Ricoeur conducts his subsequent analysis of narrative identity rests on the fundamental difference he sees between two major uses of identity, namely, identity as sameness (idem) and identity as self (ipse). "Ipseity," he argues, is not sameness, and his revised thesis is that many difficulties that obscure the question of personal identity result from not distinguishing between these two stages of the term.[70]

Building on Ricoeur, this text posits that the autobiographical and emotionally rational ontology of narrative identity is best understood via its four primary manifestations in contemporary neuropsychology: (i) metaontological, (ii) political, (iii) metanarrative, and (iv) conceptual.[71] Regarding the first, metaontological narratives are stories that moral agents use to make sense of (and act in) their lives. These "master narratives of existence" are used to define who one is and therefore serve as the precondition for knowing what one is to do. This "doing" will in turn produce new metaontological narratives and, hence, new actions insofar as the relationship between narrative identity and ontology is processual and mutually constitutive. Hence, both are conditions of one another; neither is a priori. Narrative location endows moral agents with identities – however diverse, ambiguous, ephemeral, or conflicting they may be. To have some sense of moral existence in the world requires that lives be more than assorted series of isolated events or combined variables and attributes. Metaontological narratives process events into episodes: individuals act or refrain from acting in accordance with how they understand their place in any number of given narratives.[72] This notion is echoed by Taylor's insight that insofar as human beings cannot but orient themselves to the good and thus determine their place relative to it, they must inescapably understand their lives in narrative form.[73] Regarding the second, political narratives are stories that attach to cultural or institutional formations larger than the individual, ranging from the narratives of one's family to those of the workplace (organizational myths), religious community, government, and nation.[74] Like all narratives, these "stories of sociality" have drama, plot, explanation, and selective criteria. As Taylor comments, individuals may sharply shift balance in their definition of identity, dethrone the given, historical community as a pole of identity, and relate only to the community defined by adherence to the good. Yet this does not sever their dependence on webs of interlocution; it merely changes the webs and the nature of their dependence.[75].

Regarding the third, metanarrativity refers to the "master narratives" in which individuals are embedded as contemporary actors in history as moral agents.[76] Sociological theories and concepts are encoded with aspects of metanarratives – stories of progress, decadence, industrialization, enlightenment, etc. – despite typically operating at a presuppositional level of social-science epistemology often beyond conscious awareness. These master narratives can be the epic dramas of our time (e.g., capitalism v. communism, individual v. society, barbarism v. civility) or progressive narratives of teleological unfolding (e.g., Marxism and the triumph of class struggle, liberalism and the triumph of liberty, the rise of Islam, etc.). Perhaps the most paradoxical aspect of metanarratives is their quality of denarrativization. That is, metanarratives are built on concepts and explanatory schemes (i.e., social systems, entities, and forces) that are in themselves abstractions. Regarding the fourth, conceptual narrativity and explanations are those stories individuals construct as moral agents. Insofar as neither moral action nor institution-building is solely produced through metaontological and political narratives, one's concepts and explanations must include factors of social force (e.g., market patterns, institutional practices, organizational constraints, etc.). The objective of conceptual narrativity is to devise a vocabulary that individuals can use to reconstruct and plot over time and space the metaontological narratives and relationships of historical moral agents, the political and cultural narratives that inform their lives, and the crucial intersection of these narratives with other relevant social forces. The conceptual challenge is to develop a social-analytic mechanism that can accommodate the contention that moral life, organization, action, and identity are narratively (i.e., temporally and relationally) constructed through both metaontological and public narratives grounded ultimately in autobiography and emotional rationality.[77]

## 5.2.2. Narrative Identity as Moral Education, Moral Methodology, and Moral Discourse

Even the most ardent devotee of analytical rigor can admit that most individuals, most of the time, learn the majority of what they know about morality from narratives of one kind or another. These lessons span from the tall tales of childhood, to the life stories shared among friends, to the narratives that permeate a particular culture.[78] In support of this claim, Martha Nussbaum makes the Aristotelian observation that a large part of learning takes place in experience of the concrete, which in turn requires the cultivation of perception and responsiveness, namely, the capacity to read a situation and signal out what is relevant for moral thought and action. This active task is not a technique, but rather a formula one learns through practical moral guidance.[79] Therefore, to say that narrative identity contributes to ethical decision making through the providence of a particular form of moral education is hardly a controversial claim, but it is an important one nonetheless. Defenders of an ethics-as-propositions conception of decision making argue that individuals are simply too dense to grasp, remember, or learn, and that because of this society must fall back on narrative identity as a heuristic device. But narratives are not second-best instruments for representing the content of morality in a vivid, memorable way. Rather, they are themselves that content.[80] If the enterprise of moral education is understood as a pursuit of truth in all its forms, requiring a deep and sympathetic investigation of all major ethical alternatives and the comparison of each with one's active sense of narrative identity, then it requires narrative identity and the experience of attending to it for its own completion.[81]

An antiquated method of case-based moral reasoning (i.e., casuistry) that contemporarily grounds itself in narrative identity has experienced a recent resurgence.[82] Blaise Pascal's brutal yet brilliant assault on its abuses made casuistry a term of dishonor.[83] Nevertheless, as ethicists struggled with actual cases, the case-centered approach inherent to casuistry was often employed on a variety of moral problems. In time, the restoration of a narrative-based casuistry as an intellectually respectable method of moral reasoning would gain credence.[84] As Albert Jonsen and Stephen Toulmin note, the heart of human experience does not lie in a mastery of rules and theoretical principles, however sound and well reasoned they might appear.[85] Instead, it is located in the practical wisdom (phronēsis) that comes from seeing how the ideas behind rules work out in the course of one's (evolving) narrative identity.[86] In this way, casuistry rejects the primacy of propositional moral logic: it declares that narrative details are equally important to case resolution; the principles from which agents draw is merely an interpretation of those narrative details, which are open to revision. It follows, therefore, that the rightness or wrongness of particular details is embodied in the narrative context insofar as the propositions from which agents draw are interpretations of that context. Thus, only an adequate recollection of the autobiographical and emotionally rational elements of narrative identity can equip individual agents with the tools necessary to weigh moral considerations of various kinds and resolve conflicts between those considerations.[87] Moral philosophers rarely behave as geometricians, forwarding axioms, definitions, or theorems in their moral discourse.[88] Rather, they typically tell stories of at least two genres. The first is the "philosopher's hypothetical," which is meant to make a particular point, usually about the plausibility of an assertion about ethics. Judith Jarvis Thomson's violinist[89] and Bernard Williams' traveler[90] are well known examples of the genre. These stories function either to reinforce confidence in the proposition being forwarded or to reveal its defect. If it is not obviously wrong for a woman who wakes up and finds herself attached to a violinist to want to disconnect the tubes that are keeping him alive, then neither is it obviously wrong for a woman who finds herself pregnant to want to cease supporting the fetus growing inside of her body – or so Thomson's story is meant to suggest. The second, less noted genre of narrative is intended to construct, motivate, and display the necessity of the theorist's approach. This is the method, for instance, of MacIntyre, by which he describes contemporary morality as a collection of incompatible shards of earlier moralities that were more coherent,[91] or H. Tristram Engelhardt's, by which he depicts a world in which differing conceptions of the good lead inexorably to tyranny and violence when one group attempts to impose its conception on another.[92] The success of this latter genre is due in large part to its birth in and motivation by narrative identity – that is, by stories about who individuals are, what they are like, and how they came to be in their current situation. [93] Most, if not all, moral discourse, including moral theory, is embedded in, conditioned by, and conducted through narratives.[94]

MacIntyre's description of narrative identity is both powerful and successful. It establishes his method, which is best understood as a form of philosophical history or historically situated philosophy.

MacIntyre suggests that non-narrative approaches to morality – analytic, phenomenological, and existential, to name a few – are incapable of noticing what he considers to be the central problem in contemporary morality. His outline of narratives as frequently comprised of flourishing beginnings, catastrophic middles, and final restorations are vivid and compelling, and forces attentive readers to think in novel ways about moral development, past and present.[95] While critics correctly identify that MacIntyre's conception of narrative identity overstates the moral homogeneity of the past and understates the currently available resources for resolving moral disagreement,[96] the important point is that, like Engelhardt's story of moral disagreement and political violence,[97] MacIntyre's narrative is more than a clever story. Against the Engelhardtian notion that human beings must abandon all efforts at imposing any particular conception of the good on others, MacIntyre insists that human beings must recover an intelligent, internally coherent narrative identity grounded in autobiographical and emotionally rational moral norms. Building on MacIntyre, Nussbaum posits that moral knowledge is not simply the intellectual grasp of historical data, but the analysis of complex, concrete narrative realities in a lucid and responsive way.[98] Autobiographical and emotionally rational narratives, therefore, serve to sharpen perception and enrich moral education, method, and discourse.

### 5.2.3. Narrative Identity as Ground and Object of Normative Ethical Principles

Contrary to the vision that there exists a sharp division between forms of reasoning driven by stories and those driven by principles and theories, John Arras contends that narrative is an essential supplement to ethical principles. Insisting that stories and moralizing theories are mutually interpenetrating and interdependent, he points out that moral narratives often embody a form of argument (i.e., a "moral") and that a keener awareness of the narrative elements embedded in substantive moral reasoning will permit a more reflective and robust model of moral analysis.[99] Arras' argument is twofold: first, that narrative elements are inevitably embedded in all forms of moral reasoning; and second, that individual responses to narrative are the ground out of which principles and theories develop. For the former part of the argument, Arras relies heavily on the work of Rita Charon, who passionately calls for narrative competence in bioethics while maintaining the fundamental structure of principlism within the field. Borrowing from the literary theory of

Walter Benjamin, Charon argues that sensitivity to questions of narrative identity constitute a "narrative competence" that serves as a prerequisite of "doing good ethics." On this view, narrative competence is recommended as a way to enhance the use of existing methods of moral analysis by directing their development to the particularity of individual lives.[100] Following Charon, Arras similarly interprets the significance of narrative identity as supplementary to principles, viewing narrative as the oil that lubricates the gears of normative principles, thereby enhancing their function.[101] The latter part of his argument appeals to the Rawlsian model of reflective equilibrium,[102] suggesting that few principlists would ground their theories in a way that prohibited them from being tested against considered judgments about actual circumstances.[103] For Arras and others, then, an ethics grounded in narrative identity is not a new approach, but rather a recognition and appreciation of the debt that principle-driven modes of discourse owe to stories.[104]

Against this backdrop, this text argues that a much stronger case can be made for the dependence of principles on narratives, grounded in the meaning individuals attribute to the historical development of principles.[105] Against Jonsen and Toulmin, Childress argues that it is a mistake to view the genesis of an individual's moral development as grounded in stories, with the understanding of general norms constituting a later, more sophisticated stage of development. Instead, he argues that individuals learn both moral norms and illustrative narratives from their parents, and that without norms, they would have no way to understand and classify the stories.[106] Yet it seems quite plausible that moral development occurs in precisely the opposite way. Consider, for example, how children learn to recognize the distribution of goods as fair or unfair – an activity that becomes a metaphor for the principle of justice. For instance, a child may watch his parents dividing goods in a particular way and witness the result such behavior has on his three other siblings. At a later time, the same child may watch his younger sister steal all of the goods for herself and similarly witness the effect it has on all parties. Finally, the child may be given the responsibility to divide and distribute the goods among his siblings and, perhaps with prompting, decide to emulate his parents and not his younger sister.[107] Hence, within the task of distributing goods, the child is inescapably reminded of the narratives in which he previously participated and provided the criteria by which to determine whether the new narrative produces a fair distribution.[108]

Another illuminating example is the development of the principle of respect for autonomy, which enjoys a relatively brief history in Western culture compared to other moral norms.[109] At some point, individuals began to tell stories about how others thwarted what they wished or desired to do, despite the fact that the others were at no risk of being harmed or being directly impacted by the proposed actions. Listeners to the stories believed (with good reason) that some moral wrong had occurred, even while they lacked the name of an established normative label for this type of violation. The stories were retold to others as examples of that as-yet-unnamed manifestation of moral wrong. Over time, the collection of these stories developed, and it become more readily evident that the wrong that occurred in each story was sufficiently similar to the wrong that occurred in the other stories to be worthy of inclusion under the same general heading. Eventually, the term "respect for autonomy" came to function as a shorthand label for the moral feature that this collection of narratives shared in common. In this way, one can similarly view every moral norm or principle as a label that saves the trouble of having to recount the many narratives that share a common moral feature. Beyond this, still more support for the notion that moral principles ultimately rest on narrative structures comes from a careful reading of the opening of Kant's *Fundamental Principles for the Metaphysics of Morals*.[110] Kant paints himself into a corner with an apparent paradox: on the one hand, morality requires that human beings devote themselves to the abstract moral law per se, not to any concrete practical maxim; on the other hand, those individuals then feel moved to behave morally, yet law per se – possessing no concrete link with any worldly behavior – is powerless to provide them with any practical advice on how to act. His effort to resolve this paradox is the first formulation of the categorical imperative. Here, Kant demands that individuals engage in an act of the imagination – to envision a world as it "would be" if a particular maxim were to function as a universal law. In an important sense, he is requiring that human beings write a personal novel of sorts. However, what Kant fails to explicate is that without this generated narrative identity, his imperative remains a futile guide to moral behavior.[111]

Apart from the exercise of principles, there is another area of moral activity that underscores the foundational import of narrative identity: discussions about virtue and character.[112] MacIntyre has led the way in claiming that the very notion of virtue is unintelligible without a narrative conception of what it means to live one's life. To be virtuous is, for MacIntyre, to attempt to become a particular sort of person – to live one's life in a particular way as it unfolds over time. This means not only that individuals who remain true to their narrative identity behave in certain ways, but also that they do so for certain reasons and with certain motives, and that they learn certain things from their past behavior and apply them to future behavior in certain ways. None of this makes sense in absence of a narrative conception of the self.[113] Even Arras, who is skeptical of MacIntyre's claims,[114] agrees that the only way to adequately depict, understand, and assess character is by telling and retelling stories.[115] Thomas Murray calls attention to another, often unnoticed way in which narrative can play a central role in moral-philosophical discourse.[116] Murray cites the "opening fables" in MacIntyre's *After Virtue* and Engelhardt's *Foundations of Bioethics* and notes that each writer begins with an extended narrative of either a real or a possible world, which subsequently highlights the specific problem that generates the philosophers' respective programs of ethics and situates the problem in actual human experience. One might view these "fables," Murray suggests, either as nothing but preliminary rhetorical prefaces that only prolong the actual work of philosophical analysis, or as justificatory explanations of the authors' respective projects, on the other hand. While foundational introductions undoubtedly serve to do both, there is a striking and revelatory irony that emerges in Engelhardt's case. He proceeds from his opening fable to develop a rigorously Kantian program for bioethics that would seem to exclude any role for narrative thought. However, if Murray is correct, and even someone like Engelhardt relies on narrative form and structure to perform a vital piece of moral work, then it may well be the case that even the most conventional forms of philosophical ethics are inexorably shaped by narrative and its search for meaning and identity.[117]

### 5.2.4. Rigor in Narrative Judgment and Ethical Justification

A general point of agreement among narrativists is that in order to critique a story, one needs a different story, or counter-story, with which to compare it.[118] As Margaret Urban Walker explains, the task of fully normative reflection is intrinsically comparative. In other words, when individuals ask themselves what can be said for some way of life, they are asking whether it is better or worse than some other way they know or imagine.[119] On the one hand, this observation exacerbates the problem of explaining how narrative reasoning can judge one story to be morally superior to another in a rigorous way that does not simply reimport moral principles. On the other hand, part of the attractiveness of narrative identity is its intuitive appeal: individuals can judge the coherence of an event only within the context of a story. For instance, the adult child who is faced with a proxy decision about ongoing life support for a critically ill parent will invariably ask what sort of narrative would be the most fitting final chapter of the parent's life. This, in turn, would explain why the adult child would lay special stress on any sentiments the parent had expressed vis-à-vis how life was ideally imagined as coming to a close. Walker, who views ethics primarily as an exercise in accountability and responsibility, places great stress on the significance of moral reliability.[120] For her, moral responsibility lies at the intersection of the respective narratives of relationships, identity, and values, and forms the basis for one's identity. That is, individuals should wish to be viewed by their peers as reliable, and this requires a reasonable prediction of how those individuals will behave in a wide variety of circumstances. On this view, acts are wrong only by a judgment of radical incoherence from the standpoint of the narrator.[121]

Tom Tomlinson has attacked this coherentist method, citing as its fundamental fallacy the indifference from the claim that one's life is best described by a narrative to the normative claim that life choices are best judged by the coherence within one's life narrative.[122] One's history (i.e., one's "story so far") will shape what one does, and no matter what one does, it will be intelligible within the narrative of one's life. For Tomlinson, whatever one decides as the adult child at the bedside of a critically ill parent will become the next chapter of the parent's life and will have coherence as part of an account of that life. However, saying that one's life is a narrative says nothing per se about what one should do in such circumstances. Tomlinson is correct insofar as he mentions that the coherentist position requires an additional normative step beyond viewing life as a unified narrative, but this step would be fully plausible only as a rough empirical generalization. Tomlinson further ignores the possibility of a radical disruption in one's life narrative such that coherence can no longer be found at all. There are, indeed, many ways of writing the last chapter of the life of one's critically ill parent so that it would seem a coherent continuation of a life narrative as it has unfolded until now, but that should not obscure the fact that there may be many more ways of writing that chapter which would be utterly incoherent. In the case of one's parent, two examples would include (i) petitioning a Catholic priest to administer last rites as a means to prepare for heaven's entrance and (ii) placing a stereo at bedside to loudly play rock music as a means of bringing cheer. While there is nothing intrinsically wrong with either act, and the unassuming pair may well bring comfort to some, any discomfort induced would derive only from a judgment of radical incoherence from the standpoint of the parent's life narrative. One can further expand the possibilities of radical incoherence if one thinks more globally and reminds oneself of the incredible number of different life stories that human beings may have. It therefore seems simply untrue, on the Tomlinsonian view, that any way of writing the next chapter of one's life narrative is a coherent way of continuing that narrative.[123] Nevertheless, it remains possible to employ a more exacting version of narrative coherence or incoherence as a guide to moral justification for actions.[124]

Wide reflective equilibrium, as articulated by Norman Daniels, locates justificatory power in coherence among three elements: particular moral judgments, general ethical principles, and background theories of human nature. Wide reflective equilibrium is a coherentist account of ethical justification inasmuch as no single element is immune from alteration or privileged as more foundational than another. This system presupposes that there is a large gap between general, abstract moral principles and concrete case judgments and that little, if anything, of interest to ethics resides in between. One may occasionally alter a long-standing general principle insofar as it fails to resonate with a particular case judgment. At other times, one may dismiss a case judgment because it fails to cohere to an attractive general principle or theory.[125] Circumstances determine how the equilibrium works for the best overall "fit" among the elements. Of course, any such fit is temporary, since a new case or background theory may disrupt the original reasoning.[126] Nevertheless, Daniels' theory has respectable roots in moral and political philosophy, and it is tempting to adopt for justification in an ethics of narrative identity. One would simply designate the particular case judgments as "narratives." Within this reimagined model, one need not reject normative principles, and one can acknowledge that at least some narrative judgments may be persuasive enough to overturn principle-based judgments on occasion.[127] One could then use principles when they are helpful and remain focused on particular narratives as sources of moral justification.[128] However, to maintain rigor, a "narrative equilibrium" must be more complex still.

On a fundamental level, narrative equilibrium is also coherentist.[129] That is, what is ethically justified is what most accurately hangs together with everything else, acknowledging that one can seldom, if ever, provide an algorithm for deciding what "hanging together" consists of. As suggested above, narratives do not stand alone; they depend for their meaning on broader background narratives that are often taken for granted by those who share a common society and culture. At least some of the time, then, moral judgments and moral behavior are justified on the grounds of coherence within and among one's narratives. On other occasions, when wider reflection is needed, it may be necessary to appeal to background theories of human nature or to general principles, including psychological, sociological, or anthropological aspects.[130] These elements are viewed as contained within a multilevel cluster of narratives.[131] In other words, background theories of human nature, even if apparently derived from the social sciences and quantified in statistical terms, are a sort of sociocultural background narrative, providing the story of how people in a society tend to behave and why. Narrative equilibrium is a product of human activity, and those humans function within a particular sociocultural context during a particular historical movement.[132] Thus, discerning that something is the case – whether this action is cruel or that ball is red – obviously involves subsuming the case under a concept, but it does not involve reaching a belief by invoking some generalization that seamlessly links premises to conclusion.[133] The question is not, therefore, how subjective stories can provide rigorous criticism, judgment, and justification in the context of ethical decision making, but how rigorous criticism, judgment, and justification can exist without the stories that frame one's narrative identity.[134]

### 5.3. The Requisite Unpredictability of Narrative Identity

Autobiographical and emotional history is, according to MacIntyre, an enacted narrative in which the characters also serve as co-authors. This notion suggests that human beings never start ab initio, but rather plunge in medias res, the beginnings of their stories already carved out by who and what has gone before.[135] Just as literary characters, human beings neither begin nor go on exactly where or how they please. All individuals, then, predisposed to significant segments of the narratives into which they come to be, are constrained by the actions of others and by the social settings presupposed in their actions. Understood this way, it becomes clear that the enacted narrative of one's life is not, and cannot be, predictable. This sort of unpredictability is required for the narrative structure of human life, and the empirical data unearthed by social scientists provides an understanding of human life that is compatible with this structure.[136]

### 5.3.1. The Cognitive Correlates of Narrative Values

While fMRI and PET imaging have successfully identified areas of the human brain that are charged with responding to events in one's surrounding environment, relatively little research has directly investigated the cognitive regions immediately linked to the development and maintenance of narrative identity.[137] Several scholars have speculated that the frontal lobe (responsible for language and information processing) and right hemisphere (responsible for making sense of disparate data) are the primary neurobiological mechanisms employed in narrative development.[138] Notwithstanding the scarcity of substantive literature concerning the neuroscience of human storytelling, these findings suggest the evolution of certain hardwired neural circuits recruited specifically to process social narratives.[139] One line of recent research has revealed that values that are closely linked to core personal, national, and cultural identities assume a special cognitive status.[140] Labeled "protected" or "sacred" values, they are non-negotiable and transcend the cost-benefit logic of rational models. These protected values play an important role in sustaining intractable cultural conflicts by impeding intergroup negotiations. Insofar as protected values are historically embedded, the targeted use of sacred rhetoric in narratives can facilitate beliefs achieving protected status. In this way, narratives serve as the vehicle par excellence for the creation and transmission of protected values. Gregory Berns and colleagues have conducted one of the few studies directly investigating the neural foundation of protected narrative values. The authors found that reading statements about protected values (defined as those narrative values that participants refused to abandon at any cost) led to increased engagement in the left TPJ and VMPFC, which they interpreted as related to semantic rule retrieval (inasmuch as protected values invoke fixed principles rather than calculations of costs and benefits).[141] Other work has established the involvement of affective brain systems when individuals react to tradeoffs involving protected values, identifying increased activity in the amygdala nuclei and anterior temporal cortices.[142] Although these findings are significant, they do not yet speak to the most critical aspect of protected values, namely, that they are distilled in the context of personal and cultural narratives and are therefore tied to one's fundamental narrative identity.[143]

In the very recent study of Kaplan and colleagues, participants undergoing fMRI read a series of short personal narratives culled from a large collection of Internet weblogs that described real events recorded by the individuals who experienced them.[144] The authors analyze the ways in which participants reported perceiving the values of the stories' respective protagonists. After each story, participants indicated whether they thought the protagonist could be paid any amount of money to make a different choice than the one made in the story – a strategy employed to measure participants' estimation of protagonists' protected values. When participants indicated that the values in their respective stories were nonnegotiable (i.e., protected), Kaplan and colleagues found that activation in the posterior medial cortices, medial PFC, inferior parietal lobes around the TPJ, and anterior temporal lobes were increased relative to stories in which protected values were not attributed to the protagonist. The collection of these regions forms what is typically called the default mode network. In addition to demonstrating greater signal for stories with protected values, all of the nodes mentioned showed significant increases in activity during story-reading compared with resting baseline. Moreover, network engagement multiplied from the beginning to the end of the story as it unfolded and participants were able to reach a more comprehensive understanding. In attempting to characterize the psychological operations that appear to engage this network, scholars have described them as related to social cognition,[145] internally directed processing,[146] mental time travel,[147] or self-related processing.[148] Interestingly, all of these operations are either implicated in the processing of narratives or rely on a narrative organization of information. The majority of existing studies have established that the cortical midline activations for self-related processing have centered on aspects of the autobiographical self, such as personality trait judgment rather than transient present-moment aspects of the self.[149] The autobiographical self is, in essence, a process of generating fragmentary narratives of personal life built from a multitude of recorded experiences. These same midline structures are activated upon considering the biographies of others, suggesting that the processing of narratives may be more important for activating these structures than self-relatedness.[150]

The anatomy of the default mode network is ideal for functioning as a high-level coordinator of information across sensory, motor, and memory domains in order to generate coherent narrative meaning.[151] The posterior medial cortices and the lateral inferior parietal cortices are interconnected with the rest of the brain, acting as hubs in the densely fused network structure. The posterior medial cortices, in particular, are connected bi-directionally to both association cortices and subcortical structures. The capacity to coordinate the activity of disparate brain regions is vital to the cognitive ability to integrate information – a process fundamental to understanding narratives and the deep-seated cultural values they contain. The complex emotions evoked by narrative values require the coordination of multiple brain systems involved in different aspects of affecting processing. Of particular significance is the brain architecture within which information flows in various directions, converging from low-sensory cortices up to increasingly more integrative association cortices, but also diverging as information and influence flow from the top of the hierarchy back to lower-level regions. In this view, there is a multilevel hierarchy of "convergence-divergence zones" (CDZs) at the bottom of which are mapped sensory cortices that represent specific sensory features. These features are fed through the hierarchy to CDZs that register the associations among them and, through top-down connections, have the capacity to reinstate the lower-level patterns (as during mental imagery or memory). At the apex of the hierarchy are CDZs that combine signals across sensory domains and, hence, represent the highest levels of abstraction and semantic integration. In the context of narrative comprehension, this network orchestrates the activation of various brain regions. For instance, the network may stimulate association cortices that represent abstract concepts, which in turn activates mental images in various early sensory cortices. It may similarly evoke related affective states from subcortical regions that represent the internal state of the body and connect to networks that participate in the recall of associated autobiographical memories and narrative self-knowledge.[152]

Narrative value and subjective meaning are, of course, inherently embedded within cultural contexts, and neuroscience is beginning to appreciate how culture can shape the functioning of the human brain.[153] Kaplan and colleagues found evidence for both universality of functional brain activation across cultures and also indications of cultural differences.[154] Curiously, cultural differences did not manifest in varying brain regions responding to the study task across groups; instead, the same regions responded in all three groups. Moreover, across all groups, activation of the posterior medial cortices, medial PFC, and TPJ regions was greater when participants indicated that protagonists were acting on the basis of non-negotiable values. However, the magnitude of this difference interacted with cultural group membership in the precuneus and left TPJ. This suggests that cultural concern for protected narrative values may be reflected in differential activity in this network. The data of Kaplan and colleagues indicates that individual measures of protected narrative values or moral concerns did not correlate with neural differences in the posterior medial cortices, yet group-level interactions were found. Iranian subjects, for example, were less likely to indicate that they would trade their values for money, and reported higher levels of moral concern compared with other cultural groups. Correspondingly, they demonstrated the greatest neural differences between protected and mundane narratives in the posterior medial cortices and left TPJ. Evidence for disparities in functional brain activations between East Asian and American individuals has also been accumulating, including data related to visual perception, causal attribution, attention, the experience of feelings, and self-concept.[155] A common theme in this research concerns the role of independent versus interdependent styles of narrative self-construal, which appears related to cultural differences in medial PFC activity. Asian individuals, who tend to have a more contextualized self-concept than American individuals, show smaller differences between self and other in the medial PFC.[156] Additionally, religious background can also modulate medial PFC activity,[157] and work in cultural psychology has confirmed that individuals from East Asian traditions tend to be more focused on global narrative values than are their American counterparts.[158]

### 5.3.2. The Variable and Creative Evolution of Narrative, Identity, and Agency

In a postmodern milieu, the narrative self is not a stable and enduring entity limited to or fixed in geographical location and time; it is neither the accumulation of experience nor the collective expression of neurophysiological characteristics.[159] Rather, narrative identity is rooted in the constancy of an unfolding narrative. As Richard Rorty indicates, human beings are the continuing generators of new descriptions and narratives rather than individuals accurately described in a fixed fashion. The self is an ongoing autobiography or, more precisely, a self-other multifaceted biography that individuals constantly pen and edit. The narrative self is an ever-changing expression of community narratives, a being and a becoming through language and storytelling as one continually attempts to make sense of the world and of oneself.[160] The narrative self, therefore, is always engaged in conversational becoming, constructed and reconstructed through continuous interactions and relationships. Individuals live narratives and narratives become living, just as realities become stories and stories become realities. Like past, present, and future, narratives are reflexive processes that cannot be separated. Such reflexivity provides continuity to the ongoing composition and recomposition of life.[161] As Ricoeur suggests, unlike the abstract identity of sameness, narrative identity, which is constitutive of self-constancy, includes development and evolution within the cohesion of a lifetime. Narrative agents, then, appear both as readers and writers of their own lives, as novelists such as Marcel Proust highlight. As the literary analysis of autobiography confirms, the story of life continues to be refigured by the truthful or fictive stories narrative agents tell about themselves. This refiguration makes life itself a cloth woven of stories told.[162] Morny Joy exemplifies this perpetual revision in her proposal that life is not a static narrative with a singular plot but a "dynamic mosaic" that lends cohesion and coherence to the many influences that ceaselessly threaten to overwhelm.[163]

Viewed through a postmodern lens, the problem of identity and continuity – what is commonly termed "selfhood" – concerns maintaining coherence and continuity in the narratives individuals tell about themselves.[164] Narratives of identity become a matter of forming and performing the "I" that individuals are always telling themselves and others that they are, have been, and will be. The self becomes the person or persons one's story demands and believes, regardless of whether the self becomes a hero or a victim. In this vein, Roy Schafer describes the self as an "experiential phenomenon," a set of more or less stable and emotionally felt ways of telling oneself about one's being and continuity through change.[165] Narrative theory, in this discursive sense, was an early avenue of challenge to the modern view of the self and the exploration of implications for defining the self as a storyteller – an outcome of the human process of producing meaning by linguistic activity. Nevertheless, human language possesses inextricable ambiguities. The word "self," for instance, refers to an object. On this ground, Emile Benveniste was among the first to challenge the traditional philosophical notion of self, noting that language is responsible for the notion of self, and language without personal pronouns is inconceivable.[166] As Benveniste suggests, "I" refers to the act of individual discourse in which it is pronounced, thereby designating the speaker. That is, the "I" exists in and by means of saying "I"; the "I" is not a subject or preexistent substance that speaks but is, as subject, a speaking subject. The "I" does not exist outside of language and discourse, but rather is created and maintained in language and discourse. In other words, it is in and through language that an individual constructs a personal account of the self: who one believes oneself to be is a linguistic construction. The "I," therefore, is not a preexisting subject or substance in the epistemological or metaphysical sense; it is a speaking subject.[167] Consciousness of self is only possible inasmuch as it is experienced by contrast. One uses "I" only when speaking to someone who will be "you" in one's address. It is this condition of dialogue that is constitutive of individual identity insofar as it implied that "I" (reciprocally) becomes "you" in the address of the one who, in turn, self-designates as "I."[168]

Narrative identity never represents a singular voice. Human beings are always as many actual and potential selves as are embedded in their conversations and relationships. These self-identities are not personal impulses made social, but social processes realized on-site of the personal. In attending to the construction of the "I," which is inextricably intertwined with the construction of the "you," one does not act "out of" one's own plans and desires, unrestricted by social circumstance and personal performance, but rather "in to" the opportunities offered to act. The formative nature of the "you" is a process by which individuals can, in communication with others, "in-form" one another's being.[169] It is through these relationships that individuals become narrative agents and thereby derive a sense of social or self-agency. Self-agency refers to a personal perception of competency for action. To act or take moral action indicates intentionality, which suggests the ability to behave, feel, think, and choose in a way that is liberating and reveals new possibilities. This agency is not primarily concerned with making choices, but rather with participating in the creation and expansion of possible choices. The concept of self-agency is analogous to having a voice and being free to use it. Hence, self-narratives can permit or hinder self-agency. That is, self-narratives create narrative identities that permit or hinder individuals from doing what they need or want to do, which allows them to experience the capacity to act. In this way, self-agency is inherent and self-accessed: just as individuals can only metaphorically empower one another, human beings can only participate in a process that maximizes the opportunity for self-agency to emerge. Agency, therefore, is exhibited in the twofold ability give shape or form to life while remaining rooted in culture. From an interpretive, meaning-generating perspective, change is inherent in dialogue with self and others. This change manifests through the telling and retelling of familiar stories, redescriptions that accrue through conversation, and different meanings that are conferred on past, present, and imagined future events and experiences. This change seamlessly becomes the developing future self, which unveils the capacity to conceptualize oneself as uniting with shifting experiences, relationships, conversations, views of self, self-narratives, self-identity, and self-transformation. The conduct of social life is finally based upon the right human beings assign to individuals to tell stories about themselves and their experiences, and to have what they say taken seriously.[170]

While empirically challenging, the issue of untold and fragmented (but narrative-informing) stories is a critical area for future research.[171] The primary question to address is whether untold (but narratively indispensable and treasured) and told stories are similar. A fundamental difference between constructing stories privately and publically is that public storytelling opens the self to social shaping. More specifically, the social recounting of situated stories makes the impact of the story on the self stronger than untold story construction as a result of social reinforcement, which occurs only when publically sharing one's life narrative. In other words, stories may be situated in one's mind – such as when one plans to tell a story – but the actual "storying" involves social processes that are absent or weaker in the imagined scenario. Nevertheless, private narration should also shape the self in social ways (e.g., due to the cultural expectations of stories or imagined audiences), but these differences are contingent on how individuals imagine their social worlds rather than how they actually are. There are at least three kinds of storying processes that remain underinvestigated: (i) how stories that are important and not told fit into the narrative self, particularly those that are socially negative; (ii) how partially told or fragmented stories fit into the narrative self; and (iii) how stories that are told differently in different contexts fit into the narrative self. It is unsurprising that individuals are less likely to disclose socially negative events, but the consequences of doing so suggests that while socially negative stories are likely to have an effect on the self through affective experience or behavior, not being able to tell these kinds of stories means that they are not able to be fully integrated into the self. Indeed, finding a greater usage of past tense, or distance, in disclosed events suggests that disclosed events are more integrated into one's sense of history. This is because individuals need social validation for themselves and their stories, but they also need personal validation, and giving voice to their stories is a way of providing such personal validation of the self – a way of owning even the most unsavory experiences. The discussion of untold stories must also be linked to theorizing and research on public and private narrative selves. Telling stories involves the management of both the public and private self inasmuch as both can be revised or maintained through storytelling. To witness this construction of situated stories is to understand the dynamic development of narrative, identity, and agency. Indeed, sharing stories in the principal mechanism through which individuals become narrative selves.[172]

### 5.3.3. The Developmental and Redemptive Metaphysics of Narrative Identity

Contemporary theories of narrative identity attempt to steer a middle course between the personal and the social, viewing narrative development as both an autobiographical project and a situated performance.[173] Nonetheless, narrative psychologists represent a wide range of theoretical perspectives and corresponding methodological preferences. While no single theory or research paradigm integrates all of the work being done, certain fundamental themes consistently emerge in the scholarly literature on the narrative study of lives. Across the various approaches, there appears to be general agreement on the validity of the following six metaphysical principles inherent in the development and redemption of narrative identity. First, the self is storied. As Damasio notes, consciousness begins when the brain acquires the power of storytelling.[174] From an early age, children tell stories about life, casting personal experiences into the structure of setting, character, scene, and plot. Moving into adolescence and adulthood, individuals gather together remembered episodes from the past into an autobiographical storehouse categorized in terms of lifetime periods and event-specific knowledge. In this way, autobiographical memories are highly selective and strategic. Thus understood, life stories always concern both the reconstructed past and the imagined future. Second, stories integrate lives. While stories serve to entertain, educate, inspire, motivate, concern and reveal, organize and disrupt, their most significant function is integrative. Stories often frame disparate ideas, characters, happenings, and other elements of life that were previously set apart. Psychologically, life stories provide integration in synchronic and diachronic ways. The formulation of an integrative narrative identity is a particularly salient challenge for individuals who seek personal integration within an ever-changing, contradictory, and multifaceted social world that offers no clear guidelines or consensus on how to live and what life means. Whether talking about the full life story or a personal narrative of a single event, individuals typically engage in a process of autobiographical reasoning in which they seek to derive general (semantic) meanings from particular (episodic) experiences in life. This exercise is symbolic of personal integration, of putting things together into a narrative pattern that affirms life meaning and purpose.[175]

Third, stories are told in social relationships.[176] A simple but profound truth about stories is that individuals tell them to one another. As such, stories are social phenomena, recounted in accord with societal expectations and norms. Underscoring the discursive and performative aspects of narrating one's life, many investigators argue that any expression of the self cannot be understood outside the context of its assumed listener or audience, with respect to which the story is designed to make a particular point or produce a desired effect.[177] Hence, autobiographical narrators anticipate what their audiences want to hear, and these anticipations influence what they tell and how they tell it. Individuals narrate personal events in different ways for different listeners, and they may switch back and forth between different modes of recounting. Monisha Pasupathi describes these modes as dramatic and reflective, respectively. In the dramatic mode, the storyteller makes frequent use of nonverbal signs, employs vivid quotes and dialogue, and attempts to reenact the original event in the telling. In the reflective mode, the storyteller spends relatively little time describing what happened in the event and focuses instead on what the event may mean or how the event made the storyteller feel.[178] Fourth, stories change over time. Autobiographical memory is notoriously imprecise, and while individuals typically remember with great accuracy the general content of an important life event as time passes, they frequently misremember the minute details. Factual errors in autobiographical recollection increase substantially as the temporal distance from the to-be-remembered event increases. This temporal instability, therefore, contributes to the change in story over time, yet many other processes are also at play, and several of these reflect alterations in how individuals come to terms with the social world. Most obviously, individuals accumulate new experiences over the life course, some of which prove to be so important as to lock into narrative identity. As motivations, goals, personal concerns, and social positions change, memories of important events in life and the meanings individuals attribute to them may similarly evolve. Individuals' autobiographical priorities change as well. Some events may increase in personal significance over time where others fade quickly into the background.[179] Consistent with the general idea that life stories adjust with personal development, a number of studies have documented significant associations between age, on the one hand, and various structural and content dimensions of personal narratives, on the other hand.[180]

Fifth, stories are cultural texts.[181] Life narratives mirror the culture within which they are created, sustained, and told. Stories live in culture: they are born, grow, proliferate, and eventually die according to the norms, rules, and traditions that prevail in a given society and its implicit understanding of what counts as a "tellable" life. Before one can formulate a convincing life story, one must become acquainted with the culture's concept of biography. Indeed, much of what human beings remember as part of their life stories is shared cultural knowledge about the specific course of life in which they are collectively embedded. Within any society, however, different stories invariably compete for dominance and acceptance. It is painfully clear that life stories echo gender and class constructions in society and reflect prevailing patterns of hegemony in the economic, political, and cultural contexts in which they are situated. Power elites give credence to certain life stories over others. At the same time, some may resist dominant cultural narratives, give voice to suppressed discourses, and struggle to bring marginalized ways of imagining and recounting lives to the cultural fore. The construction of such counter-narratives is especially salient among minorities, the economically disadvantaged, and other marginalized social groups. Finally, some stories are better than others. Life narratives always suggest moral perspectives insofar as human characters are intentional agents whose actions can always be construed from the standpoint of what is "good" and "bad" in a given society. Moreover, stories themselves can be evaluated as relatively good or bad from a moral-psychological standpoint, though these evaluations also suggest particular perspectives that reflect the values and norms of the society within which a narrative is evaluated. The past decade has witnessed an escalated interest among narrative researchers in the constitution of a good life story,[182] examining narrative coherence and complexity, as well as the extent to which certain features are associated with psychological maturity, mental health, and professional satisfaction. A growing number of clinical and counseling psychologists are beginning to view psychotherapy as fundamentally a process of story reformulation and repair.[183] This form of therapy aims guide individuals in transforming their faulty narratives into new stories that affirm growth, health, and adaptation.[184]

Finding positive meanings in negative events is the cardinal theme that runs through the redemptive metaphysics of narrative identity.[185] In a series of nomothetic and idiographic studies conducted over the past twenty years, McAdams and colleagues have consistently identified that midlife American adults who score highly on self-report measures of generativity – suggesting a strong commitment to promoting the well-being of future generations and the world in which they live – tend to see their lives as narratives of redemption.[186] Compared to their less generative American counterparts, highly generative adults frequently construct life stories that feature redemption sequences in which the protagonist is delivered from suffering to an enhanced status or state. Moreover, highly generative individuals are more likely than their less generative peers to construct narratives in which the protagonist (i) enjoys special advantage or blessing early in life; (ii) expresses sensitivity to the suffering of others or social injustice as a child; (iii) establishes a clear and strong value system in adolescence that remains of source of unwavering conviction through adult life; (iv) experiences significant conflicts between desires for agency or power and desires for communion or love; and (v) looks to achieve goals to benefit society in the future. Taken together, these themes articulate a narrative prototype that many highly generative adults employ to make sense of their lives. For these individuals, the redemptive self is a model of the good life. This narrative describes how generative adults seek to give back to society in gratitude for the advantages they have received. While generativity is undoubtedly difficult and frustrating work in any life, the construction of a narrative identity in which the protagonist's suffering in the short term gives rise to reward later on is better equipped to sustain the conviction that seemingly thankless investments today will pay off for future generations. In this way, redemptive life narratives support the kind of life strivings that highly generative individuals are likely to set forth.[187]

### 5.3.4. The Narrative Objectivity of Subjective Identity

In light of the developmental and redemptive metaphysics of selfhood delineated above, what is required for an objective analysis of subjective identity is a sophisticated model that can mediate and synthesize the diverse and heterogeneous aspects of life.[188] That model is requisitely narrative. Of particular import is the capacity of narrative identity to coordinate different orders of time. Drawing on Heidegger's phenomenology, Ricoeur argues that individuals are essentially practical beings who are necessarily oriented to life in terms of what they expect to become.[189] This orientation gives self-consciousness a fundamentally temporal and primarily future-oriented character. That is, one's life exists across a stretch of cosmological time from birth to death, but that linear time-span is experienced in terms of an interplay of past-present-future orientations within phenomenological time. The Ricoeurean practicality of being is argued cogently in Christine Korsgaard's account of autonomy,[190] which emphasizes the role of self-reflection in ascribing normative weight to moral reasons. Self-consciousness is herein defined as the reflective capacity to call into question one's beliefs, desires, and motives, and to ask whether these constitute reasons. Reasons, in turn, are impulses, perceptions, and desires that have withstood objective scrutiny. The capacity for reflection is therefore the source of normativity: it provides a choice about how one should believe, decide, and act. When individuals reflect upon their situations and look for reasons to guide actions, they appeal to those beliefs, feelings, and ideas that serve to motivate them. The candidates that function as reasons for acting are the beliefs, feelings, and ideas that have practical significance, and such reasons emerge directly from one's concept of self – one's practical narrative identity. As Korsgaard highlights, practical narrative identity is not a theoretical view of the self, but a description under which one values one's life to be worth living and one's actions to be worth undertaking.[191] When individuals act against or disavow their reflectively endorsed reasons, they prompt questions of who they are by way of questioning what they did and why they so acted. In this way, narrative objectivity and subjective identity mesh in a reflective conception of individual and social moral existence.[192]

The capacity to form reasons in support of particular actions and to behave in accordance with those reasons turns on critical self-reflection.[193] This requires a minimal level of objective physical, psychological, and social coherence that is only offered by a subjective experience of narrative identity. It is not enough that one's psychological states are induced in one way or another, or that there is a high degree of overlap. Rather, what finally matters is that individuals are able to take their psychological states – including beliefs, feelings, and convictions – as their own and endorse them as the source of obligation with which they feel compelled to act in conformity. That is, individuals must assume a position vis-à-vis themselves in order to serve as narrative agents. This Lockean self-constitution requires that a self possesses both constancy and variability insofar as the circumstances in which they are required to act vary greatly. Individuals who exhibit compartmentalization of disjointed personalities, for example, are limited to narrow situation-specific control. Lacking global self-knowledge, their actions tend to be directed by circumstances rather than by a coherent narrative identity. Within an integrated personality, however, characterological strands unite the disparate elements of the authentic self. These characterological strands ground the reasons that govern the autonomous individual's conduct. Korsgaard has argued that the moral life arises from the narrative first-person perspective inasmuch as the normative force of reasons is tied to the position of an agent on which morality is making a difficult claim.[194] This suggests that morality is essential a matter of Aristotelian phronēsis, and that the grounds from which human beings reason or deliberate are constituted by personal identity. The conclusion that practical narrative identity underpins the normative authority of reasons turns on the notion of constituted ends. Unless the reasons that direct actions are grounded in valued core attributes of the self – that is, unless ends (goods) are embodied in self-constitution – actions and beliefs will remain arbitrary and pointless. Nevertheless, constitutive ends need not be deterministic. Through critical self-reflection, self-knowledge, and self-direction, one can emphasize an existing trait or work to replace it. The latter is achieved by situating oneself in circumstances that promote the acquisition of the trait desired. This can only occur, however, through the exercise of first-person perspective – by taking one's narrative traits as reasons for acting. By these means, the moral life is shaped by both social experiences and personal direction – in narrative terminology, by the coordination of first, second, and third-person perspectives through self-constancy. Hence, the model required to make sense of oneself as a moral being takes the form of practical subjective identity rooted firmly in a rigorous narrative objectivity.[195]

To be identified as this or that person (Thomas, Mary, or Robert), to be the object of various attributes (kind, honest, or intense), and to be self-referential ("I said," "I went," or "I saw") is to be realized in language.[196] It is largely through discourse that individuals achieve a sense of individuated selves with particular attributes and self-referential capacities. With the discursive construction of the self foregrounded, there are significant ways in which subjective identity is importantly fashioned through narrative objectivity. From the beginning, the very categories by which individual identities are understood is primarily a byproduct of discursive elaboration. The sense of an individual self possessing and exercising the capacity for rational and objective deliberation, for instance, can be traced back to the texts (and conversations) of educated seventeenth and eighteenth century societies.[197] An appreciation of the capacity for genius and inspiration, the recognition of sincerely held passions, and the suspicion of deep-seated disturbance were made possible by the discourses (in the form of both arts and letters) of the nineteenth century romanticists.[198] Moreover, these concepts of the person – as rational, passionate, inspired, and the like – are embedded within broader narratives: they are not simply names for existing entities, but discursive creations requiring extensive narration. To paraphrase Ludwig Wittgenstein, the limits of one's narrative traditions mark the limits of one's identity.[199] In this context, it is helpful to consider the process of autobiographical memory – one's means of identifying oneself through reports of personal history. Insofar as there exist no means of identifying a particular psychological state associated with or responsible for producing various actions that are publicly indexed as manifestations of autobiographical memory, the conditions for ascribing memory are not signaled by the existence of a mental event but are designated socially. That is, only under circumscribed conditions do human beings collectively treat certain actions as "remembering." Suggesting that autobiographical memory is a discursive achievement is also to propose that "possessing a memory" is to participate in a cultural narrative tradition. To speak of one's past is to enter into a tradition for which the rules of well-formed narration are apposite. One cannot speak of last week's adventure, for instance, by recalling a random patch of color, a breeze, a word, and a bee sting. Rather, one must identify oneself as a particular identity, moving through time in certain directions with certain prominent endpoints. To "remember properly," therefore, is to generate a story replete with all the subjective qualities of an objective narrative.[200]

Narrative objectivity is vital to both the creation and sustenance of value, and to the achievement of a robust subjective identity. To the extent that historical consciousness is inherently consciousness of narrative, the conversational realities created by historical accounts inevitably perform certain functions within cultures.[201] They can be valuable constituents of long-standing cultural traditions, serving to construct a particular tradition, invest it with honor, and articulate a rationale for its future. In this way, historical narration is inexorably linked to cultural values and normative morality. To lend intelligibility to a given tradition is to lend silent affirmation to the notion of good it embodies. Just as subjective identity is configured or implicated in objective narratives, the achievement of moral being is sustained or impeded by historical accounts. For better or worse, human beings live within and are constructed by particular historical narratives that serve as a foreground for achieving moral identity. The logic may be delineated as such: if individuals with a particular narrative history do not engage in X and instead uphold the moral ideals of Y, then in choosing Y over X those individuals participate in their narrative history and become more fully a member of their particular culture. The capacity to achieve a robust moral identity, then, is inextricably linked to relationships with narratives of the past. Applied in the context of religious belief, debating the "facts" of a particular historical account – for instance, the Christian resurrection – including its magnitude and extensity, has been a delicate endeavor. From a constructionist standpoint, one may view this difficulty as problematically grounded but essential in implication. The controversy is problematic in its realist premises, namely, that an examination of the "facts of the case" will ultimately reveal the truth of history. There are myriad means of depicting "what actually happened," each felicitous within its own community of intelligibility. Further, each history will inevitably select "the facts" necessary to sustain its existence as an intelligible story. There is, in this sense, no impartial history, no story that transcends community, context, and discursive tradition. Insofar as the conversations become increasingly remote from the indexed particulars necessary to sustain their objective truth, there are few ways by which the sense of objective grounding can be secured. In effect, the quest for the "pure historical truth" of certain religious beliefs can never, in this narrow sense, achieve satisfactory closure. Nevertheless, the significance of history for ongoing religious life is not thereby extinguished and cannot therefore be underestimated.[202] To tell the story is to participate and sustain this tradition and, in terms of social pragmatics, to achieve moral identity within its terms. Moral identity is similarly at stake with regard to one's place within the narrative implicature.[203] In the broadest sense, then, historical accounts are only manifestly about the past. The creation of this past gains its cardinal significance via its contribution to present cultural life and the range of values that it initiates. It marks the manner in which human beings achieve a sense of the good that establishes the objective fore-structure of subjective intelligibility by which individuals determine their collective future identities.[204]

### 5.4. The Threat of Manipulation to Narrative, Authenticity, and Ethical Decision Making

Narrative theorists from John Locke to Charles Taylor underscore that in the creation of autobiographical structures upon which self-understanding hinges,[205] one is able to participate in the process of selecting particular memories, based on important life incidents and themes,[206] which in turn produce one's sense of existence in the world.[207] This process marks the means by which individuals interpret, make sense of, and extract meaning from life events. Regaining and reforming a systematized and consistent narrative after a traumatic event remains, then, vital to the larger reconstruction of one's autobiographical narrative of authenticity in the effort to make sense of trauma and to identify meaning within it.[208] In this way, autobiographical memory is equally critical to one's ability to persist through time, retain moral agency, and maintain moral responsibility.[209] If who one depends to a greater or lesser extent upon what one does, then it follows that what one does depends to a greater or lesser extent upon what one remembers – or, put more precisely, what one remembers in light of one's narrative identity. The conclusion gleaned is that who one is depends to a greater or lesser extent upon what one remembers, and, more immediately, who one remembers oneself to be.

### 5.4.1. Manipulation as Splintering of the Narrative Self

Through the disintegration of autobiographical memory, which in turn degenerates emotional rationality, MM acutely decays narrative identity, thereby splintering the self. In his *Genealogy of Morals*, Nietzsche comments that human beings are the "animal[s] that can promise."[210] In this (second) essay on guilt and conscience, Nietzsche attempts to convey that individuals possess an efficient memory that does not forget what was once said and done, and that therefore enables them to project their wishes and purposes into the future. On the one hand, this observation presumes an inner censor, commonly known as conscience, which keeps a close watch on one's promises and obligations lest they be violated. On the other hand, Nietzsche's primary concern is whether this kind of historical memory bars human beings from happiness. Lesser animals, by living unhistorically (knowing relatively little about the past and having no concept of the future), live in simple happiness. Contrarily, human beings, who are naturally unable to intentionally forget, are never able to achieve this hollow form of pleasure. Nietzsche's doubts have been confirmed by Freud's pessimism regarding the progress of culture: the moral constraints that culture imposes on individuals necessarily conflicts with their quest for purely hedonic happiness, resulting in various forms of neuroticism.[211] Along the same lines, Ricoeur places the essence of individualism within the temporal relationship individuals have with themselves.[212] For Ricoeur, human beings are not just things in the world that are characterized by remaining the same over time, but rather individuals who act and speak and, in so doing, prove to be self-understanding beings. This is why Ricoeur suggests, individuals ask who they are rather than what they are. The answer to the question of who one is lends another quality to narrative singularity and sameness: it is only oneself who can be responsible for one's actions in the past; and one remains oneself by being faithful to one's promises in the future. The "ipseity," or selfhood, of the individual opens up the sphere of responsibility and faithfulness instituted by accepted values and norms, and thus establishes the historical continuity essential for personal narrative identity.[213]

Both responsibility and promise make clear that the concept of narrative identity is essentially related to others to whom one speaks and is accountable.[214] There exists an inner witness in most actions and intentions to which one is able to give an account of what one did and is doing with others. This leads to a narrative conception of the self as posited in the writings of MacIntyre and Ricoeur. Narrative identity implies a meaningful coherence of the personal past, present, and future that is similar to the unity of an individually-narrated story. Individuals live immersed in narrative, recounting and reassessing the meaning of their past actions, anticipating the outcome of their future projects, and situating themselves at the intersection of several stories not yet completed. This narrative only makes sense, however, for a real or an implicit other.

MacIntyre therefore comments that the unity of an individual life is the unity of a coherent narrative enacted through the course of that life – or, at minimum, the unity of a narrative quest that one is striving to make coherent and bring to a meaningful conclusion.[215] Yet such narrative structuring does not suggest that narratives are finished products of reflective thought. Rather, stories are lived before they are told as enacted narratives. In each of one's actions (and the actions of others), one assumes a fundamentally intelligible, meaningful course of goals and means (i.e., beginnings, middles, and endings), and understands the acting person to be the agent or author of this sequence. Hence, one's everyday dealings with others already imply a narrative "pre-understanding" that rarely requires communication through an explicit story. Moreover, the concept of narrative identity does not signal that narratives are solitary works of isolated writers and their individual stories. Instead, personal identities are constituted by a complex interaction between first-, second-, and third-person perspectives. Others are not only the implicit auditors and witnesses of an individual's narrative identity, but also the co-authors of that individual's life story. If recent hermeneutical theory is correct, and temporality, narrativity, and coherence of identity are closely and inextricably intertwined, then narrative identity is essentially grounded in the capacity of the individual to integrate contradictory aspects and tendencies into a coherent, overarching sense and view of the self. This indispensable capacity, which marks the very possibility of human meaning-making, is radically fragmented in the crippling webs of manipulation.[216]

Beyond these detriments, manipulation also compromises the goal structure of the narrative self, which serves a vital function in both the encoding and retrieval of autobiographical knowledge.[217] Indeed, as several scholars have commented, autobiographical memories are primarily records of success or failure in narrative goal attainment.[218] There is now substantial evidence demonstrating that memories are not only directly related to narrative goals, but that broad subgroups of similar goals may selectively raise the accessibility of groups of goal-related memories. Major findings in the study of Jefferson Singer and Peter Salovey included the observation that autobiographical memories associated with feelings of happiness and pride were strongly linked with narrative goal attainment and the successful execution of personal plans.[219] In contrast, autobiographical memories associated with feelings of sadness and anger were linked to the progressive failure to achieve narrative goals. The authors suggest that each subject possessed a set of "self-defining" memories that contained critical knowledge of progress toward the attainment of long-term narrative goals. Goals such as attaining independence, intimacy, and mastery were likely adopted as solutions to dominant self-discrepancies arising from childhood experience. Indeed, for recalled negative memories, the content is even more strongly related to narrative attachment. Mario Mikulincer, for instance, investigated the autobiographical memories of anger experiences in individuals with secure, anxious-ambivalent, and avoidant narrative-attachment styles. In stark contrast to individuals with anxious-ambivalent and avoidant attachment styles, individuals with secure styles recalled memories of anger experiences that revealed a capacity for functional anger with characteristics of rational appraisal of the experience, lack of intense urges to punish the anger provoker, and no hostile attributions to the other.[220] Taken together, the foregoing patterns indicate that knowledge concerning narrative goals permeates autobiographical memory. The goals of the working self determine access to the knowledge base, and this occurs in the generation of retrieval models used to guide the narrative search process. On the one hand, these models facilitate access by setting constraints in a way that benefits the search-elaborate-evaluative retrieval cycle. On the other hand, retrieval models may aim to attenuate or prevent access by setting constraints that the search processes cannot satisfy or by prohibiting the recall of destabilizing knowledge such as highly emotional materials or attachment memories, the recall of which would increase self-discrepancies and reactivate dysfunctional attachment behavior and feelings. Relatedly, Avril Thorne found that study participants were unaware of the narrative goals their autobiographical memories so clearly expressed, leading her to conclude that motives were automatically but nonconsciously encoded in long-term memory.[221] This lack of reflexiveness and insight into narrative goal aspects of memories, or anosognosia for past narrative goals, likely arises insofar as specific autobiographical memories do not directly represent narrative goals (in retrievable form) but rather represent the outcomes of plans generated to attain narrative goals.[222]

As suggested above, narrative identity requires the constant labor of temporal integration.[223] This work entails remembering and adhering to autobiographical obligations and responsibilities. That is, one must accept what has happened as a meaningful part of one's biography, even at price of painful feelings of guilt, remorse, or sorrow. Insofar as one is wisely constrained by one's values, convictions, and decisions, MM – which fragments the narrative self – creates a present that is disconnected from a past and a future: an isolated now. The more one plunges into the present, the more one can identity momentary impulses and get rid of the burden of past doubts. Viewed in this light, manipulation is beyond notions of good and evil, innocence and guilt: it nefariously sentences to death the narrative self whose truths have become inconvenient. On the one hand, the temporal splintering of the narrative self excludes past and future as dimensions of self constancy, bonding, commitment, responsibility, and guilt. Hence, it avoids the necessity to tolerate the threatening ambiguity and uncertainty of longer-term interpersonal relationships. On the other hand, this fragmentation does not spawn innocent bliss but even greater suffering, not from neurotic repression but from reality itself – above all, from others who refuse to comply with this extreme and remain available whenever it proves fitting. Suffering results from the inability to develop authentic attachments, from inner emptiness and loss of meaning, and from being overwhelmed by excessive feelings of anxiety, shame, sorrow, and other emotions that currently replace the feelings associated with one's past. In the end, inner emptiness, numbness, and alienation are the result of being split from the narrative self. Experiences that cannot be integrated into a historical coherence of life, but are continually abandoned for the sake of new ones, can leave only emptiness behind, however intense they may be. Such experiences remain meaningless insofar as they can only gain meaning against the background of an overarching narrative concept or direction of life. The loss of time as a continuum that extends into the past and the future creates a current reality without depth.[224] One is finally faced, therefore, with the question of which suffering and happiness is most narratively genuine and morally dignified: the hopeful and redemptive suffering implied in the arduous but joyful labor of weaving the threads of life into a meaningful pattern, or the shallow and inauthentic happiness siphoned from having these (frequently) sorrowful but (perpetually) veracious threads continually extinguished, leaving only unintelligible fragments behind.

### 5.4.2. Manipulation as Corrosive of Narrative Authenticity

Through the splintering of the self, MM corrodes narrative authenticity. The moral notion of narrative authenticity is something relatively new and peculiar to contemporary culture.[225] Born at the end of the eighteenth century, it builds on earlier forms of individualism, such as the individualism of disengaged rationality, originally pioneered by Descartes, which demands individuals to think self-responsibly, or the political individualism of Locke, which seeks to make the individual or the subjective will prior to social obligation. However, as a child of the Romantic period, which was critical of disengaged rationality and an atomism that failed to recognize the ties of community, narrative authenticity also conflicts with these earlier forms. One way of describing its development is to witness its starting point in the eighteenth-century notion that human beings are endowed with a moral sense that reveals itself through the intuitive sense of right and wrong. The notion of narrative authenticity develops out of a displacement of the idea that morality is, in a genuine sense, a voice within. On an original view, the inner voice is significant insofar as it informs individuals of the right thing to do in a given set of circumstances. Being in touch with moral sentiments matters inasmuch as that they serve as the narrative means to the authentic end of acting rightly. Thus understood, the expulsion of the moral accent on narrative authenticity comes about when being in touch with these sentiments engenders independent and critical moral significance. Hence, authenticity comes to be something individuals must attain to be true and full human beings.[226] Jean-Jacques Rousseau frequently presented the issue of morality as following a (narrative) voice of nature within.[227] This voice is frequently drowned out by the passions induced by one's dependence on others, of which the principal culprit is pride (amour propre). One's moral salvation, therefore, springs from recovering authentic narrative contact with oneself. Rousseau goes on to attribute a name to this intimate contact with oneself, more fundamental than any particular moral view, which serves as a source of joy and contentment: "le sentiment de l'existence."[228]

Self-determining freedom is a notion with immense power in American political life.[229] In Rousseau's work, it takes political form via a social contract state founded on a general will that, due to the common freedom in which it is grounded, can brook no opposition in the name of freedom. Following Rousseau, Johann Gottfried von Herder posited the idea that each human being has an original way of being human. As von Herder writes, all individuals have their own "measure"[230] – an idea that has subsequently entered deeply into modern consciousness. Prior to the late eighteenth century, the notion of moral differences between human beings was largely absent and therefore insignificant. There is a certain way of being narratively authentic that is uniquely one's own way. One is called upon to live one's life in such an authentic way, and not in imitation of anyone else. This axiom gives critical significance to being true (i.e., narratively authentic) to oneself. As Taylor puts it, "if I am not [true to my authentic self], I miss the point of my life, I miss what being human is for *me*."[231] The powerful moral ideal of narrative authenticity accords vital import to a form of self-contact with one's own inner nature, which it views as chronically endangered, partly through the pressures toward outward conformity, but also because in assuming an instrumental stance to oneself, one necessarily sacrifices the capacity to listen to this inner voice. In turn, this greatly increases the significance of this self-contact by introducing the principle of originality: every individual's voice has something of its own to say. Not only should one refrain from conforming one's life to the demands of the external community in this way; one cannot even begin locate a model by which to live outside of oneself. Being true to oneself, therefore, means being true to one's own originality, and that is something only discovered and articulated within a framework of narrative authenticity. In so doing, one realizes a potentiality that is properly one's own.[232] Such is the background that gives moral force to the culture of authenticity, including its most degraded, absurd, and trivialized forms via the pseudo-freedom promised by manipulation proponents.

If narrative authenticity is being true to oneself and recovering one's own "sentiment de l'existence," then narrative authenticity can only be integrally achieved to the extent that it connects one to a wider whole.[233] Reasoning in moral matters is always reasoning with somebody, which implies its fundamentally dialogical character. Human beings become full moral agents, capable of understanding themselves and, hence, creating and defining a narrative identity, through their acquisition of rich languages of authentic expression. Individuals do not acquire the languages needed for self-definition on their own. They are introduced to them, rather, through exchanges with others who matter to them. In this sense, the genesis of the human mind is not monological (exercised individually), but dialogical (exercised with others). It is not simply that individuals learn narrative languages in dialogue and then individually utilize them to achieve personal ends. Human beings are expected to develop their own opinions, outlooks, and social stances largely through solitary reflection. However, this is typically not how important issues are properly approached, including the definition of narrative identity. Individuals requisitely approach (and often struggle against) this question in dialogue with the narrative identities their respective interlocutors recognize in them. Even when individuals outgrow some of these roles – the moral influence of parents, for instance – the conversation with them continues to the extent that they live in accord with narrative authenticity. Insofar as narrative identity is understood as who one is and who one is becoming in relation to the diverse platforms from which one is constantly emerging, it serves as the background against which one's tastes, desires, opinions, and aspirations make sense. Hence, if some of the things one values most are accessible only in relation to the others one loves, then those others become internal to one's narrative identity.[234]

While narrative credibility depends, in part, on the plausibility of a chain of objective events that can be corroborated, narrative authenticity aims to couch these events within subjective events that cannot be contradicted.[235] Autobiographical remembering is a similarly subjective event. While remembering itself is an unobservable and therefore unverifiable mental state, a thought cast as remembered is presented as true. The word "remember" is a factual mental verb that presupposes the certainty of a proposition. Thus, the notion that one "remembers X" presupposes that one actually experienced X. Narrative recollection, then, is an authenticating act: narrative agents publicly claim to have brought to conscious awareness a state, event, or condition that is real in their eyes and believed to be true. In this sense, acts of narrative authenticity are attempts to seize authority with respect to a topic of concern. For the presupposed truths to become recognized as such, however, these acts require validation by others. Defining oneself entails finding what is significant in one's difference from others. In the context of manipulation, the value of narrative authenticity is contaminated by its lack of connection with another leading ideal: self-determining freedom. Only to the extent that one begins to define oneself by the ability to articulate the truths of narrative authenticity is one thereby able to enter the domain of recognizable self-definitions. This suggests that inasmuch as autobiographical memories take on significance only against a background of narrative intelligibility, the suppression or denial (through manipulation) of the horizons against which memories are able to take on significance is incompatible with the endeavor to define oneself authentically. Put simply, one can define one's identity only against the background of things that matter. Hence, to bracket out history, nature, society, the demands of solidarity, and everything but what one finds in oneself would be to eliminate all candidates for what matters. Only if one exists in a world in which history, the demands of nature, the needs of others, the duties of citizenship, the call of God, or something else of this order matters crucially can one define an authentic narrative identity for oneself that is not trivial.[236] Insofar as the tension between certainty and doubt drives narrative activity in pursuit of an authentically remembered self,[237] narrative authenticity is not the enemy of demands that emanate from beyond the self; rather, it presupposes them.[238]

### 5.4.3. Manipulation as Destructive of Narrative Self-Control, Narrative Growth, and Narrative Responsibility

Through the corrosion of narrative authenticity, MM destroys narrative self-control, narrative growth, and narrative responsibility. Philosophers use the term "self-control" in a variety of ways.[239] Daniel Dennett, for instance, employs it to communicate what most other philosophers mean by "autonomy."[240] Following Jeanette Kennett,[241] this text distinguishes among three senses of narrative self-control, each of which is terminated via manipulation: (i) a form of control required for intentional human action; (ii) a form of control specifically manifested in careful reflection about one's desires and which action would best serve those desires overall in circumstances wherein those desires are in danger of being forgotten or wherein a lesser immediate satisfaction looks promising; and (iii) a form of control that requires a commitment by the agent to the rule of reason. One current philosophical use of self-control is linked to Aristotle, who comments that self-control (enkrateia) and its contrary (akrasia) "are concerned with that which is in excess of the state characteristic of most men; for the [self-controlled] man abides by his resolutions more and the [akratic] man less than most men can."[242] The philosophical project of understanding narrative self-control, construed broadly along Aristotelian lines, has its roots in ancient efforts to explain intentional actions. Hence, in *De Motu Animalium*, Aristotle asks: "But how is it that thought is sometimes followed by action, sometimes not; sometimes by movement, sometimes not? There the end is the truth seen."[243] A proper narrative answer requires understanding, among other things, how it happens that individuals sometimes act in accordance with the conclusions of their reasoning about what is best to do, on the whole, and sometimes fail to do so, instead pursuing courses of action at odds with those narrative conclusions. Relevant judgments include those in which the notion of "best" is relativized to options envisioned by the agent at the time. In instances of strict akratic action, moral agents do not need to judge that a course of action is narratively superior to the alternatives envisioned. If, however, in the absence of compulsion, moral agents do not choose to act with narrative authenticity and instead pursue one of the envisioned alternatives, they thereby act akratically. It is a truism that narratively self-controlled agents cannot act akratically insofar as doing so would require the abandonment of narrative identity. Hence, narrative self-control presupposes a robust narrative identity in accordance with which one is free to participate or depart (akrasia). To remove this narrative commonality through manipulation is to remove the very possibility self, control, and freedom, and therefore the freedom of narrative self-control.[244]

Descartes' "cogito ergo sum" presupposes that every individual is a self, immortal beyond change and time.[245] Yet the "I" whose existence Descartes proves is not the bodily "I," nor the entity that manifests itself in agency in the public world. The "I" that thinks is a totally inner entity, disengaged from the body and its actions in the world. However, Descartes' argument works only on a particular occasion of use: it says nothing about what happens to the "I" in between, nor does it express anything about the ontological framework in which the "I" develops. Without intending to, some psychoanalysts have taken Descartes' axiom to mean that the mind is unitary, forever unknowable, unchanging, and hidden behind the behavior of the body and its antics.[246] Responding to Descartes, Hume argued that his own experience yielded no such thing as a subject of experiences but only the experiences themselves.[247] Nevertheless, an adequate view of the self necessarily anticipates narrative growth, one aspect of which is reaching goals intentionally set for oneself. In absence of one's autobiographical memory, however, setting goals is impossible and growing upon their achievement is more impossible still. In this way, the narrative self grows to the extent that it comes to know itself in the process of growing. Thus understood, the narrative self is not a timeless, ineffable, Cartesian something behind appearances, as Descartes suggests. To lead a narrative life means to possess something of a conception of life – of one's own life, projects, goals, and moral values. Narrative growth is, therefore, not merely a description, but also an evaluation. A narrative entity who is true and reliable to self others is focused on the present, enriched by the past, and directed toward the future. The product of narrative growth is an autobiographical integrity that unites one with a coherent set of values in accordance with which one is able to act freely. An accomplished narrative self is capable of self-knowledge and global valuation – that is, self-aware assessment in which one's different values have been weighed and bound together such that they allow for a singular value to guide behavior. This process begins when individuals release the fantasies that defend them from past anxieties that erect defensive screens between themselves and the world. Such "narrative emergence" is what distinguished authentic from inauthentic narrative growth and, hence, self-awareness.[248]

Among the chief promises of manipulation are narrative self-deception, denial, dissociation, and repression, each of which entails the division of the narrative self and the inhibition of its potential for growth. Philosophers typically regard narrative self-deception as paradoxical, questioning the possibility of simultaneously believing and not believing the same proposition.[249] In interpersonal narrative deception, A believes that P, but wants B to believe that not-P. A sets about to persuade B of what A believes P is. If A is successful, then B comes to believe that not-P. For instance, when Iago seeks to deceive Othello about Desdemona's chastity, Iago believes she is chaste, but wants Othello to believe that she is not. To treat narrative self-deception on this model, one ends up trying to find a way in which two minds can be one, or one mind two. Contrary to interpersonal deception, narrative self-deception involves both A and B selves being consciously aware of the secret. The phenomenon that decisively reveals the singular character of self-deception is known as intrapsychic conflict. The induction of intrapsychic conflict via manipulation does not attempt to satisfy multiple desires but dissolve sets of anxiety, each of which is presumed to exaggerate the other. Dostoyevsky's Raskolnikoff – a complex, self-loathing character – hates his landlady and resolves to murder her.[250] The desire is buried in other desires and anxieties that make Raskolnikoff feel guilty and remain opaque due to his failed effort to understand them. Raskolnikoff kills the landlady and instead of dissolving his anxiety, the murder increases it. He justifies his action by thinking of her as a vermin who deserved to be killed, yet he does not really believe this, and it only serves to increase his guilty sense that he is the vermin. As Dostoyevsky's vivid portrait demonstrates well, the trouble with narrative self-deception, repression, and disassociation is that it is not a simple matter of changing one's mind, nor of successfully veiling what it is one wants to deny. One inescapably remains aware of the haunting truths of narrative reality, even if only dimly.[251]

For a moral agent to be responsible for action A, it seems necessary for the agent to bear some sort of narrative relation to A.[252] This fundamental requirement of moral judgment is vitiated through manipulation. In order to understand the relation above, most philosophers begin with "the platitude," which entails the slogan that moral responsibility presupposes personal identity – that is, in order for someone to be morally responsible for A, that person must be the same person as the agent of A. If the entailment obtains, then one ought to be able to provide an analysis of the narrative relation between the agent and A by offering the correct theory of personal identity as the proper account of what makes an action one's own. In other words, if some past action is one's own by virtue of the fact that one is the same person who performed that action, then whatever makes one identical to that past person is what makes one's actions one's own (for purposes of responsibility). In order to illuminate a critical component of moral responsibility, then, the nature of narrative responsibility must be broached. Thus, to say that one is morally responsible for one's actions means that one is eligible for judgments of narrative attributability: responsibility for those actions that ultimately emanate from the authentic self – constituted by one's various evaluative commitments – with which one is narratively identified. In this way, the platitude is less about personal than practical identity. Insofar as the relevant form of identification involved in one's practical identity is not restricted by the considerations of numerical or narrative identity, it remains possible that one may be identified with the actions of other individuals – a point that raises the question of collective responsibility. Viewed in light of the relationship between moral responsibility and narrative identity, the platitude can be expanded in (at least) two ways. First, it may incorporate not only actions but attitudes: inasmuch as moral attitudes flow forth from the narrative self, the expanded platitude could account for the judgments of narrative responsibility one has in response to them. Second, a version of the platitude restricted to responsibility-as-attributability might advance a stronger claim, namely, that one is (prima facie) morally responsible for all and only one's narrative actions.[253]

### 5.4.4. Manipulation as Disabling of Narrative-Based Ethical Decision Making

Through the destruction of narrative self-control, narrative growth, and narrative responsibility, MM disables one to seek, identify, and act on the good, thereby rendering narrative-based ethical decision making impossible. The morality of particular actions are identified by invoking both implicit and explicit contexts.[254] The intentions of moral agents are placed in causal and temporal order with reference to their role in individual narrative histories and in the larger narrative context of the setting or settings to which they belong. In determining what causal efficacy particular moral intentions previously had in one or more directions and how short-term intentions succeeded or failed to be constitutive of long-term intentions, individuals continue to write next chapters in their narrative histories. By these means, narrative history becomes the basic and essential genre for characterizing human actions. Hence, narrative is not the arcane work of poets, dramatists, and novelists reflecting on events that have no narrative order prior to that imposed by the writer; narrative form is neither disguised nor decorated in this way. Rather, human beings dream, remember, anticipate, hope, despair, believe, doubt, plan, revise, construct, learn, hate, and love by narrative, and this engages moral agents not merely as actors but also as authors. What agents are able to do or communicate intelligibly as actors is deeply affected by the fact that they are never more (and sometimes less) than the co-authors of their own narratives. Only in fantasy do moral agents live the story they please. In the authenticity of narrative life, individuals are always under certain constraints: they enter upon a stage that they did not design and find themselves part of an action that was not of their own making, each remaining a main character in an individual drama while simultaneously playing subordinate parts in the drama of others. This unpredictability coexists with the teleological characteristic of all narratives: individuals live out their lives, both individually and in relationship with others, in light of particular conceptions of a shared future in which certain possibilities beckon and others repel, some foreclosed and others inevitable.[255]

There exists no present that is not informed by an image of some future introducing itself via a telos toward which one is either moving or failing to move in the moment.[256] As suggested above, the requisite unpredictability of narrative identity and life goals coexist as part of life: like characters in a fictional narrative, individuals do not know what will happen next, but nonetheless their lives have a certain form which projects itself toward the future. Thus, the narratives that human beings live out have both an unpredictable and partially teleological character: if the narratives of their individual and social lives are to continue intelligibly, it is both the case that there are constraints on how those stories can continue and, within those constraints, indefinitely many ways in which they can continue. Here, a central thesis begins to emerge. It concerns the notion that human beings are, in both their actions and their fictions, essentially story-telling creatures. That is, through one's narrative history, one becomes a teller of stories that aspire to truth. It thus becomes clear that one can only identify what one is to do – that is, what is the right, the good, or perhaps the least worse thing to do in a given instance – if one first identifies the narrative or narratives of which one finds oneself a part. One enters human society, in other words, with one or more characters imputed – roles into which one has been drafted – and one must learn what they are in order to understand how others are to respond and how one's responses to others are appropriately construed. To be the subject of an authentic narrative that runs from one's birth to one's death and possesses both autobiographical and emotional integrity is to be accountable for the actions and experiences that compose a "narratable" life. Put simply, it is to be open to give a particular account of what one did, experienced, and witnessed at any earlier point in life than the time at which the question is posed.[257]

It is through hearing stories about wicked stepmothers, good but misguided kings, and prodigal sons in need of mercy that children learn or mislearn what a child and what a parent is, what the cast of characters may be in the drama into which they have been born, and what the ways of the world are.[258] Hence, there is no way to provide an understanding of any society, including one's own, except through the volumes of stories that constitute its initial dramatic resources. This suggests the correlative nature of narrative selfhood: individuals are not just held to narrative account; they can also demand an account from others. Each individual is part of the stories of others, and the stories of others are part of the individual's own story. The narrative of any one life is part of an interlocking set of narratives, and this asking for and giving of accounts plays a cardinal role in constituting narratives. Asking what others did and why, explaining what one did and why, and pondering the differences between the same pursuit of the true, beautiful, and good are the simplest and barest forms of narrative-based ethical decision making. Thus, without the accountability of the narrative self, those trains of events that constitute all but the simplest and barest narratives could not occur. Moreover, without that same accountability, narratives would lack the continuity required to make them (and the actions that constitute them) intelligible. To the inevitable question of what comprises the unity of an individual life, the answer concerns the unity of a narrative embodied in that life. To ask what is good for oneself is to ask how best one might live out that unity and bring it to completion. To ask what is good for humanity is to ask what all answers to the former question must have in common. The unity of human life, therefore, is the unity of the narrative quest. It is important to emphasize that the systematic asking of these two fundamental questions and the attempt to answer them in word and deed is what provides the moral life with its unity. The unity of a human life is the unity of narrative questions. Like any quest, this basal narrative endeavor sometimes fails, is frustrated, abandoned, or scatters into distraction; and human lives may, in all of these ways, similarly fail. Yet the only criteria for moral success or failure in human life as a whole are the criteria for success or failure in a narrated or to-be narrated quest.[259]

Against this backdrop, one is unable to seek, identify, and act on the good merely qua individual.[260] This is partly because narrative-based ethical decision making varies from one circumstance to another even when it is one and the same conception of the good life and one and the same set of virtues that are being embodied in a particular instance. What the good life is for a fifth-century general will not be the same as what it is for a medieval nun or an eighteenth-century farmer. However, it is not simply that different individuals live in different social circumstances; it is also that all human beings approach their own circumstances as bearers of a particular narrative identity. One is always someone's son or daughter, someone else's cousin or uncle, a member of this or that clan, tribe, or nation. Hence, what is morally good for oneself has to be good for one who inhabits these narrative roles. As such, one inherits from the past of one's family, city, tribe, and nation a variety of narrative debts, inheritances, rightful expectations, and moral obligations. These constitute the given of one's life – one's ethical starting point. It is precisely this that gives one's life its own particularity. What one is, therefore, is in large part what one inherits: a specific past that is present to some degree in one's current reality. One finds oneself part of a narrative history, which means that whether or not one likes or recognizes it, one is a bearer of a tradition. It is vital to recognize that the characterization of any narrative-based ethical decision making process enjoys a history, and that at any given moment what a decision yields depends on a mode of understanding that has been transmitted through multiple and diverse generations. Thus, insofar as narrative identity sustains the relationships required for ethical practices, they must sustain relationships to the past (and future) as well as in the present. The presupposition of narrative-based moral objectivity is that individuals can understand the notion of "good for X" and cognate notions in terms of some conception of the unity of X's life. What is better or worse for X depends on the character of that intelligible narrative that provides X's life with its unity. Unsurprisingly, it is the lack of any such unifying conception of a human life that underlies modern denials of the factual character of moral judgments, especially those that ascribe virtues or vices to individual ethical decisions.[261]

## 5.5. Conclusion

As the third and fourth links in the book's logic chain, this chapter endeavored to (i) underscore the moral significance of narrative identity for ethical decision-making and to thereby make the case against direct and intentional narrative decaying via MM, and (ii) delineate the crippling effect of narrative disintegration via MM, namely, the disability to pursuit seek, identify, and act on the good. It suggested, first, that insofar as cultural narratives affect the ways in which individuals experience things, there can be no pure, raw, immediately-given experiences, the narrative interpretation of which would necessarily be a matter of retrospective distortion. Rather, human beings are always entangled in stories, wearing personal narratives in a dialogical relation to cultural narratives, both of which are objects of unremitting reinterpretation. Hence, individual existence cannot be separated from the stories that comprise the individual. Second, the chapter addressed narrative identity as the product of autobiographical memory and emotional rationality. For better or worse, the convergence of one's autobiographical memory and emotional rationality produces one's narrative identity – one's conception of self. The notion of the "narrative self" centers on the innate effort of human beings to understand and interpret the world through storytelling. Thus, building on the claim that human beings are "embodied conversations," and that the unity of conversation serves to support human existence, it is evident that individuals experience and comprehend life as a series of narratives that possess various beginnings, middles, and ends.

Third, the chapter addressed requisite unpredictability of narrative identity. The narrative self is not a stable and enduring entity limited to or fixed in geographical location and time; it is neither the accumulation of experience nor the collective expression of neurophysiological characteristics. Rather, narrative identity is rooted in the constancy of an unfolding narrative. As an ongoing autobiography – or, more precisely, a self-other multifaceted biography that individuals constantly pen and edit – the narrative self is always engaged in conversational becoming, constructed and reconstructed through continuous interactions and relationships. Finally, the chapter addressed the threat of MM to narrative, authenticity, and ethical decision making. Through the disintegration of autobiographical memory, which in turn degenerates emotional rationality, MM acutely decays narrative identity, thereby splintering the self. As a result, this self-splintering corrodes narrative authenticity, which subsequently destructs narrative self-control, narrative growth, and narrative responsibility. Consequently, through the destruction of narrative responsibility, MM disables one to seek, identify, and act on the good, thereby rendering narrative-based ethical decision making impossible.

Against this backdrop, at least three conclusions can be drawn. First, narrative identity serves as the necessary condition for moral agency. At some point, specific episodic memories must be integrated into narrative form that references a self whose story it is. Tragically, the practice of MM actively and intentionally decays the narrative means by which moral agency is made possible. Second, narratives do not stand alone; they depend for their meaning on broader background narratives that are often taken for granted by those who share a common society and culture. The question is not, therefore, how subjective stories can provide rigorous criticism, judgment, and justification in the context of ethical decision making, but how rigorous criticism, judgment, and justification can exist without the stories that frame one's narrative identity. Hence, when narrative identity is acutely blunted through MM, so too is one's capacity to identify the good and work to pursue it. Finally, narrative identity has a function beyond serving as the source of continuity required to make moral actions intelligible; it also sustains the relationships required for ethical practices. One is always something to someone – a son or daughter, cousin or uncle, member of this tribe or that nation. Hence, what is morally good for oneself has to be good for one who inhabits these narrative roles. To ask what is good for oneself is to ask how best one might live out narrative unity and bring it to completion; but to ask what is good for humanity is to ask what all answers to the former question must have in common.

## 5.6. Notes

1. Dan P. McAdams, "Personal Narratives and the Life Story," in *Handbook of Personality: Theory and Research*, 3rd ed., ed. Oliver P. John, Richard W. Robins, and Lawrence A. Pervin (New York: The Guilford Press, 2008), 242-62; see especially p. 243.

2. McAdams, "Personal Narratives," 242-62; see especially p. 243.

3. McAdams, "Personal Narratives," 242-62; see especially p. 243.

4. Howard Brody, *Stories of Sickness*, 2nd ed. (New York: Oxford University Press, 2003), 5-22; see especially p. 8.

5. Kathryn Montgomery Hunter, "'There Was This One Guy . . . ': The Use of Anecdotes in Medicine," *Perspectives in Biology and Medicine* 29 (1986): 619-30.

6. Hunter, "'There Was This One Guy,'" 619-30.

7. Brody, *Stories of Sickness*, 5-22; see especially pp. 8-9.

8. James Hillman, "The Fiction of Case History: A Round," in *Religion as Story*, ed. J. B. Wiggins (New York: Harper and Row, 1975), 123-73; see especially p. 124.

9. Eugene B. Brody and Judith F. Tormey, "Clinical Psychoanalytic Knowledge—An Epistemological Perspective," *Perspectives in Biology and Medicine* 24 (1980): 143-59; see especially p. 148.

10. Brody and Tormey, " Clinical Psychoanalytic Knowledge," 143-59; see especially p. 149.

11. Dan P. McAdams, "The Role of Narrative in Personality Psychology Today," *Narrative Inquiry* 16, no. 1 (2006): 11-18; see especially p. 14.

12. McAdams, " The Role of Narrative," 11-18; see especially p. 14.

13. Brody, *Stories of Sickness*, 5-22; see especially p. 12.

14. Kathryn Montgomery Hunter, *Doctors' Stories: The Narrative Structure of Medical Knowledge* (Princeton, NJ: Princeton University Press, 1991).

15. See, for instance, M. Faith McLellan and Anne Hudson Jones, "Why Literature and Medicine?" *The Lancet* 348 (1996): 109-11.

16. Hilde Lindemann Nelson, ed., *Stories and Their Limits: Narrative Approaches to Bioethics* (New York: Routledge, 1997).

17. Trisha Greenhalgh and Brian Hurwitz, eds., *Narrative Based Medicine: Dialogue and Discourse in Clinical Practice* (London: BMJ Books, 1998).

18. Anne Hunsaker Hawkins, *Reconstructing Illness: Studies in Pathography*, 2nd ed. (West Lafayette, IN: Purdue University Press, 1999).

19. McAdams, " The Role of Narrative," 11-18; see especially p. 14.

20. McAdams, "The Role of Narrative," 11-18; see especially p. 14.

21. McAdams, "The Role of Narrative," 11-18; see especially p. 14.

22. Hanna Meretoja, "Narrative and Human Experience," *New Literary History* 45 (2014): 89-109; see especially p. 89.

23. Meretoja, "Narrative and Human Experience," 89-109; see especially pp. 89-96.

24. Charles Taylor, *Philosophical Papers 1: Human Agency and Language* (Cambridge, UK: Cambridge University Press, 1985), 45-76; see especially p. 45.

25. See Paul Ricoeur, *Time and Narrative*, vols. 1-3, trans. Kathleen McLaughlin and David Pellauer (Chicago: University of Chicago Press, 1984).

26. See Paul Ricoeur, *Oneself as Another*, trans. Kathleen Blamey (Chicago: The University of Chicago Press, 1992).

27. Meretoja, "Narrative and Human Experience," 89-109; see especially p. 96.

28. Paul Ricoeur, "L'histoire Comme Récit et Comme Pratique," *Esprit* 54, no. 6 (June 1981): 155-65; see especially p. 156.

29. Meretoja, "Narrative and Human Experience," 89-109; see especially p. 96.

30. Jerome Bruner, "Life as Narrative," *Social Research* 54, no. 1 (2004): 11-32; see especially p. 31.

31. Meretoja, "Narrative and Human Experience," 89-109; see especially pp. 97-98.

32. Meretoja, "Narrative and Human Experience," 89-109; see especially p. 98.

33. Paul Ricoeur, "The Creativity of Language," in *A Ricoeur Reader: Reflection and Imagination,* ed. Mario Valdés (New York: Harvester Wheatsheaf, 1991), 463-81; see especially p. 468.

34. See Edmund Husserl, *The Crisis of the European Sciences and Transcendental Phenomenology* (Exeter, UK: Imprint Academic, 1970).

35. Grant R. Gillett, "The Subjective Brain, Identity, and Neuroethics," *American Journal of Bioethics* 9, no. 9 (2009): 5-13; see especially p. 7.

36. Gillett, "The Subjective Brain," 5-13; see especially p. 7.

37. Gillett, "The Subjective Brain," 5-13; see especially p. 12.

38. Charles Taylor, *Sources of the Self* (Cambridge, MA: Harvard University Press, 1992), 25-52; see especially p. 48.

39. Gillett, "The Subjective Brain," 5-13; see especially p. 12.

40. Ronald A. Carson, "Medical Ethics as Reflexive Practice," in *Philosophy of Medicine and Bioethics: A Twenty-Year Retrospective and Critical Analysis,* ed. Ronald A. Carson and Chester R. Burns (Dordrecht: Kluwer Academic Publishers, 1997), 181-92; see especially p. 185.

41. Daniel K. Lapsley, "Moral Agency, Identity and Narrative in Moral Development," *Human Development* 53 (2010): 87-97; see especially p. 88.

42. Monisha Pasupathi and Cecilia Wainryb, "Developing Moral Agency through Narrative," *Human Development* 53 (2010): 55-80.

43. Pasupathi and Wainryb, "Developing Moral Agency," 55-80.

44. Pasupathi and Wainryb, "Developing Moral Agency," 55-80; see especially pp. 62-63.

45. Lapsley, "Moral Agency," 87-97; see especially pp. 88-89.

46. Lapsley, "Moral Agency," 87-97; see especially p. 91.

47. Grazyna Kochanska, "Mutually Responsive Orientation Between Mothers and Their Young Children: A Context for the Early Development of Conscience," *Current Directions in Psychological Science* 11 (2002): 191-95.

48. Lapsley, "Moral Agency," 87-97; see especially p. 94.

49. Michael Bamberg, "*Who am I?* Narration and its Contribution to Self and Identity," *Theory and Psychology* 2, no. 1 (2010): 1-22; see especially p. 4.

50. Bamberg, "*Who am I?*" 1-22; see especially p. 4.

51. See Taylor, *Sources of the Self,* at various locations; e.g., p. 292.

52. Bamberg, "*Who am I?*" 1-22; see especially p. 5.

53. See Erik H. Erikson, *Childhood and Society* (New York: W. W. Norton & Company, 1963).

54. See Jerome Bruner, *Actual Minds, Possible Worlds* (Cambridge, MA: Harvard University Press, 1986).

55. See Donald E. Polkinghorne, *Narrative Knowing and the Human Sciences* (New York: State University of New York Press, 1988).

56. See Theodore R. Sarbin, "The Poetics of My Identities," in *Narrative Identities: Psychologists Engaged in Self-Construction,* eds. George Yancy and Susan Hadley (London: Jessica Kingsley Publishers, 2003), 13-35.

57. See Elliot G. Mishler, "The Analysis of Interview-Narratives," in *Narrative Psychology: The Storied Nature of Human Conduct,* ed. Theodore R. Sarbin (New York: Praeger Press, 1986), 233-55.

58. Jack J. Bauer, Dan P. McAdams, and Jennifer L. Pals, "Narrative Identity and Eudaimonic Well-Being," *Journal of Happiness Studies* 9 (2008): 81-104; see especially p. 96.

59. Bauer et al., "Narrative Identity," 81-104; see especially pp. 96-97.

60. Walter R. Fisher, "Technical Logic, Rhetorical Logic, and Narrative Rationality," *Argumentation* 1 (1987): 3-21; see especially p. 16.

61. This means that coherent narratives do not possess altered facts or neglect pertinent details, and have considered possible alternative interpretations to ensure veracity. See Deanna D. Sellnow, *The Rhetorical Power of Popular Culture: Considering Mediated Texts* (Thousand Oaks, CA: SAGE Publications, 2010), 37-49; see especially p. 38.

62. Sellnow, *The Rhetorical Power,* 37-49; see especially pp. 38-39.

63. See Sellnow, *The Rhetorical Power,* 37-49; see especially p. 37.

64. See Martin Heidegger, *Existence and Being* (Chicago: Henry Regnery Company, 1949); see especially p. 278.

65. Fisher, "Technical Logic," 3-21; see especially p. 24.

66. This notion is attributed to and was first employed by Fisher in 1984. See Walter R. Fisher, "Narration as a Human Communication Paradigm: The Case of Public Moral Argument," *Communication Monographs* 51 (1984): 1-22.

67. Sellnow, *The Rhetorical Power,* 37-49; see especially p. 38.

68. Fisher, "Technical Logic," 3-21; see especially p. 16.

69. See Ricoeur, *Time and Narrative,* vols. 1-3; see especially vol. 3.

70. Paul Ricoeur, "L'identité Narrative," *Esprit* 7-8 (1988): 295-304.

71. See Margaret R. Somers, "The Narrative Constitution of Identity: A Relational and Network Approach," *Theory and Society* 23 (1994): 605-49; see especially p. 618.

72. Somers, "The Narrative Constitution," 605-49; see especially p. 618.

73. Taylor, *Sources of the Self,* 25-52; see especially pp. 51-52.

74. Somers, "The Narrative Constitution," 605-49; see especially p. 619.

75. Taylor, *Sources of the Self,* 25-52; see especially p. 39.

76. Somers, "The Narrative Constitution," 605-49; see especially p. 619.

77. Somers, "The Narrative Constitution," 605-49; see especially pp. 619-20.

78. See Thomas H. Murray, "What Do We Mean By 'Narrative Ethics'?" in *Stories and Their Limits: Narrative Approaches to Bioethics,* ed. Hilde Lindemann Nelson (New York: Routledge, 1997), 3-17; see especially pp. 6-7.

79. Martha C. Nussbaum, *Love's Knowledge: Essays on Philosophy and Literature* (New York: Oxford University Press, 1990), 3-53; see especially p. 26.

80. Murray, "What Do We Mean By 'Narrative Ethics'?" 3-17; see especially p. 7.

81. Nussbaum, *Love's Knowledge,* 3-53; see especially p. 26.

82. Murray, "What Do We Mean By 'Narrative Ethics'?" 3-17; see especially pp. 7-8.

83. Murray, "What Do We Mean By 'Narrative Ethics'?" 3-17; see especially p. 8.

84. This trail was blazed by Albert Jonsen and Stephen Toulmin, whose method would gain credence by the late 1980s. See Murray, "What Do We Mean By 'Narrative Ethics'?" 3-17; see especially pp. 8-9.

85. Albert R. Jonsen and Stephen Toulmin, *The Abuse of Casuistry: A History of Moral Reasoning* (Los Angeles: University of California Press, 1988), 304-31; see especially p. 314.

86. In particular, seeing more precisely what is involved in accepting (or rejecting) this or that rule in one or another set of circumstances. See Jonsen and Toulmin, *The Abuse of Casuistry,* 304-31; see especially p. 314.

87. Murray, "What Do We Mean By 'Narrative Ethics'?" 3-17; see especially pp. 8-9.

88. Murray, "What Do We Mean By 'Narrative Ethics'?" 3-17; see especially p. 9.

89. Judith Jarvis Thomson, "A Defense of Abortion," *Philosophy and Public Affairs* 1, no. 1 (1971): 67-95.

90. Bernard Williams, "A Critique of Utilitarianism," in *Utilitarianism: For and Against,* ed. J. J. C. Smart and Bernard Williams (New York: Cambridge University Press, 1973), 77-150; see especially pp. 93-150.

91. Alasdair MacIntyre, *After Virtue: A Study in Moral Theory,* 3rd ed. (Notre Dame, IN: Notre Dame University Press, 2007), 1-22.

92. See H. Tristram Engelhardt Jr., *The Foundations of Bioethics,* 2nd ed. (New York: Oxford University Press, 1996).

93. See Murray, "What Do We Mean By 'Narrative Ethics'?" 3-17; see especially p. 10.

94. Murray, "What Do We Mean By 'Narrative Ethics'?" 3-17; see especially p. 10.

95. See MacIntyre, *After Virtue*, 1-5; see especially pp. 1-2.

96. Murray, "What Do We Mean By 'Narrative Ethics'?" 3-17; see especially pp. 12-13.

97. See Engelhardt Jr., *The Foundations of Bioethics;* see especially its preface and ch. 2.

98. Nussbaum, *Love's Knowledge*, 148-67; see especially p. 152.

99. At the same time, Arras is skeptical about whether narrative identity can lead directly to moral knowledge. See John D. Arras, "Nice Story, But So What? Narrative and Justification in Ethics," in *Stories and Their Limits: Narrative Approaches to Bioethics*, ed. Hilde Lindemann Nelson (New York: Routledge, 1997), 65-88.

100. See Rita Charon, "Narrative Contributions to Medical Ethics: Recognition, Formulation, Interpretation, and Validation in the Practice of the Ethicist," in *A Matter of Principles? Ferment in U.S. Bioethics*, ed. Edwin R. DuBose, Ron Hamel, and Laurence J. O'Connell (Valley Forge, PA: Trinity Press International, 1994), 260-83.

101. See Arras, "Nice Story, But So What?" 65-88; see especially pp. 69-70.

102. On this view, most principlists are moral coherentists in the sense of relying on particular, considered judgments as a necessary means to test general theories. See Arras, "Nice Story, But So What?" 65-88; see especially p. 72.

103. Arras points out that the cases that give rise to these considered judgments are themselves revelatory of some narrative identity: they contain either micro-narratives that describe what it means for a particular person to behave in a particular way, or macro-narratives that describe the history of a particular behavior and its particular social benefits or burdens. See Arras, "Nice Story, But So What?" 65-88; see especially p. 72.

104. Brody, *Stories of Sickness*, 172-92; see especially p. 179.

105. Brody, *Stories of Sickness*, 172-92; see especially p. 179.

106. See James F. Childress, "Narrative(s) Versus Norm(s): A Misplaced Debate in Bioethics," in *Stories and Their Limits: Narrative Approaches to Bioethics*, ed. Hilde Lindemann Nelson (New York: Routledge, 1997), 252-71; see especially p. 256.

107. This can be understood as a new (and perhaps first) instance in which the child invokes moral judgment. See Brody, *Stories of Sickness*, 172-92; see especially pp. 179-80.

108. Brody, *Stories of Sickness*, 172-92; see especially pp. 179-80.

109. Brody, *Stories of Sickness*, 172-92; see especially p. 180.

110. See Immanuel Kant, *Fundamental Principles of the Metaphysics of Morals*, trans. Thomas Kingsmill Abbott (Mineola, NY: Dover Publications, Inc., 2005); see especially secs. 1-2.

111. Brody, *Stories of Sickness*, 172-92; see especially p. 180.

112. Brody, *Stories of Sickness*, 172-92; see especially pp. 180-81.

113. See MacIntyre, *After Virtue*, 181-225.

114. That is, claims concerning whether truth can be identified through narrative. See Arras, "Nice Story, But So What?" 65-88; see especially p. 73.

115. Brody, *Stories of Sickness,* 172-92; see especially p. 181.

116. See Murray, "What Do We Mean By 'Narrative Ethics'?" 3-17.

117. Brody, *Stories of Sickness,* 172-92; see especially p. 181.

118. Brody, *Stories of Sickness,* 193-237; see especially p. 218.

119. See Margaret Urban Walker, *Moral Understandings: A Feminist Study of Ethics* (New York: Routledge, 1998), 3-34; see especially p. 13.

120. Walker, *Moral Understandings,* 109-36; see especially p. 115.

121. Brody, *Stories of Sickness,* 193-237; see especially pp. 218-19.

122. See Tom Tomlinson, "Perplexed about Narrative Ethics," in *Stories and Their Limits: Narrative Approaches to Bioethics,* ed. Hilde Lindemann Nelson (New York: Routledge, 1997), 123-33.

123. Tomlinson, "Perplexed about Narrative Ethics," 123-33; see especially p. 130.

124. Brody, *Stories of Sickness,* 193-237; see especially pp. 219-20.

125. See Norman Daniels, "Wide Reflective Equilibrium and Theory Acceptance in Ethics," *Journal of Philosophy* 79 (1979): 256-82.

126. Brody, *Stories of Sickness,* 193-237; see especially pp. 229-30.

127. See Brody, *Stories of Sickness,* 193-237; see especially p. 230.

128. Brody, *Stories of Sickness,* 193-237; see especially p. 230.

129. See Brody, *Stories of Sickness,* 193-237; see especially pp. 230-31.

130. See Brody, *Stories of Sickness,* 193-237; see especially p. 231.

131. See Brody, *Stories of Sickness,* 193-237; see especially pp. 231-32.

132. Brody, *Stories of Sickness,* 193-237; see especially pp. 231-32.

133. Margaret O. Little, "Moral Generalities Revisited," in *Moral Particularism,* ed. Brad Hooker and Margaret O. Little (New York: Oxford University Press, 2000), 276-304; see especially p. 292.

134. Brody, *Stories of Sickness,* 193-237; see especially p. 235.

135. MacIntyre, *After Virtue,* 204-225; see especially p. 215.

136. MacIntyre, *After Virtue,* 204-225; see especially p. 215.

137. Rachel D. Godsil and Brianna Goodale, *Telling Our Own Story: The Role of Narrative in Racial Healing* (East Battle Creek, MI: W. K. Kellogg Foundation, June 2013), 1-10; see especially p. 7.

138. See, for instance, Raymond A. Mar, "The Neuropsychology of Narrative: Story Comprehension, Story Production and Their Interrelation," *Neuropsychologia* 42 (2004): 1414-43.

139. Godsil and Goodale, *Telling Our Own Story,* 1-10; see especially p. 7.

140. Jonas T. Kaplan, Sarah I. Gimbel, Morteza Dehghani, Mary Helen Immordino-Yang, Kenji Sagae, Jennifer D. Wong, Christine M. Tipper, Hanna Damasio, Andrew S. Gordon, and Antonio Damasio, "Processing Narratives Concerning Protected Values: A Cross-Cultural Investigation of Neural Correlates," *Cerebral Cortex* 26, no. 1 (January 2016): 1-11; see especially pp. 1-2.

141. See Gregory S. Berns, Emily Bell, C. Monica Capra, Michael J. Prietula, Sara Moore, Brittany Anderson, Jeremy Ginges, and Scott Atran, "The Price of Your Soul: Neural Evidence for the Non-Utilitarian Representation of Sacred Values," *Philosophical Transactions of the Royal Society B* 367 (2012): 754-62.

142. See Corinne Duc, Martin Hanselmann, Peter Boesiger, and Carmen Tanner, "Sacred Values: Trade-Off Type Matters," *Journal of Neuroscience, Psychology, and Economics* 6, no. 4 (December 2013): 252-63.

143. Kaplan et al., "Processing Narratives," 1-11; see especially p. 2.

144. See Kaplan et al., "Processing Narratives," 1-11.

145. See Rogier B. Mars, Franz-Xaver Neubert, MaryAnn P. Noonan, Jerome Sallet, Ivan Toni, and Matthew F. S. Rushworth, "On the Relationship between the 'Default Mode Network' and the 'Social Brain,'" *Frontiers in Human Neuroscience* 6 (June 2012): 1-9.

146. See Mary Helen Immordino-Yang, Joanna A. Christodoulou, and Vanessa Singh, "Rest Is Not Idleness: Implications of the Brain's Default Mode for Human Development and Education," *Perspectives on Psychological Science* 7, no. 4 (2012): 352-64.

147. See Ylva Østby, Kristine B. Walhovd, Christian K. Tamnes, Håkon Grydeland, Lars Tjelta Westlye, and Anders M. Fjell, "Mental Time Travel and Default-Mode Network Functional Connectivity in the Developing Brain," *Proceedings of the National Academy of Sciences of the United States of America* 109, no. 42 (2012): 16800-04.

148. See Pengmin Qin and Georg Northoff, "How is Our Self Related to Midline Regions and the Default-Mode Network?" *NeuroImage* 57, no. 3 (August 2011): 1221-33.

149. See, for instance, Georg Northoff, Alexander Heinzel, Moritz de Greck, Felix Bermpohl, Henrik Dobrowolny, and Jaak Panksepp, "Self-Referential Processing in Our Brain—A Meta-Analysis of Imaging Studies on the Self," *NeuroImage* 31 (2006): 440-57.

150. Kaplan et al., "Processing Narratives," 1-11; see especially p. 5.

151. Kaplan et al., "Processing Narratives," 1-11; see especially p. 6.

152. Kaplan et al., "Processing Narratives," 1-11; see especially p. 6.

153. See, for instance, Hyemin Han, Gary H. Glover, and Changwoo Jeong, "Cultural Influences on the Neural Correlate of Moral Decision Making Processes," *Behavioural Brain Research* 259 (February 2014): 215-28.

154. Kaplan et al., "Processing Narratives," 1-11; see especially p. 7.

155. Regarding the latter, see Joan Y. Chiao, Tokiko Harada, Hidetsugu Komeda, Zhang Li, Yoko Mano, Daisuke Saito, Todd B. Parrish, Norihiro Sadato, and Tetsuya Iidaka, "Dynamic Cultural Influences on Neural Representations of the Self," *Journal of Cognitive Neuroscience* 22, no. 1 (January 2010): 1-11.

156. Kaplan et al., "Processing Narratives," 1-11; see especially p. 8.

157. See, for instance, Yanhong Wu, Cheng Wang, Xi He, Lihua Mao, and Li Zhang, "Religious Beliefs Influence Neural Substrates of Self-Reflection in Tibetans," *Social Cognitive and Affective Neuroscience* 5, nos. 2-3 (June/September 2010): 324-31.

158. See, for instance, Richard E. Nisbett and Yuri Miyamoto, "The Influence of Culture: Holistic versus Analytic Perception," *Trends in Cognitive Sciences* 9, no. 10 (October 2005): 467-73.

159. Harlene Anderson, *Conversation, Language, and Possibilities: A Postmodern Approach to Therapy* (New York: BasicBooks, 1997), 211-34; see especially pp. 215-16.

160. See Richard Rorty, *Philosophy and the Mirror of Nature* (Princeton, NJ: Princeton University Press, 1979).

161. Anderson, *Conversation, Language, and Possibilities*, 211-34; see especially p. 216.

162. See Ricoeur, *Time and Narrative*, vol. 3, 241-74; see especially p. 246.

163. Morny Joy, "Feminism and the Self," *Theory and Psychology* 3 (1993): 275-302; see especially pp. 296-97.

164. Anderson, *Conversation, Language, and Possibilities*, 211-34; see especially p. 217.

165. See Roy Schafer, *Language and Insight* (New Haven, CT: Yale University Press, 1978).

166. See Emile Benveniste, "Subjectivity in Language," in *Problems in General Linguistics*, trans. Mary Elizabeth Meek (Miami: University of Miami Press, 1971), 223-230.

167. Anderson, *Conversation, Language, and Possibilities*, 211-34; see especially p. 217-19.

168. See Benveniste, "Subjectivity in Language," 223-30.

169. Anderson, *Conversation, Language, and Possibilities*, 211-34; see especially p. 227-28.

170. Anderson, *Conversation, Language, and Possibilities*, 211-34; see especially pp. 230-34.

171. Kate C. McLean, Monisha Pasupathi, and Jennifer L. Pals, "Selves Creating Stories Creating Selves: A Process Model of Self-Development," *Personality and Social Psychology Review* 11, no. 3 (August 2007): 262-78; see especially p. 274.

172. McLean et al., "Selves Creating Stories," 262-78; see especially pp. 274-75.

173. McAdams, "Personal Narratives," 242-62; see especially p. 243.

174. Antonio Damasio, *The Feeling of What Happens: Body and Emotion in the Making of Consciousness* (New York: Harcourt, Inc., 1999), 1-32; see especially p. 10.

175. McAdams, "Personal Narratives," 242-62; see especially pp. 244-45.

176. McAdams, "Personal Narratives," 242-62; see especially p. 245.

177. See, for instance, Monisha Pasupathi, "The Social Construction of the Personal Past and its Implications for Adult Development," *Psychological Bulletin* 127 (2001): 651-72.

178. See Monisha Pasupathi, "Silk from Sows' Ears: Collaborative Construction of Everyday Selves in Everyday Stories," in *Identity and Story: Creating Self in Narrative*, ed. Dan P. McAdams, Ruthellen Josselson, and Amia

Lieblich (Washington, DC: American Psychological Association, 2006), 129-50.

179. McAdams, "Personal Narratives," 242-62; see especially pp. 245-46.

180. See, for instance, Jack J. Bauer, Dan P. McAdams, and April R. Sakaeda, "Interpreting the Good Life: Growth Memories in the Lives of Mature, Happy People," *Journal of Personality and Social Psychology* 88, no. 1 (2005): 203-17.

181. McAdams, "Personal Narratives," 242-62; see especially p. 246.

182. See, for instance, Laura A. King, "The Hard Road to the Good Life: The Happy, Mature Person," *Journal of Humanistic Psychology* 41 (2001): 51-72.

183. See, for instance, Jefferson A. Singer, *Personality and Psychotherapy: Treating the Whole Person* (New York: Guilford Press, 2005).

184. McAdams, "Personal Narratives," 242-62; see especially pp. 246-48.

185. McAdams, "Personal Narratives," 242-62; see especially p. 255

186. See, for instance, McAdams et al., "When Bad Things Turn Good and Good Things Turn Bad," 472-83.

187. McAdams, "Personal Narratives," 242-62; see especially pp. 252-56.

188. Kim Atkins, "Narrative Identity, Practical Identity, and Ethical Subjectivity," *Continental Philosophy Review* 37 (2004): 341-66; see especially p. 347.

189. See Ricoeur, *Time and Narrative*, vols. 1-3; see especially vol. 3.

190. See, for instance, Christine M. Korsgaard, "Autonomy and the Second Person Within: A Commentary on Stephen Darwall's The Second-Person Standpoint," *Ethics* 118, no. 1 (2007): 8-23.

191. See Christine M. Korsgaard, *The Sources of Normativity* (Cambridge, UK: Cambridge University Press, 1996), 7-48; see especially pp. 14–16.

192. Atkins, "Narrative Identity," 341-66; see especially pp. 362-63.

193. Atkins, "Narrative Identity," 341-66; see especially p. 363.

194. See Korsgaard, *The Sources of Normativity*, 7-48; see especially p. 14.

195. Atkins, "Narrative Identity," 341-66; see especially pp. 363-64.

196. Kenneth J. Gergen, "Narrative, Moral Identity, and Historical Consciousness: A Historical Constructionist Account," in *Narration, Identity, and Historical Consciousness*, ed. Jürgen Straub (New York: Berghahn Books, 2005), 99-119; see especially p. 110.

197. See John O. Lyons, *The Invention of the Self: The Hinge of Consciousness in the Eighteenth Century* (Carbondale, IL: Southern Illinois University Press, 1978).

198. See Kenneth J. Gergen, *The Saturated Self* (New York: Basic Books, 1991).

199. See Ludwig Wittgenstein, *Tractatus Logico-Philosophicus* (New York: Cosimo, Inc., 2007).

200. Gergen, "Narrative," 99-110; see especially pp. 110-12.

201. Gergen, "Narrative," 99-110; see especially p. 116.

202. For theists and non-theists alike, this history has largely been negotiated in terms of a moral account. In the Christian story, there is the loss (or negation) of a highly valued endpoint, as represented in the crucifixion of Jesus of

Nazareth. In mourning the tragic loss while celebrating the hopeful promise, the value is vindicated. Hence, the retelling of the narrative represents the continuation of a long-standing cultural tradition. See Gergen, "Narrative," 99-110; see especially p. 117.

203. For Christians, the resurrection plays a moral role in a salvation history to which their present-day identity is inextricably wed. To question this history is to challenge the intelligibility, identity, and standing of one's moral being. See Gergen, "Narrative," 99-110; see especially p. 117.

204. Gergen, "Narrative," 99-110; see especially pp. 116-18.

205. See John Locke, *An Essay Concerning Human Understanding,* ed. Roger Woolhouse (London: Penguin Books, 1997).

206. See Charles Taylor, *The Ethics of Authenticity* (Cambridge, MA: Harvard University Press, 1991).

207. For a helpful analysis of the ways in which temporal remoteness biases memories related to moral autobiography, see Jessica R. Escobedo and Ralph Adolphs, "Becoming a Better Person: Temporal Remoteness Biases Autobiographical Memories for Moral Events," *Emotion* 10, no. 4 (August 2010): 511-18.

208. Jennifer Bell, "Propranolol, Post-Traumatic Stress Disorder and Narrative Identity," *Journal of Medical Ethics* 34, no. 11 (2008): e23.

209. Glannon, "Psychopharmacology and Memory," 74-78; see especially p. 78.

210. Friedrich Nietzsche, *The Genealogy of Morals,* trans. Horace B. Samuel (New York: Dover Publications, Inc., 2003), essay II; see especially p. 34.

211. Thomas Fuchs, "Fragmented Selves: Temporarily and Identity in Borderline Personality Disorder," *Psychopathology* 40 (2007): 379-87; see especially pp. 379-80.

212. See Ricoeur, *Oneself as Another;* see especially pp. 140-68.

213. Fuchs, "Fragmented Selves," 379-87; see especially p. 380.

214. Fuchs, "Fragmented Selves," 379-87; see especially p. 380.

215. See MacIntyre, *After Virtue,* 204-25; see especially p. 219.

216. Fuchs, "Fragmented Selves," 379-87; see especially p. 380.

217. Martin A. Conway and Christopher W. Pleydell-Pearce, "The Construction of Autobiographical Memories in the Self-Memory System," *Psychological Review* 107, no. 2 (2000): 261-88; see especially p. 266.

218. See, for instance, Lawrence W. Barsalou, "The Content and Organization of Autobiographical Memories," in *Remembering Reconsidered: Ecological and Traditional Approaches to the Study of Memory,* ed. Ulric Neisser and Eugene Winograd (New York: Cambridge University Press, 1988), 193-243.

219. See Jefferson A. Singer and Peter Salovey, *The Remembered Self: Emotion and Memory in Personality* (New York: Free Press, 1993).

220. See Mario Mikulincer, "Adult Attachment Style and Individual Differences in Functional versus Dysfunctional Experiences of Anger," *Journal of Personality and Social Psychology* 74 (1998): 513-24.

221. See Avril Thorne, "Developmental Truths in Memories of Childhood and Adolescence," *Journal of Personality* 63, no. 2 (1995): 139-63.

222. Conway and Pleydell-Pearce, "The Construction of Autobiographical Memories," 261-88; see especially pp. 266-69.

223. Fuchs, "Fragmented Selves," 379-87; see especially p. 385.

224. Fuchs, "Fragmented Selves," 379-87; see especially pp. 385-86.

225. Taylor, *The Ethics of Authenticity*, 25-29; see especially p. 25.

226. Taylor, *The Ethics of Authenticity*, 25-29; see especially pp. 25-27.

227. See Jean-Jacques Rousseau, *Les Rêveries du Promeneur Solitaire*, in *Oeuvre Complètes*, vols. 1-5 (Paris: Gallimard, 1959); see especially vol. 1.

228. Rousseau, *Les Rêveries du Promeneur Solitaire*; see especially p. 1047.

229. Taylor, *The Ethics of Authenticity*, 25-29; see especially p. 28.

230. "Jeder Mensch haat ein eigenes Mass, gleichsam eine eigne Stimmung aller seiner sinnlichen GefSee Johann Gottfried von Herderühle zu einander." See Johann Gottfried von Herder, *Ideen*, in *Herders Sämtliche Werke*, ed. Bernhard Suphan, vols. I-XV (Berlin: Weidmann, 1877-1913); see especially vol. XIII, p. 291.

231. Taylor, *The Ethics of Authenticity*, 25-29; see especially p. 29; emphasis his.

232. Taylor, *The Ethics of Authenticity*, 25-29; see especially pp. 28-29.

233. Taylor, *The Ethics of Authenticity*, 81-91; see especially p. 91.

234. Taylor, *The Ethics of Authenticity*, 31-41; see especially pp. 32-34.

235. Lisa Capps, "Narrative Authenticity," *Journal of Narrative and Life History* 7, nos. 1-4 (1997): 83-89; see especially p. 83.

236. Taylor, *The Ethics of Authenticity*, 31-41; see especially pp. 35-41.

237. Capps, "Narrative Authenticity," 83-89; see especially p. 88.

238. Taylor, *The Ethics of Authenticity*, 31-41; see especially p. 41.

239. Alfred R. Mele, "Self-Control in Action," in *The Oxford Handbook of The Self*, ed. Shaun Gallagher (New York: Oxford University Press, 2013), 465-86; see especially p. 465.

240. See Daniel Dennett, *Elbow Room: The Varieties of Free Will Worth Wanting* (Cambridge, MA: The MIT Press, 1984).

241. See Jeanette Kennett, *Agency and Responsibility: A Common Sense Moral Psychology* (Oxford, UK: Oxford University Press, 2001), 119-53; see especially pp. 126-29.

242. See Aristotle, *Nicomachean Ethics*, in *The Complete Works of Aristotle*, ed. Jonathan Barnes, vols. 1-2, trans. W. D. Ross, rev. J. O. Urmson (Princeton, NJ: Princeton University Press, 1984; see especially vol. 2, pp. 1729-1867 at 1152a25-27.

243. Aristotle, *Movement of Animals*, in *The Complete Works of Aristotle*, ed. Jonathan Barnes, vols. 1-2, trans. A. S. L. Farquharson (Princeton, NJ: Princeton University Press, 1984; see especially vol. 1, pp. 1087-96 at 701a7-8.

244. Mele, "Self-Control in Action," 465-86; see especially pp. 465-66.

245. Maria Cavell, "The Self: Growth, Integrity, and Coming Apart," in *The Oxford Handbook of The Self*, ed. Shaun Gallagher (New York: Oxford University Press, 2013), 592-605; see especially p. 595.

246. See, for instance, Heinz Kohut, *The Restoration of the Self* (Chicago: The University of Chicago Press, 1977).

247. See Hume, *A Treatise of Human Nature*.

248. Cavell, "The Self," 592-605; see especially p. 595-601.

249. Cavell, "The Self," 592-605; see especially p. 601-02.

250. See Fyodor Dostoyevsky, *Crime and Punishment,* trans. David McDuff (New York: Penguin Books, 2002).

251. Cavell, "The Self," 592-605; see especially p. 602-03.

252. David Shoemaker, "Moral Responsibility and the Self," in *The Oxford Handbook of The Self,* ed. Shaun Gallagher (New York: Oxford University Press, 2013), 487-518; see especially p. 488.

253. Shoemaker, "Moral Responsibility and the Self," 487-518; see especially pp. 488-515.

254. MacIntyre, *After Virtue,* 204-225; see especially p. 208.

255. MacIntyre, *After Virtue*, 204-225; see especially pp. 208-15.

256. MacIntyre, *After Virtue*, 204-225; see especially pp. 215-16.

257. MacIntyre, *After Virtue*, 204-225; see especially pp. 216-18.

258. MacIntyre, *After Virtue*, 204-225; see especially p. 216.

259. MacIntyre, *After Virtue*, 204-225; see especially pp. 218-19.

260. MacIntyre, *After Virtue*, 204-225; see especially p. 220.

261. MacIntyre, *After Virtue*, 204-225; see especially pp. 220-25.

# 6.
# CONCLUSION: THE TERMINAL NORMLESSNESS OF MEMORY MANIPULATION

In light of the contemporary possibilities of MM, this text examined the respective relationships shared between autobiographical memory, emotional rationality, and narrative identity. It posited the argument that MM cannot be justified as a morally licit biomedical practice insofar as it disables ones to seek, identify, and act on the good. To secure the justification of this thesis, it drew from the logic chain of four evidential effects: (i) MM disintegrates autobiographical memory; (ii) the disintegration of autobiographical memory degenerates emotional rationality; (iii) the degeneration of emotional rationality decays narrative identity; and (iv) the decay of narrative identity disables one to seek, identify, and act on the good. In pursuit of its end, the present effort has been successful. However, a final case must be made against MM vis-à-vis the glaring and terminal normlessness of its endeavors.

## 6.1. The Case Against MM: The Normative Demands of Proportionate Reason

For reasons beyond the arguments posited in chapters 2-5, this text contends that even LUMM cannot be justified as a morally licit biomedical practice within the confines of a comprehensive normative ethical framework. The application of Richard McCormick's threefold criteria of proportionate reason can serve to illuminate the normative astigmatism of LUMM and the text's corresponding endeavor to correct it.[1] However, before an adequate moral assessment of an action's proportionality can be made, its effect on all ends and values must first be weighed. Moral values can be considered and a final decision made only after all values have been compared.[2] It is this systematic weighing of moral values, for instance, that has made noncombatant immunity a virtually exceptionless moral rule. The strength of moral norms touching concrete conduct is an elaboration of what is judged – within a particular culture, with a particular history, based on a particular experience – to be proportionate or disproportionate. Proportionality is always the criterion where actions cause damage, and MM involves both personal and social damage. The corrective vision provided herein attempts to halt the narrowly-conceived notion of proportionality embedded in the arguments of LUMM proponents.[3]

If there exist norms that are teleologically established and yet are virtually exceptionless,[4] the remaining task is to clarify the metaethical assertions in view of which those norms are held as exceptionless. This task includes nondemonstrable calculations – prudential judgments based on the certainties of history and the uncertainties of the future. The sense of what individuals ought and ought not to do is therefore informed by past experience and agnosticism with regard to future behavior and its long-term effects. This suggests that where norms are viewed as virtually exceptionless, it is because of the prudential validity technically referred to as lex lata in praesumptione periculi communis: a law established on the presumption of common or universal danger.[5] The notion of presumed universal danger is frequently associated with positive law. That is, even if the action in question does not threaten the individual personally (though LUMM does), there remains the further presumption that to allow individuals to make that decision for themselves poses a threat to the common good (which LUMM also does). Hence, the ethical impetus to retain autobiographically accurate, emotional rational, and narratively authentic memories can be viewed in a way analogous to the exceptionless character of norms such as noncombatant immunity. The risk in alternative policies is simply too great. In the context of moral development, autobiographical, emotional, and narrative memories are enormous goods at stake. Past experience of human failure, inconsistency, and frailty, along with uncertainty regarding long-term effects of such irreversible actions as dampening or erasing human memories, suggests that societies should continue to hold some norms - such as authentic self-knowledge – as virtually exceptionless. That is the conclusion of prudence in the face of dangers too grave to make risk tolerable.

For McCormick, proportionate reason (for permitting the occurrence of harm otherwise judged as illicit within a normative moral calculus) means three things: (i) the value at stake is at least equal to the value being sacrificed; (ii) there is no less harmful way to protect the value here and now; and (iii) the means used to protect the value (here and now) will not undermine it in the long run.[6]

Conversely, an action is disproportionate if (i) a lesser value is preferred to a more important one; (ii) harm is unnecessarily caused in the protection of a greater good; or (iii) in the circumstances, the manner of protecting the good will undermine it in the long run. To determine if an action involving harm is proportionate in the circumstances, one must judge whether the specific choice is the best possible service to all values in the difficult and, in the context of PTSD and substance addiction, tragic circumstances. What constitutes the best promotion of all values in the circumstances will, of course, depend on how one defines and understands the circumstances. An adequate account of the circumstances indicates not simply how much quantitative good can be salvaged from an individual conflict of values, but also the weight and balance of social implications and reverberating aftereffects insofar as they can be foreseen. This account will test generalizability, consider cultural climate, draw from historical wisdom, seek guidance from others, and distance itself from self-interested tendencies. In sum, the criterion of proportionality is found within the ordo bonorum, which determines the good one ought to do and serves as the objectively licit character of one's activity. So informed and organized, individuals do all that can be expected of them.[7]

Before applying McCormick's first criterion of proportionality, namely, that the value at stake is at least equal to the value being sacrificed – to the action of LUMM,[8] it is necessary, first, to identify both the value at stake and the value being sacrificed in the circumstances of LUMM. The strongest proponents of LUMM are likely to define the value at stake as individual (and, in turn, social) health, well-being, and safety (which has been compromised by disproportionate psychosocial conditions including, but not limited to, PTSD and substance addiction). This text defines the value being sacrificed as individual (and social) autobiographical memory, emotional rationality, and narrative identity – elements that together comprise the ability to seek, identify, and act on the good. There is no doubt that individual (and social) health, well-being, and safety are massively significant values. However, as LUMM proponents fail to discern, these values depend almost exclusively for existence on some manifestation of the interplay between memory, emotion, and identity. If overall health, well-being, and safety are largely contingent on the capacities to understand (employing memory), evaluate (employing emotion), and reason (employing identity), then LUMM does not meet

McCormick's first criterion of proportionality. Indeed, the value at stake is not equal to the value being sacrificed.

McCormick's second criterion of proportionality requires that there exists a no less harmful way to protect the value here and now.[9] Applied in the context of LUMM, the criterion becomes thus: there exists no less harmful way to protect the value of individual (and social) health, well-being, and safety than to annihilate the memories that are presumed to be responsible for perpetuating disproportionate psychosocial conditions (including, but not limited to, PTSD and substance addiction). This claim is an enormous stretch, even in light of most aggressive and debilitating psychosocial conditions. Moreover, it is based on the false premises that (i) memories are solely or even primarily responsible for perpetuating psychosocial conditions, and (ii) memories are independently and immutably morally charged. Neither of these claims carries any neurobiological or neuroethical merit. Lastly, if the value being scarified is defined as individual (and social) autobiographical memory, emotional rationality, and narrative identity – elements that, again, collectively comprise the ability to seek, identify, and act on the good – then it is difficult to imagine a more harmful way to protect the value here and now than to utilize LUMM to do so. In light of the abilities of intensive psychotherapy and the proportionate use of rebalancing pharmacologicals (when utilized to their fullest potential),[10] LUMM does not meet McCormick's second criterion of proportionality. Indeed, there exists a drastically less harmful (and impermanent) way to protect the value here and now.

McCormick's third criterion of proportionality requires that in the circumstances, the means used to protect the value here and now will not undermine it in the long run.[11] Applied in the context of LUMM, the criterion becomes thus: in the circumstances of threatening psychosocial conditions, the means of LUMM used (here and now) to protect the value of individual (and social) health, well-being, and safety will not undermine it (i.e., individual [and social] health, well-being, and safety) in the long run. Here, the neuroethical astigmatism of LUMM proponents is most acutely evident. Proponents of LUMM frequently couch their arguments in what they believe would result from the redemptive practice of LUMM: the restoration and reinstallation of autonomy once plundered by debilitating and disproportionate psychosocial conditions. By itself, the restoration and reinstallation of autonomy is a noble and desirable goal for any treatment. As mentioned above, if the consequence of certain psychosocial conditions involves the loss of autonomy, then the conditions(s) must be viewed to somehow undercut the individual (and social) ability to pursue goals. This presumes, however, that the pursuit of individual (and social) goals – which depends for existence on the capacities to (i) recall (through memory) and (ii) act (through informed and intentioned will) on them – will somehow be resurrected by annihilating the very faculty (i.e., memory) that makes the pursuit possible and, more importantly, morally responsible. However, goals are always specific; they cannot exist apart from individuals and societies who espouse them, and they depend for definition, therefore, on the characteristics those individuals and societies possess – characteristics that are, no doubt, very different from other individuals and societies.

A standard understanding of autonomy refers to the freedom individuals (and societies) ought to enjoy to choose their own way in life and to make their own decisions within moral limits. Proponents of LUMM contend there is no more immediate and humane way, in extreme circumstances, to restore and reinstate lost freedom than to dampen or erase the memories that seem to imprison it. For them, LUMM marks a bridge for individuals (and societies) bound by the devastating effects of psychosocial disorders to once again be free to act freely. A first blush, this position is tempting and intoxicating, but it ultimately proves astigmatic and impossible. Societies have long nuanced the notion of pure autonomy to preclude the freedom of its members to simply do as they wish. This is exemplified in the rejection, for instance, of strictly utilitarian calculi and purely consequentialist logistics. Against the notion of pure autonomy, this text suggests that a more adequate understanding of autonomy is not fundamentally concerned with the freedom to do as one wants, but with the freedom to do as one ought in light of one's moral responsibilities.

If this nuanced notion of autonomy is persuasive, then it raises the question of how individuals can decipher what they have a moral responsibility to do. As the central thesis of this text holds, the answer is determined by one's narrative identity, the sum of one's autobiographical memory and emotional rationality. The "autonomous ought" can, therefore, exist only in light of an idiosyncratic narrative. Hence, only within an idiosyncratic narrative structure can one determine one's future "oughtness" – the pull of moral responsibility grounded in and determined by the story of one's life and the values and commitments that comprise it. Conceived of concretely, the "autonomous ought" is what separates an individual's lack of desire to help the child in front of her from the responsibility to help the child in front of her because this child is her child and she is this child's parent. This structure of identity, which serves as the basis for determining the good and one's responsibility to act on it, is irreparably damaged by memory manipulation, even in its most limited forms and exceptional applications. For this reason, LUMM does not meet McCormick's third criterion of proportionality. Indeed, in the circumstances, the means used (here and now) to protect the value ultimately undermine it in the long run.

### 6.2. Implications

The implications of this text are significant. To be sure, the reality of manipulation extremism is a genuine and growing fear. However, rather than serving as de facto deterrents of neuroscientific research and medical progress, the foregoing arguments serve as wise constraints and prudent reminders that while the prospective benefits of neurocognitive technologies are important, how and how far they are utilized is more important still. Hence, the impetus to continue the exploration of narratively complimentary and cognitively non-debilitating means of therapy to assist individuals in coping with trauma and addiction should remain a priority of contemporary biomedical research.

Giambattista Vico, the seventeenth century philosopher who never left Naples yet observed much about the world, formulated a maxim that was often adopted by early scholastic philosophers: verum ipsum factum. The phrase can be translated as "the true is the made," and its logic offers significant wisdom to the future of memory research. According to Vico's maxim, individuals can understand only insofar as they can imitate or replicate. Although Vico was doubtlessly commenting on scientific discovery, its translation to the present context – in particular, to the ethical significance of autobiographical memory, emotional rationality, and narrative identity – is utterly clear: human beings can pursue the good to the extent that they can autobiographically recollect against emotionally rational moral norms embedded in robust and rooted narrative identities. Vico took this maxim from theology and moved it to secular philosophy. In theology, the maxim was used as an argument that only God can understand the world because God created it. On similar grounds, this text has suggested that the mechanism of memory can be comprehensively understood only inasmuch as the sanctity of its autobiographical, emotional, and narrative origins are first appreciated and finally preserved.

### 6.3. Notes

1. See Richard A. McCormick, "Ambiguity in Moral Choice," in *Doing Evil to Achieve Good: Moral Choice in Conflict Situations,* ed. Richard A. McCormick and Paul Ramsey (Lanham, MD: University Press of America, 1985), 7-53; see especially pp. 45-46.

2. Charles Curran, *A New Look at Christian Morality* (Notre Dame, IN: Fides Publishers, 1970); see especially p. 239.

3. McCormick, "Ambiguity in Moral Choice, 7-53; see especially p. 44.

4. For instance, the direct destruction of noncombatants in warfare.

5. McCormick, "Ambiguity in Moral Choice, 7-53; see especially p. 44.

6. McCormick, "Ambiguity in Moral Choice, 7-53; see especially pp. 45-46.

7. McCormick, "Ambiguity in Moral Choice, 7-53; see especially p. 46.

8. McCormick, "Ambiguity in Moral Choice, 7-53; see especially p. 45.

9. McCormick, "Ambiguity in Moral Choice, 7-53; see especially p. 45.

10. This oversimplified statement is meant to include participation by those with addiction in the many historically-successful 12-step recovery programs.

11. McCormick, "Ambiguity in Moral Choice, 7-53; see especially pp. 45-46.

# BIBLIOGRAPHY

Addis, Donna Rose, Alana T. Wong, and Daniel L. Schacter. "Remembering the Past and Imagining the Future: Common and Distinct Neural Substrates During Event Construction and Elaboration." *Neuropsychologia* 45, no. 7 (April 2007): 1363-77.

———. "Age-Related Changes in the Episodic Simulation of Future Events." *Psychological Science* 19, no. 1 (2008): 33-41.

Amodio, David M., and Chris D. Frith. "Meeting of Minds: The Medial Frontal Cortex and Social Cognition." *Nature Reviews Neuroscience* 7 (April 2006): 268-77.

Anderson, Harlene. *Conversation, Language, and Possibilities: A Postmodern Approach to Therapy.* New York: BasicBooks, 1997.

Aristotle. *Movement of Animals.* In *The Complete Works of Aristotle,* edited by Jonathan Barnes, 1087-96. Vols. 1-2. Translated by A. S. L. Farquharson. Princeton, NJ: Princeton University Press, 1984.

———. *Nicomachean Ethics.* In *The Complete Works of Aristotle,* edited by Jonathan Barnes, 1729-1867. Vols 1-2. Translated by W. D. Ross. Revised by J. O. Urmson. Princeton, NJ: Princeton University Press, 1984.

Arndt, Jason. "The Role of Memory Activation in Creating False Memories of Encoding Context." *Journal of Experimental Psychology* 36, no. 1 (2010): 66-79.

Arras, John D. "Nice Story, But So What? Narrative and Justification in Ethics." In *Stories and Their Limits: Narrative Approaches to Bioethics,* edited by Hilde Lindemann Nelson, 65-88. New York: Routledge, 1997.

Atkins, Kim. "Narrative Identity, Practical Identity, and Ethical Subjectivity." *Continental Philosophy Review* 37 (2004): 341-66.

Baker, Dana Lee, Leonard, Brandon. *Neuroethics in Higher Education Policy.* Palgrave MacMillan, 2016.

Balfour, David J. K. "The Neurobiology of Tobacco Dependence: A Preclinical Perspective on the Role of the Dopamine Projections to the Nucleus Accumbens." *Nicotine & Tobacco Research* 6, no. 6 (2004): 899-912.

Bamberg, Michael. "*Who am I?* Narration and its Contribution to Self and Identity." *Theory and Psychology* 2, no. 1 (2010): 1-22.

Barsalou, Lawrence W. "The Content and Organization of Autobiographical Memories." In *Remembering Reconsidered: Ecological and Traditional Approaches to the Study of Memory,* edited by Ulrich Neisser and Eugene Winograd, 193-243. New York: Cambridge University Press, 1988.

Bartlett, Francis E. *Remembering: A Study in Experimental and Social Psychology.* Cambridge, UK: Cambridge University Press, 1932.

Bassok, Miriam, and Laura R. Novick. "Problem Solving." In *The Oxford Handbook of Thinking and Reasoning,* edited by Keith J. Holyoak and Robert G. Morrison, 413-32. New York: Oxford University Press, 2012.

Bauer, Jack J., Dan P. McAdams, and Jennifer L. Pals. "Narrative Identity and Eudaimonic Well-Being." *Journal of Happiness Studies* 9 (2008): 81-104.

Bauer, Jack J., Dan P. McAdams, and April R. Sakaeda. "Interpreting the Good Life: Growth Memories in the Lives of Mature, Happy People." *Journal of Personality and Social Psychology* 88, no. 1 (2005): 203-17.

Baum, Matthew L. The *Neuroethics of Biomarkers. What the Development of Bioprediction Means for Moral Responsibility, Justice, and the Nature of Mental Disorder.* New York: Oxford University Press, 2017.

Baumeister, Roy F. "Ego Depletion and Self-Control Failure: An Energy Model of the Self's Executive Function." *Self and Identity* 1, no. 2 (2002): 129-36.

Baumeister, Roy F., Arlene Stillwell, and Sara R. Wotman. "Victim and Perpetrator Accounts of Interpersonal Conflict: Autobiographical Narratives About Anger." *Journal of Personality and Social Psychology* 59, no. 5 (November 1990): 994-1005.

Baylis, Françoise. "'I Am Who I Am': On the Perceived Threats to Personal Identity from Deep Brain Stimulation." *Neuroethics* 6 (2013): 513-26.

Bechara, Antoine, Antonio R. Damasio, Hanna Damasio, and Steven W. Anderson. "Insensitivity to Future Consequences Following Damage to Human Prefrontal Cortex." *Cognition* 50 (1994): 7-15.

Bechara, Antoine, Hanna Damasio, and Antonio R. Damasio. "Emotion, Decision Making and the Orbitofrontal Cortex." *Cerebral Cortex* 10 (March 2000): 295-307.

———. "Role of the Amygdala in Decision-Making." *Annals of the New York Academy of Sciences* 985 (2003): 356-69.

Bell, Jennifer. "Propranolol, Post-Traumatic Stress Disorder and Narrative Identity." *Journal of Medical Ethics* 34, no. 11 (2008): e23.

Ben-Ze'ev, Aaron. "The Thing Called Emotion." In *The Oxford Handbook of Philosophy of Emotion,* edited by Peter Goldie, 41-62. New York: Oxford University Press, 2010.

Benoit, Roland G., Karl K. Szpunar, and Daniel L Schacter. "Ventromedial Prefrontal Cortex Supports Affective Future Simulation by Integrating Distributed Knowledge." *Proceedings of the National Academy of Sciences of the United States of America* 111, no. 46 (November 2014): 16550-55.

Benveniste, Emile. "Subjectivity in Language." In *Problems in General Linguistics*, 223-30. Translated by Mary Elizabeth Meek. Miami: University of Miami Press, 1971.

Berns, Gregory S., Emily Bell, C. Monica Capra, Michael J. Prietula, Sara Moore, Brittany Anderson, Jeremy Ginges, and Scott Atran. "The Price of Your Soul: Neural Evidence for the Non-Utilitarian Representation of Sacred Values." *Philosophical Transactions of the Royal Society B* 367 (2012): 754-62.

Bernstein, Daniel M., Nicole L. M. Pernat, and Elizabeth F. Loftus. "The False Memory Diet: False Memories Alter Food Preferences." In *Handbook of Behavior, Food and Nutrition*, edited by Victor R. Preedy, Ronald Ross Watson, and Colin R. Martin, 1645-63. New York: Springer, 2011.

Bickle, John, ed. *The Oxford Handbook of Philosophy and Neuroscience*. New York: Oxford University Press, 2013.

Bird, Amy, and Elaine Reese. "Emotional Reminiscing and the Development of an Autobiographical Self." *Developmental Psychology* 42, no. 4 (July 2006): 613-26.

Blank, Robert H. *Intervention in the Brain: Politics, Policy and Ethics*. Cambridge, MA: MIT Press, 2013.

Blasi, Augusto. "Emotions and Moral Motivation." *Journal for the Theory of Social Behaviour* 29, no. 1 (March 1999): 1-19.

Bloom, Daniel. "The Self-Awareness of Reason in Plato." *Journal of Cognition and Neuroethics* 2, no. 1 (2014): 95-103.

Bluck, Susan, Jacqueline M. Baron, Sarah A. Ainsworth, Amanda N. Gesselman, and Kim L. Gold. "Eliciting Empathy for Adults in Chronic Pain through Autobiographical Memory Sharing." *Applied Cognitive Psychology* 27, no. 1 (January/February 2013): 81-90.

Bluck, Susan, and Karen Z. H. Li. "Predicting Memory Completeness and Accuracy: Emotion and Exposure in Repeated Autobiographical Recall." *Applied Cognitive Psychology* 15, no. 2 (March/April 2001): 145-58.

Boals, Adriel, and Darnell Schuettler. "A Double-Edged Sword: Event Centrality, PTSD and Posttraumatic Growth." *Applied Cognitive Psychology* 25, no. 5 (September/October 2011): 817-22.

Bolla, K. I., D. A. Eldreth, E. D. London, K. A. Kiehl, M. Mouratidis, C. Contoreggi, J. A. Matochik, V. Kurian, J. L. Cadet, A. S. Kimes, F. R. Funderburk, and M. Ernst. "Orbitofrontal Cortex Dysfunction in Abstinent Cocaine Abusers Performing A Decision-Making Task." *NeuroImage* 19, no. 3 (July 2003): 1085-94.

Boritz, Tali, Lynne Angus, Georges Monette and Laurie Hollis-Walker. "An Empirical Analysis of Autobiographical Memory Specificity Subtypes in Brief Emotion-Focused and Client-Centered Treatments of Depression." *Psychotherapy Research* 18, no. 5 (September 2008): 584-93.

Boritz, Tali Zweig, Lynne Angus, Georges Monette, Laurie Hollis-Walker, and Serine Warwar. "Narrative and Emotion Integration in

Psychotherapy: Investigating the Relationship Between Autobiographical Memory Specificity and Expressed Emotional Arousal in Brief Emotion-Focused and Client-Centered Treatments of Depression." *Psychotherapy Research* 21, no. 1 (January 2011): 16-26.

Bower, Gordon R. "A Brief History of Memory Research." In *The Oxford Handbook of Memory*, edited by Endel Tulving and Fergus I. M. Craik, 3-32. New York: Oxford University Press, 2000.

Buchanan, Allen. *Beyond Humanity? The Ethics of Biomedical Enhancement.* New York: Oxford University Press, 2011.

Brody, Eugene B., and Judith F. Tormey. "Clinical Psychoanalytic Knowledge—An Epistemological Perspective." *Perspectives in Biology and Medicine* 24 (1980): 143-59.

Brody, Howard. *Stories of Sickness.* 2nd ed. New York: Oxford University Press, 2003.

Bruner, Jerome. *Actual Minds, Possible Worlds.* Cambridge, MA: Harvard University Press, 1986.

———. "Life as Narrative." *Social Research* 54, no. 1 (2004): 11-32.

Brunet, Alain, Scott P. Orr, Jacques Tremblay, Kate Robertson, Karim Nader, and Roger K. Pitman. "Effect of Post-Retrieval Propranolol on Psychophysiologic Responding During Subsequent Script-Driven Traumatic Imagery in Post-Traumatic Stress Disorder." *Journal of Psychiatric Research* 42, no. 6 (May 2008): 503-06.

Bryant, R. A. "Early Interventions Following Psychological Trauma." *CNS Spectrum* 7, no. 9 (2002): 650-54.

Bryant, Richard A. "Psychological Interventions for Trauma Exposure and PTSD." In *Post-Traumatic Stress Disorder*, edited by Dan J. Stein, Matthew J. Friedman, and Carlos Blanco, 171-202. Hoboken, NJ: Wiley-Blackwell, 2011.

Cahill, Larry, Ralf Babinsky, Hans J. Markowitsch, and James L. McGaugh. "The Amygdala and Emotional Memory." *Nature* 377 (September 1995): 295-96.

Cahill, Larry, Bruce Prins, Michael Weber, and James L. McGaugh. "$\beta$-Adrenergic Activation and Memory for Emotional Events." *Nature* 371 (October 1994): 702-04.

Calvo, Manuel G., and David Beltrán. "Recognition Advantage of Happy Faces: Tracing the Neurocognitive Processes." *Neuropsychologia* 51, no. 11 (September 2013): 2051-61.

Camille, Nathalie, Giorgio Coricelli, Jerome Sallet, Pascale Pradat-Diehl, Jean-René Duhamel, and Angela Sirigu. "The Involvement of the Orbitofrontal Cortex in the Experience of Regret." *Science* 304 (May 2004): 1167-70.

Canli, Turhan, and Zenab Amin. "Neuroimaging of Emotion and Personality: Ethical Considerations." In *Neuroethics: An Introduction with Readings*, edited by Martha J. Farah, 147-54. Cambridge, MA: The MIT Press, 2010.

Caplan, Arthur. "Denying Autonomy in Order to Create It: The Paradox of Forcing Treatment on Addicts." *Addiction* 103, no. 12 (December 2008): 1919-21.

Capps, Lisa. "Narrative Authenticity." *Journal of Narrative and Life History* 7, nos. 1-4 (1997): 83-89.

Carruthers, Mary J. *The Book of Memory: A Study of Memory in Medieval Culture.* Cambridge, UK: Cambridge University Press, 1990.

Carson, Ronald A. "Medical Ethics as Reflexive Practice." In *Philosophy of Medicine and Bioethics: A Twenty-Year Retrospective and Critical Analysis,* edited by Ronald A. Carson and Chester R. Burns, 181-92. Dordrecht: Kluwer Academic Publishers, 1997.

Carter, Sid, and Marcia Smith Pasqualini. "Stronger Autonomic Response Accompanies Better Learning: A Test of Damasio's Somatic Marker Hypothesis." *Cognition and Emotion* 18, no. 7 (November 2004): 901-911.

Cavell, Maria. "The Self: Growth, Integrity, and Coming Apart." In *The Oxford Handbook of The Self,* edited by Shaun Gallagher, 592-605. New York: Oxford University Press, 2013.

Cialdini, Robert B., Stephanie L. Brown, Brian P. Lewis, Carol Luce, and Steven L. Neuberg. "Reinterpreting the Empathy-Altruism Relationship: When One Into One Equals Oneness." *Journal of Personality and Social Psychology* 73, no. 3 (1997): 481-94.

Chandler, Jennifer A., Alexandra Mogyoros, Tristana Martin Rubio, and Eric Racine. "Another Look at the Legal and Ethical Consequences of Pharmacological Memory Dampening: The Case of Sexual Assault." *Journal of Law, Medicine & Ethics* 41, no. 4 (Winter 2013): 859-71.

Charland, Louis C. "Cynthia's Dilemma: Consenting to Heroin Prescription." *American Journal of Bioethics* 2, no. 2 (Spring 2002): 37-47.

Charon, Rita. "Narrative Contributions to Medical Ethics: Recognition, Formulation, Interpretation, and Validation in the Practice of the Ethicist." In *A Matter of Principles? Ferment in U.S. Bioethics,* edited by Edwin R. DuBose, Ron Hamel, and Laurence J. O'Connell, 260-83. Valley Forge, PA: Trinity Press International, 1994.

Chater, Nick, and Mike Oaksford. "Normative Systems: Logic, Probability, and Rational Choice." In *The Oxford Handbook of Thinking and Reasoning,* edited by Keith J. Holyoak and Robert G. Morrison, 11-21. New York: Oxford University Press, 2012.

Chatterjee, Anjan, Farah, Martha J., eds. *Neuroethics in Practice.* New York: Oxford University Press, 2013.

Chiao, Joan Y., Tokiko Harada, Hidetsugu Komeda, Zhang Li, Yoko Mano, Daisuke Saito, Todd B. Parrish, Norihiro Sadato, and Tetsuya Iidaka. "Dynamic Cultural Influences on Neural Representations of the Self." *Journal of Cognitive Neuroscience* 22, no. 1 (January 2010): 1-11.

Childress, James F. " Narrative(s) Versus Norm(s): A Misplaced Debate in Bioethics." In *Stories and Their Limits: Narrative Approaches to Bioethics,* edited by Hilde Lindemann Nelson, 252-71. New York: Routledge, 1997.

Clausen, Jens, Levy, Neil. *Handbook of Neuroethics.* New York: Springer, 2014.

Clifasefi, Seema L., Daniel M. Bernstein, Antonia Mantonakis, and Elizabeth F. Loftus. "'Queasy Does It': False Alcohol Beliefs and Memories May Lead to Diminished Alcohol Preferences." *Acta Psychologica* 143 (2013): 14-19.

Clore, Gerald L. "Psychology and the Rationality of Emotion." *Modern Theology* 27, no. 2 (April 2011): 325-38.

Clore, Gerald L., and Justin Storbeck. "Affect as Information about Liking, Efficacy, and Importance." In *Hearts and Minds: Affective Influences on Social Cognition and Behavior,* edited by Joseph Forgas, 123-42. New York: Psychology Press, 2006.

Conrad, Peter, and Joseph W. Schneider. *Deviance and Medicalization: From Badness to Sickness.* St. Louis: The C. V. Mosby Company, 1980.

Conway, Martin A., and Christopher W. Pleydell-Pearce. "The Construction of Autobiographical Memories in the Self-Memory System." *Psychological Review* 107, no. 2 (2000): 261-88.

Coricelli, Giorgio, Hugo D. Critchley, Mateus Joffily, John P. O'Doherty, Angela Sirigu, and Raymond J. Dolan. "Regret and Its Avoidance: A Neuroimaging Study of Choice Behavior." *Nature Neuroscience* 8 (2005): 1255-62.

Crisp, Thomas M., et al., eds. *Neuroscience and the Soul. The Human Person, in Philosophy, Science and Theology.* Eerdmans, 2016.

Curran, Charles. *A New Look at Christian Morality.* Notre Dame, IN: Fides Publishers, 1970.

D'Argembeau, Arnaud, Helena Cassol, Christophe Phillips, Evelyne Balteau, Eric Salmon, and Martial Van der Linden. "Brain Creating Stories of Selves: The Neural Basis of Autobiographical Reasoning." *Social Cognitive and Affective Neuroscience* 9 (2014): 646-52.

D'Argembeau, Arnaud, and Martial Van der Linden. "Phenomenal Characteristics Associated with Projecting Oneself Back into the Past and Forward into the Future: Influence of Valence and Temporal Distance." *Consciousness and Cognition* 13, no. 4 (2004): 844-58.

Damasio, Antonio. *Descartes' Error: Emotion, Reason, and the Human Brain.* New York: Grosset/Putnam Group, 1994.

———. *The Feeling of What Happens: Body and Emotion in the Making of Consciousness.* New York: Harcourt, Inc., 1999.

———. "Neuroscience and Ethics—Intersections." *American Journal of Bioethics—AJOB Neuroscience* 7, no. 1 (2007): 3-7.

Daniels, Norman. "Wide Reflective Equilibrium and Theory Acceptance in Ethics." *Journal of Philosophy* 79 (1979): 256-82.

Davidov, Maayan, and Joan E. Grusec. "Untangling the Links of Parental Responsiveness to Distress and Warmth to Child Outcomes." *Child Development* 77, no. 1 (January/February 2006): 44-58.

Davis, Mark H. *Empathy: A Social Psychological Approach*. Madison, WI: Westview Press, 1996.

De Brigard, Felipe and Kelly S. Giovanello. "Influence of Outcome Valence in the Subjective Experience of Episodic Past, Future, and Counterfactual Thinking." *Consciousness and Cognition* 21 (2012): 1085-96.

de Waal, Frans B. M., and Pier Francesco Ferrari. "Towards a Bottom-Up Perspective on Animal and Human Cognition." *Trends in Cognitive Sciences* 14, no. 5 (May 2010): 201-07.

Decety, Jean, and Stephanie Cacioppo. "The Speed of Morality: A High-Density Electrical Neuroimaging Study." *Journal of Neurophysiology* 108 (September 2012): 3068-72.

Delistraty, Cody C. "The Ethics of Erasing Bad Memories." *The Atlantic*. 15 May 2014.

Deng, Zhi-De, Shawn M. McClintock, Nicodemus E. Oey, Bruce Luber, and Sarah H. Lisanby. "Neuromodulation for Mood and Memory: From the Engineering Bench to the Patient Bedside." *Current Opinion in Neurobiology* 30 (2015): 38-43.

Denke, Claudia, Michael Rotte, Hans-Jochen Heinze, and Michael Schaefer. "Belief in a Just World is Associated with Activity in Insula and Somatosensory Cortices as a Response to the Perception of Norm Violations." *Social Neuroscience* 9, no. 5 (2014): 514-21.

Dennett, Daniel. *Elbow Room: The Varieties of Free Will Worth Wanting*. Cambridge, MA: The MIT Press, 1984.

Denny, Bryan T., Hedy Kober, Tor D. Wager, and Kevin N. Ochsner. "A Meta-analysis of Functional Neuroimaging Studies of Self- and Other Judgments Reveals a Spatial Gradient for Mentalizing in Medial Prefrontal Cortex." *Journal of Cognitive Neuroscience* 24, no. 8 (2012): 1742-52.

Descartes, René. *The Philosophical Work of Descartes*. Vols. 1-2. Translated by Elizabeth S. Haldane and G. R. T. Ross. Cambridge, UK: Cambridge University Press, 1911.

Dhar, Benulal. "The Phenomenology of Value-Experience: Some Reflections on Scheler and Hartmann." *Indian Philosophical Quarterly* 26, no. 2 (April 1999): 183-97.

Dilthey, Wilhelm. "Über Eine Beschreibende und Zergliedernde Psychologie." Vol. 5 of *Die geistige Welt: Einleitung in die Philosophie des Lebens*, edited by Wilhelm Dilthey, 139-240. Stuttgart, Göttingen: Teubner, Vandenhoeck & Ruprecht, 1895.

Dodson, Geran F. *Free Will, Neuroethics, Psychology and Theology*. Wilmington, Delaware: Vernon Press, 2017.

Donovan, Elsie. "Propranolol Use in the Prevention and Treatment of Posttraumatic Stress Disorder in Military Veterans." *Perspectives in Biology and Medicine* 53, no. 1 (Winter 2010): 61-74.

Dolcos, Florin, Kevin S. LaBar, and Roberto Cabeza. "Interaction Between the Amygdala and the Medial Temporal Lobe Memory System Predicts Better Memory for Emotional Events." *Neuron* 42 (June 2004): 855-63.

———. "Remembering One Year Later: Role of the Amygdala and the Medial Temporal Lobe Memory System in Retrieving Emotional Memories." *Proceedings of the National Academy of Sciences of the United States of America* 102, no. 7 (February 2005): 2626-31.

Dostoyevsky, Fyodor. *Crime and Punishment.* Translated by David McDuff. New York: Penguin Books, 2002.

Duc, Corinne, Martin Hanselmann, Peter Boesiger, and Carmen Tanner. "Sacred Values: Trade-Off Type Matters." *Journal of Neuroscience, Psychology, and Economics* 6, no. 4 (December 2013): 252-63.

Duncan, Jhodie R., and Andrew J. Lawrence. "Molecular Neuroscience and Genetics." In *Addiction Neuroethics: The Ethics of Addiction Neuroscience Research and Treatment*, edited by Adrian Carter, Wayne Hall, and Judy Illes, 27-54. San Diego: Academic Press, 2012.

Dunlop, William L., and Jessica L. Tracy. "The Autobiography of Addiction: Autobiographical Reasoning and Psychological Judgment in Abstinent Alcoholics." *Memory* 21, no 1 (2013): 64-78.

Drummond, John J. "Moral Objectivity: Husserl's Sentiments of the Understanding." *Husserl Studies* 12, no. 2 (1995): 165-83.

Ebbinghaus, Hermann. Urmanuskript "Über das Gedächtnis." Passau, Germany: Passavia Universitätsverlag, 1983.

Elliott, Carl. "Who Holds the Leash?" *American Journal of Bioethics* 2, no. 2 (Spring 2002): 48.

Elster, Jon. "Emotional Choice and Rational Choice." In *The Oxford Handbook of Philosophy of Emotion*, edited by Peter Goldie, 263-81. New York: Oxford University Press, 2010.

Engelhardt Jr., H. Tristram. *The Foundations of Bioethics.* 2nd ed. New York: Oxford University Press, 1996.

Erikson, Erik H. *Childhood and Society.* New York: W. W. Norton & Company, 1963.

Erk, Susanne, Markus Kiefer, Jo Grothe, Arthur P. Wunderlich, Manfred Spitzer, and Henrik Walter. "Emotional Context Modulates Subsequent Memory Effect." *NeuroImage* 18, no. 2 (February 2003): 439-47.

Erk, Susanne, Sonja Martin, and Henrik Walter. "Emotional Context During Encoding of Neutral Items Modulates Brain Activation Not Only During Encoding But Also During Recognition." *NeuroImage* 26, no. 3 (July 2005): 829-38.

Ersche, Karen D., and Barbara J. Sahakian. "The Neuropsychology of Amphetamine and Opiate Dependence: Implications for Treatment." *Neuropsychology Review* 17, no. 3 (September 2007): 317-36.

Escobedo, Jessica R., and Ralph Adolphs. "Becoming a Better Person: Temporal Remoteness Biases Autobiographical Memories for Moral Events." *Emotion* 10, no. 4 (August 2010): 511-18.

Farah, Martha J., Judy Illes, Robert Cook-Deegan, Howard Gardner, Eric Kandel, Patricia King, Erik Parens, Barbara Sahakian, and

Paul Root Wolpe. "Neurocognitive Enhancement: What Can We Do and What Should We Do?" In *Neuroethics: An Introduction with Readings*, edited by Martha J. Farah, 30-41. Cambridge, MA: The MIT Press, 2010.

Fingarette, Herbert. *Heavy Drinking: The Myth of Alcoholism as a Disease*. Berkeley, CA: University of California Press, 1988.

Fisher, Walter R. "Narration as a Human Communication Paradigm: The Case of Public Moral Argument." *Communication Monographs* 51 (1984): 1-22.

———. "Technical Logic, Rhetorical Logic, and Narrative Rationality." *Argumentation* 1 (1987): 3-21.

Fivush, Robyn, Tilmann Habermas, Theodore E. A. Waters, and Widaad Zaman. "The Making of Autobiographical Memory: Intersections of Culture, Narratives and Identity." *International Journal of Psychology* 46, no. 5 (2011): 321-45.

Fivush, Robyn, Catherine A. Haden, and Elaine Reese. "Elaborating on Elaborations: Role of Maternal Reminiscing Style in Cognitive and Socioemotional Development." *Child Development* 77, no. 6 (November/December 2006): 1568-88.

Fivush, Robyn, and Jessica McDermott Sales. "Coping, Attachment, and Mother-Child Narratives of Stressful Events." *Merrill-Palmer Quarterly* 52, no. 1 (January 2006): 125-50.

Flavell, John H., and Henry M. Wellman. "Metamemory." In *Perspectives on the Development of Memory and Cognition*, edited by Robert V. Kail and John W. Hagen, 3-34. Hillsdale, NJ: Lawrence Erlbaum Associates, Inc., Publishers, 1977.

Ford, Jaclyn Hennessey, Donna Rose Addis, and Kelly S. Giovanello. "Differential Effects of Arousal in Positive and Negative Autobiographical Memories." *Memory* 20, no. 7 (2012): 771-78l.

Frankfurt, Harry. "Freedom of the Will and the Concept of a Person." *Journal of Philosophy* 68 (1971): 5-20.

Fraser, Louisa M., Ronan E. O'Carroll, and Klaus P. Ebmeier. "The Effect of Electroconvulsive Therapy on Autobiographical Memory: A Systematic Review." *Journal of ECT* 24, no. 1 (March 2008): 10-17.

Freud, Sigmund, and Joseph Breuer. *Studies in Hysteria*. Translated by Nicola Luckhurst. New York: Penguin Books, 2004.

Fuchs, Thomas. "Fragmented Selves: Temporarily and Identity in Borderline Personality Disorder." *Psychopathology* 40 (2007): 379-87.

Gaesser, Brendan. "Constructing Memory, Imagination, and Empathy: A Cognitive Neuroscience Perspective." *Frontiers in Psychology* 3 (January 2013): 1-6.

Gaesser, Brendan, Daniel C. Sacchetti, Donna Rose Addis, and Daniel L. Schacter. "Characterizing Age-Related Changes in Remembering the Past and Imagining the Future." *Psychology and Aging* 26, no. 1 (March 2011): 80-84.

Gahr, Maximilian, Carlos Schönfeldt-Lecuona, Manfred Spitzer, and Heiko Graf. "Electroconvulsive Therapy and Posttraumatic Stress Disorder: First Experience With Conversation-Based Reactivation of Traumatic Memory Contents and Subsequent ECT-mediated Impairment of Reconsolidation." *Journal of Neuropsychiatry and Clinical Neurosciences* 26, no. 3 (Summer 2014): E38-39.

Gardiner, John M., and Alan Richardson-Klavehn. "Remembering and Knowing." In *The Oxford Handbook of Memory*, edited by Endel Tulving and Fergus I. M. Craik, 229-44. New York: Oxford University Press, 2000.

Gergen, Kenneth J. *The Saturated Self.* New York: Basic Books, 1991.

———. "Narrative, Moral Identity, and Historical Consciousness: A Historical Constructionist Account." In *Narration, Identity, and Historical Consciousness*, edited by Jürgen Straub, 99-119. New York: Berghahn Books, 2005.

Gerlach, Kathy D., David W. Dornblaser, and Daniel L. Schacter. "Adaptive Constructive Processes and Memory Accuracy: Consequences of Counterfactual Simulations in Young and Older Adults." *Memory* 22, no. 1 (January 2014): 1-26.

Gilhooly, Kenneth J., and P. Murphy. "Differentiating Insight from Noninsight Problems." *Thinking and Reasoning* 11, no. 3 (2005): 279-302.

Gillett, Grant R. "The Subjective Brain, Identity, and Neuroethics." *American Journal of Bioethics* 9, no. 9 (2009): 5-13.

Gino, Francesca, and Sreedhari D. Desai. "Memory Lane and Morality: "How Childhood Memories Promote Prosocial Behavior." *Journal of Personality and Social Psychology* 102, no. 4 (April 2012): 743-58.

Giordano, James J., Gordijn, Bert., eds. *Scientific and Philosophical Perspectives in Neuroethics.* New York: Cambridge University Press, 2010.

Glannon, Walter. "Psychopharmacology and Memory." *Journal of Medical Ethics* 32, no. 2 (2006): 74-78.

———. *Brain, Body, and Mind: Neuroethics with a Human Face.* New York: Oxford University Press, 2011.

Gligorov, Nasda. *Neuroethics and the Scientific Revision of Common Sense.* New York: Springer, 2016.

Godsil, Rachel D., and Brianna Goodale. *Telling Our Own Story: The Role of Narrative in Racial Healing.* East Battle Creek, MI: W. K. Kellogg Foundation, June 2013.

Goldstein, Rita Z., and Nora D. Volkow. "Drug Addiction and Its Underlying Neurobiological Basis: Neuroimaging Evidence for the Involvement of the Frontal Cortex." *American Journal of Psychiatry* 159, no. 10 (October 2002): 1642-52.

Gray, Kevin M., Noreen L. Watson, Matthew J. Carpenter, and Steven D. LaRowe. "N-Acetylcysteine (NAC) in Young Marijuana Users: An Open-Label Pilot Study." *American Journal of Addiction* 19, no. 2 (March 2010): 187-89.

Greenberg, Leslie S., and Lynne E. Angus. "The Contributions of Emotional Processes to Narrative Change in Psychotherapy: A Dialectical Constructivist Approach." In *The Handbook of Psychotherapy*, edited by Lynne E. Angus and John McLeod, 330-50. Thousand Oaks, CA: SAGE Publications, 2004.

Greene, Joshua D., Sylvia A. Morelli, Kelly Lowenberg, Leigh E. Nystrom, and Jonathan D. Cohen. "Cognitive Load Selectively Interferes with Utilitarian Moral Judgment." *Cognition* 107, no. 3 (June 2008): 1144-54.

Greene, Joshua D., Leigh E. Nystrom, Andrew D. Engell, John M. Darley, and Jonathan D. Cohen. "The Neural Bases of Cognitive Conflict and Control in Moral Judgment." *Neuron* 44 (October 2004): 389-400.

Greenhalgh, Trisha, and Brian Hurwitz, eds. *Narrative Based Medicine: Dialogue and Discourse in Clinical Practice*. London: BMJ Books, 1998.

Greenspan, Patricia. "Practical Reasoning and Emotion." In *The Oxford Handbook of Rationality*, edited by Alfred R. Mele and Piers Rawling, 206-21. New York: Oxford University Press, 2004.

———. "Learning Emotions and Ethics." In *The Oxford Handbook of Philosophy of Emotion*, edited by Peter Goldie, 539-59. New York: Oxford University Press, 2010.

Grillon, Christian, Jeremy Cordova, Charles Andrew Morgan III, Dennis S. Charney, and Michael Davis. "Effects of the Beta-blocker Propranolol on Cued and Contextual Fear Conditioning in Humans." *Psychopharmacology* 175, no. 3 (September 2004): 342-52.

Grühn, Daniel, Kristine Rebucal, Manfred Diehl, Mark Lumley, and Gisela Labouvie-Vief. "Empathy Across the Adult Lifespan: Longitudinal and Experience-Sampling Findings." *Emotion* 8, no. 6 (December 2008): 753-65.

Gui, Dan-Yang, Tian Gan, and Chao Liu. "Neural Evidence for Moral Intuition and the Temporal Dynamics of Interactions Between Emotional Processes and Moral Cognition." *Social Neuroscience* [Epub ahead of print] (August 2015): 1-15.

Habermas, Tilmann. "Autobiographical Reasoning: Arguing and Narrating from a Biographical Perspective." *New Directions for Child and Adolescent Development* 131 (Spring 2011): 1-17.

Habermas, Tilmann, and Susan Bluck. "Getting a Life: The Emergence of the Life Story in Adolescence." *Psychological Bulletin* 126, no. 5 (September 2000): 748-69.

Haidt, Jonathan. "The New Synthesis in Moral Psychology." *Science* 316 (May 2007): 998-1002.

Hall, Wayne, and Adrian Carter. "Debunking Alarmist Objections to the Pharmacological Prevention of PTSD." *American Journal of Bioethics* 7, no. 9 (2007): 23-24.

Han, Hyemin, Gary H. Glover, and Changwoo Jeong. "Cultural Influences on the Neural Correlate of Moral Decision Making Processes." *Behavioural Brain Research* 259 (February 2014): 215-28.

Han, Sanghoon, Akira R. O'Connor, Andrea N. Eslick, and Ian G. Dobbins. "The Role of Left Ventrolateral Prefrontal Cortex During Episodic Decisions: Semantic Elaboration or Resolution of Episodic Interference." *Journal of Cognitive Neuroscience* 24, no. 1 (January 2012): 223-34.

Hariz, Marwan, Patric Blomstedt, and Ludvic Zrinzo. "Future of Brain Stimulation: New Targets, New Indications, New Technology." *Movement Disorders* 28, no. 13 (2013): 1784-92.

Harris, Richard Jackson, Steven J. Hoekstra, Christina L. Scott, Fred W. Sanborn, Laura A. Dodds, and Jason Dean Brandenburg. "Autobiographical Memories for Seeing Romantic Movies on Date: Romance Is Not Just for Women." *Media Psychology* 6, no. 3 (August 2004): 257-84.

Hart, J. T. "Memory and the Feeling-of-Knowing Experience." *Journal of Educational Psychology* 56, no. 4 (August 1965): 208-16.

Hartmann, Nicolai. *Ethics.* Translated by Stanton Colt. London: George Allen & Unwin Ltd., 1958.

Hauser, Marc D. "The Liver and the Moral Organ." *Social Cognitive and Affective Neuroscience* 1, no. 3 (December 2006): 214-20.

Hawkins, Anne Hunsaker. *Reconstructing Illness: Studies in Pathography.* 2nd ed. West Lafayette, IN: Purdue University Press, 1999.

Hays, Sean, et al., eds. *Nanotechnology, the Brain, and the Future.* New York: Springer, 2013.

Heidegger, Martin. *Existence and Being.* Chicago: Henry Regnery Company, 1949.

———. *Being and Time.* Translated by John Macquarrie and Edward Robinson. Cambridge, MA: Blackwell Publishers Ltd., 1962.

Hein, Grit, and Robert T. Knight. "Superior Temporal Sulcus—It's My Area: Or Is It?" *Journal of Cognitive Neuroscience* 20, no. 12 (December 2008): 2125-36.

Helm, Bennett W. "Emotions and Motivation: Reconsidering Neo-Jamesian Accounts." In *The Oxford Handbook of Philosophy of Emotion,* edited by Peter Goldie, 303-23. New York: Oxford University Press, 2010.

Hemmer, Pernille, and Kimele Persaud. "Interaction Between Categorical Knowledge and Episodic Memory Across Domains." *Frontiers in Psychology* 5 (June 2014): 1-5.

Henderson, Michael B., Alan I. Green, Perry S. Bradford, David T. Chau, David W. Roberts, and James C. Leiter. "Deep Brain Stimulation of the Nucleus Accumbens Reduces Alcohol Intake in Alcohol-Preferring Rats." *Neurosurgical Focus* 29, no. 2 (August 2010): 1-7.

Henry, Michael, Jennifer R. Fishman, and Stuart J. Younger. "Propranolol and the Prevention of Post-Traumatic Stress Disorder: Is

It Wrong to Erase the 'Sting' of Bad Memories?" *American Journal of Bioethics* 7, no. 9 (2007): 12-20.

Heyman, Gene M. *Addiction: A Disorder of Choice*. Cambridge, MA: Harvard University Press, 2009.

Higgins, Stephen T., Alan J. Budney, Warren K. Bickel, Florian E. Foerg, Robert Donham, and Gary J. Badger. "Incentives Improve Outcome in Outpatient Behavioral Treatment of Cocaine Dependence." *Archives of General Psychiatry* 51, no. 7 (1994): 568-76.

Hildt, Elisabeth, Franke, Andreas G. Franke, eds. *Cognitive Enhancement: An Interdisciplinary Perspective*. New York: Springer, 2013.

Hillman, James. "The Fiction of Case History: A Round." In *Religion as Story*, edited by J. B. Wiggins, 123-73. New York: Harper and Row, 1975.

Hintzman, Douglas L. "Memory Judgments." In *The Oxford Handbook of Memory*, edited by Endel Tulving and Fergus I. M. Craik, 165-77. New York: Oxford University Press, 2000.

Hoffmann, Janina A., Bettina von Helversen, and Jörg Rieskamp. "Pillars of Judgment: How Memory Abilities Affect Performance in Rule-Based and Exemplar-Based Judgments." *Journal of Experimental Psychology: General* 143, no. 6 (2014): 2242-61.

Holland, Alisha C., and Elizabeth A. Kensinger. "Emotion and Autobiographical Memory." *Physics of Life Reviews* 7, no. 1 (March 2010): 88-131.

Horberg, Elizabeth J., Christopher Oveis, and Dacher Keltner. "Emotions as Moral Amplifiers: An Appraisal Tendency Approach to the Influences of Distinct Emotions upon Moral Judgment." *Emotion Review* 3, no. 3 (July 2011): 237-44.

Hu, Jiangyuan, Larissa Ferguson, Kerry Adler, Carole A. Farah, Margaret H. Hastings, Wayne S. Sossin, and Samuel Schacher. "Selective Erasure of Distinct Forms of Long-Term Synaptic Plasticity Underlying Different Forms of Memory in the Same Postsynaptic Neuron." *Current Biology* 12 (July 2017): 1-12.

Huebner, Bryce, Susan Dwyer, and Marc Hauser. "The Role of Emotion in Moral Psychology." *Trends in Cognitive Sciences* 13, no. 1 (January 2009): 1-6.

Hume, David. *A Treatise of Human Nature*. Edited by P. H. Nidditch. Oxford: Clarendon Press, 1978.

———. "Of the Standard of Taste." In *Essays: Moral, Political, and Literary*, edited by Eugene F. Miller, 226-49. Indianapolis: Liberty Classics, 1985.

Hunter, Kathryn Montgomery. "'There Was This One Guy . . .': The Use of Anecdotes in Medicine." *Perspectives in Biology and Medicine* 29 (1986): 619-30.

———. *Doctors' Stories: The Narrative Structure of Medical Knowledge*. Princeton, NJ: Princeton University Press, 1991.

Hurlemann, R., H. Walter, A. K. Rehme, J. Kukolja, S. C. Santoro, C. Schmidt, K. Schnell, F. Musshoff, C. Keysers, W. Maier, K. M.

Kendrick, and O. A. Onur. "Human Amygdala Reactivity is Diminished by the β-noradrenergic Antagonist Propranolol." *Psychological Medicine* 27 (2010): 1-10.

Husserl, Edmund. *The Crisis of the European Sciences and Transcendental Phenomenology.* Exeter, UK: Imprint Academic, 1970.

Hyman Ira E., and F. James Billings Jr. "Individual Differences and the Creation of False Childhood Memories." *Memory* 6, no. 1 (1998): 1-20.

Illes, Judy, Sahakian, Barbara J., eds. *The Oxford Handbook of Neuroethics.* New York: Oxford University Press, 2013.

Illes, Judy, ed. *Neuroethics: Anticipating the Future.* New York: Oxford University Press, 2017.

Immordino-Yang, Mary Helen, Joanna A. Christodoulou, and Vanessa Singh. "Rest Is Not Idleness: Implications of the Brain's Default Mode for Human Development and Education." *Perspectives on Psychological Science* 7, no. 4 (2012): 352-64.

Inbar, Yoel, David A. Pizarro, Joshua Knobe, and Paul Bloom. "Disgust Sensitivity Predicts Intuitive Disapproval of Gays." *Emotion* 9, no. 3 (2009): 435-39.

Jaggar, Alison M. "Love and Knowledge: Emotion in Feminist Epistemology." *Inquiry* 32, no. 2 (1989): 151-76.

James, William. *The Principles of Psychology.* Vols. 1-2. Cambridge, MA: Harvard University Press, 1890.

Johnson, L. Syd M., Rommelfanger, Karen S., eds. *The Routledge Handbook of Neuroethics.* Routledge, 2017.

Jones, Gregory V., and Maryanne Martin. "Primacy of Memory Linkage in Choice Among Valued Objects." *Memory & Cognition* 34, no. 8 (2006): 1587-97.

Jonsen, Albert R., and Stephen Toulmin. *The Abuse of Casuistry: A History of Moral Reasoning.* Los Angeles: University of California Press, 1988.

Joy, Morny. "Feminism and the Self." *Theory and Psychology* 3 (1993): 275-302.

Kalberg, Stephen. "Max Weber's Types of Rationality: Cornerstones for the Analysis of Rationalization Processes in History." *The American Journal of Sociology* 85, no. 5 (March 1980): 1145-79.

Kant, Immanuel. *The Metaphysics of Morals.* Translated by M. J. Gregor. Cambridge: Cambridge University Press, 1996.

\_\_\_\_. *Fundamental Principles of the Metaphysics of Morals.* Translated by Thomas Kingsmill Abbott. Mineola, NY: Dover Publications, Inc., 2005.

Kaplan, Jonas T., Sarah I. Gimbel, Morteza Dehghani, Mary Helen Immordino-Yang, Kenji Sagae, Jennifer D. Wong, Christine M. Tipper, Hanna Damasio, Andrew S. Gordon, and Antonio Damasio. "Processing Narratives Concerning Protected Values: A Cross-Cultural Investigation of Neural Correlates." *Cerebral Cortex* 26, no. 1 (January 2016): 1-11.

Keltner, Dacher, Phoebe C. Ellsworth, and Kari Edwards. "Beyond Simple Pessimism: Effects of Sadness and Anger on Social Perception." *Journal of Personality and Social Psychology* 64, no. 5 (1993): 740-52.

Kennett, Jeanette. *Agency and Responsibility: A Common Sense Moral Psychology.* Oxford: Oxford University Press, 2001.

Kensinger, Elizabeth A. "Negative Emotion Enhances Memory Accuracy: Behavioral and Neuroimaging Evidence." *Current Directions in Psychological Science* 16, no. 4 (2007): 213-18.

Kensinger, Elizabeth A., and Suzanne Corkin. "Two Routes to Emotional Memory: Distinct Neural Processes for Valence and Arousal." *Proceedings of the National Academy of Sciences of the United States of America* 101, no. 9 (March 2004): 3310-15.

Kensinger, Elizabeth A., and Daniel L. Schacter. "When the Red Sox Shocked the Yankees: Comparing Negative and Positive Memories." *Psychonomic Bulletin & Review* 13, no. 5 (2006): 757-63.

Kiefer, Markus, Stefanie Schuch, Wolfram Schenck, and Klaus Fiedler. "Emotion and Memory: Event-Related Potential Indices Predictive for Subsequent Successful Memory Depend on the Emotional Mood State." *Advances in Cognitive Psychology* 3, no. 3 (2007): 363-73.

Kilpatrick, Lisa, and Larry Cahill. "Amygdala Modulation of Parahippocampal and Frontal Regions During Emotionally Influenced Memory Storage." *NeuroImage* 20, no. 4 (December 2003): 2091-99.

Kindt, Merel, Marieke Soeter, and Bram Vervliet. "Beyond Extinction: Erasing Human Fear Responses and Preventing the Return of Fear." *Nature Neuroscience* 12 (2009): 266-58.

King, Laura A. "The Hard Road to the Good Life: The Happy, Mature Person." *Journal of Humanistic Psychology* 41 (2001): 51-72.

Klein, Stanley B. "The Complex Act of Projecting Oneself into the Future." *Wiley Interdisciplinary Reviews: Cognitive Science* 4, no. 1 (January/February 2013): 63-79.

Klein, Stanley B., Judith Loftus, and John F. Kihlstrom. "Memory and Temporal Experience: The Effects of Episodic Memory Loss On An Amnesic Patient's Ability to Remember the Past and Imagine the Future." *Social Cognition* 20, no. 5 (2002): 353-79.

Kochanska, Grazyna. "Mutually Responsive Orientation Between Mothers and Their Young Children: A Context for the Early Development of Conscience." *Current Directions in Psychological Science* 11 (2002): 191-95.

Koenigs, Michael, Liane Young, Ralph Adolphs, Daniel Tranel, Fiery Cushman, Marc Hauser, and Antonio Damasio. "Damage to the Prefrontal Cortex Increases Utilitarian Moral Judgements [sic]." *Nature* 446, no. 7138 (April 2007): 908-11.

Kohut, Heinz. *The Restoration of the Self.* Chicago: The University of Chicago Press, 1977.

Kolber, Adam J. "Therapeutic Forgetting: The Legal and Ethical Implications of Memory Dampening." *Vanderbilt Law Review* 59, no. 5 (2006): 1561-1626.

Korsgaard, Christine M. *The Sources of Normativity.* Cambridge, UK: Cambridge University Press, 1996.

———. "Autonomy and the Second Person Within: A Commentary on Stephen Darwall's The Second-Person Standpoint." *Ethics* 118, no. 1 (2007): 8-23.

Kramer, Peter D. *Listening to Prozac: A Psychiatrist Explores Antidepressant Drugs and the Remaking of the Self.* New York: Viking Press, 1993.

Kraemer, Felicitas. "Me, Myself and My Brain Implant: Deep Brain Stimulation Raises Questions of Personal Authenticity and Alienation." *Neuroethics* 6 (2013): 483-97.

Kroes, Marijn C. W., Indira Tendolkar, Guido A. van Wingen, Jeroen A. van Waarde, Bryan A. Strange, and Guillén Fernández. "An Electroconvulsive Therapy Procedure Impairs Reconsolidation of Episodic Memories in Humans." *Nature Neuroscience* 17, no. 2 (February 2014): 204-08.

Kuhn, Jens, Theo O. J. Gründler, Robert Bauer, Wolfgang Huff, Adrian G. Fischer, Doris Lenartz, Mohammad Maarouf, Christian Bührle, Joachim Klosterkötter, Markus Ullsperger, and Volker Sturm. "Successful Deep Brain Stimulation of the Nucleus Accumbens in Severe Alcohol Dependence is Associated with Changed Performance Monitoring." *Addiction Biology* 16, no. 4 (October 2011): 620-23.

Kuhn, Jens, Doris Lenartz, Wolfgang Huff, SunHee Lee, Athanasios Koulousakis, Joachim Klosterkötter, and Volker Sturm. "Remission of Alcohol Dependency Following Deep Brain Stimulation of the Nucleus Accumbens: Valuable Therapeutic Implications?" *Journal of Neurology, Neurosurgery & Psychiatry* 78 (2007): 1152-53.

LaBar, Kevin S., and Roberto Cabeza. "Cognitive Neuroscience of Emotional Memory." *Nature Reviews Neuroscience* 7, No. 1 (January 2006): 54-64.

Laible, Deborah J., Gustavo Carlo, and Scott C. Roesch. "Pathways to Self-Esteem in Late Adolescence: The Role of Parent and Peer Attachment, Empathy, and Social Behaviors." *Journal of Adolescence* 27, no. 6 (December 2004): 703-16.

Lambie, John A. "On the Irrationality of Emotion and the Rationality of Awareness." *Consciousness and Cognition* 17 (2007): 946-71.

Lapsley, Daniel K. "Moral Agency, Identity and Narrative in Moral Development." *Human Development* 53 (2010): 87-97.

Lavazza, Andrea, ed. *Frontiers in Neuroethics.* England: Cambridge Scholars Publishing, 2016.

LeBoeuf, Robyn A., and Eldar Shafir. "Decision Making." In *The Oxford Handbook of Thinking and Reasoning,* edited by Keith J. Holyoak and Robert G. Morrison, 301-21. New York: Oxford University Press, 2012.

LeDoux, Joseph. *The Emotional Brain: The Mysterious Underpinnings of the Emotional Life.* New York: Touchstone, 1998.

Lerner, Jennifer S., and Dacher Keltner. "Fear, Anger, and Risk." *Journal of Personality and Social Psychology* 81, no. 1 (July 2001): 146-59.

Leshner, Alan. "Science-Based Views of Drug Addiction and Its Treatment." *Journal of the American Medical Association* 282 (1999): 1314-16.

Levy, Dino J., and Paul W. Glimcher. "The Root of All Value: A Neural Common Currency for Choice." *Current Opinion in Neurobiology* 22 (2012): 1027-38.

Levy, Neil. "Autonomy, Responsibility and the Oscillation of Preference." In *Addiction Neuroethics: The Ethics of Addiction Neuroscience Research and Treatment,* edited by Adrian Carter, Wayne Hall, and Judy Illes, 139-51. San Diego: Academic Press, 2012.

Liao, S. Matthew, ed. *Moral Brains: The Neuroscience of Morality.* New York: Oxford University Press, 2016.

Little, Margaret O. "Moral Generalities Revisited." In *Moral Particularism,* edited by Brad Hooker and Margaret O. Little, 276-304. New York: Oxford University Press, 2000.

Locke, John. *An Essay Concerning Human Understanding.* Edited by Roger Woolhouse. London: Penguin Books, 1997.

Lombardo, Michael V., Jennifer L. Barnes, Sally J. Wheelwright, and Simon Baron-Cohen. "Self-Referential Cognition and Empathy in Autism." *PLoS ONE* 2, no. 9 (September 2007): e883.

Long, Roderick T. *Reason and Value: Aristotle versus Rand.* Washington, DC: The Atlas Society, 2000.

Lyons, John O. *The Invention of the Self: The Hinge of Consciousness in the Eighteenth Century.* Carbondale, IL: Southern Illinois University Press, 1978.

Lyubomirsky, Sonja, Lorie Sousa, and Rene Dickerhoof. "The Costs and Benefits of Writing, Talking, and Thinking About Life's Triumphs and Defeats." *Journal of Personality and Social Psychology* 90, no. 4 (2006): 692-708.

MacIntyre, Alasdair. *After Virtue: A Study in Moral Theory.* 3rd ed. Notre Dame, IN: Notre Dame University Press, 2007.

Mar, Raymond A. "The Neuropsychology of Narrative: Story Comprehension, Story Production and Their Interrelation." *Neuropsychologia* 42 (2004): 1414-43.

Maratos Elizabeth J., and Michael D. Rugg. "Electrophysiological Correlates of the Retrieval of Emotional and Non-Emotional Context." *Journal of Cognitive Neuroscience* 13, no. 7 (October 2001): 877-91.

Marder, Eve, and Jean-Marc Goaillard. "Variability, Compensation and Homeostasis in Neuron and Network Function." *Nature Reviews Neuroscience* 7 (July 2006): 563-74.

Markowitsch, Hans J. "Neuroanatomy of Memory." In *The Oxford Handbook of Memory,* edited by Endel Tulving and Fergus I. M. Craik, 465-84. New York: Oxford University Press, 2000.

Mars, Rogier B., Franz-Xaver Neubert, MaryAnn P. Noonan, Jerome Sallet, Ivan Toni, and Matthew F. S. Rushworth. "On the Relationship between the 'Default Mode Network' and the 'Social Brain.'" *Frontiers in Human Neuroscience* 6 (June 2012): 1-9.

Martínez-Galindo, Joyce Graciela, and Selene Cansino. "Positive and Negative Emotional Contexts Unevenly Predict Episodic Memory." *Behavioural Brain Research* 291 (2015): 89-102.

Masten, Carrie L., Sylvia A. Morelli, and Naomi I. Eisenberger. "An fMRI Investigation of Empathy for 'Social Pain' and Subsequent Prosocial Behavior." *NeuroImage* 22 (2011): 381-88.

Mathur, Vani A., Tokiko Harada, Trixie Lipke, and Joan Y. Chiao. "Neural Basis of Extraordinary Empathy and Altruistic Motivation." *NeuroImage* 51 (2010): 1468-75.

Mazzoni, Giuliana, Andrew Clark, and Robert A. Nash. "Disowned Recollections: Denying True Experiences Undermines Belief in Occurrence But Not Judgments of Remembering." *Acta Psychologica* 145 (January 2014): 139-46.

Mazzoni, Giuliana, Alan Scoboria, and Lucy Harvey. "Nonbelieved Memories." *Psychological Science* 21 (August 2010): 1334-40.

McAdams, Dan P. "The Role of Narrative in Personality Psychology Today." *Narrative Inquiry* 16, no. 1 (2006): 11-18.

———. "Personal Narratives and the Life Story." In *Handbook of Personality: Theory and Research,* 3rd ed., edited by Oliver P. John, Richard W. Robins, and Lawrence A. Pervin, 242-62. New York: The Guilford Press, 2008.

McAdams, Dan P., Jeffrey Reynolds, Martha Lewis, Allison H. Patten, and Phillip J. Bowman. "When Bad Things Turn Good and Good Things Turn Bad: Sequences of Redemption and Contamination in Life Narrative and their Relation to Psychosocial Adaptation in Midlife Adults and in Students." *Personality and Social Psychology Bulletin* 27, no. 4 (April 2001): 472-83.

McCabe, Allyssa, Carole Peterson, And Dianne M. Connors. "Attachment Security and Narrative Elaboration." *International Journal of Behavioral Development* 30, no. 5 (2006): 8-19.

McCormick, Richard A. "Ambiguity in Moral Choice." In *Doing Evil to Achieve Good: Moral Choice in Conflict Situations,* edited by Richard A. McCormick and Paul Ramsey, 7-53. Lanham, MD: University Press of America, 1985.

McCullough, Andrew M., and Andrew P. Yonelinas. "Cold-Pressor Stress After Learning Enhances Familiarity-Based Recognition Memory in Men." *Neurobiology of Learning and Memory* 106 (November 2013): 11-17.

McLean, Kate C., and Cade D. Mansfield. "To Reason or Not to Reason: Is Autobiographical Reasoning Always Beneficial?" *New Di-*

rections for Child and Adolescent Development 131 (Spring 2011): 85-97.

McLean, Kate C., Monisha Pasupathi, and Jennifer L. Pals. "Selves Creating Stories Creating Selves: A Process Model of Self-Development." *Personality and Social Psychology Review* 11, no. 3 (August 2007): 262-78.

McLellan, M. Faith, and Anne Hudson Jones. "Why Literature and Medicine?" *The Lancet* 348 (1996): 109-11.

Mele, Alfred R. "Self-Control in Action." In *The Oxford Handbook of The Self*, edited by Shaun Gallagher, 465-86. New York: Oxford University Press, 2013.

Mendez, Mario, Eric Anderson, and Jill S. Shapira. "An Investigation of Moral Judgement [sic] in Frontotemporal Dementia." *Cognitive and Behavioral Neurology* 18, no. 4 (December 2005): 193-97.

Menzies, Robin. "Propranolol Treatment of Traumatic Memories." *Advances in Psychiatric Treatment* 15 (2009): 159-60.

———. "Propranolol, Traumatic Memories, and Amnesia: A Study of 36 Cases." *The Journal of Clinical Psychiatry* 73, no. 1 (January 2012): 129-30.

Meretoja, Hanna. "Narrative and Human Experience." *New Literary History* 45 (2014): 89-109.

Messer, Neil. *Theological Neuroethics: Christian Ethics Meets the Science of the Human Brain.* New York: T&T. Clark, Bloomsbury, 2017.

Metcalfe, Janet. "Metamemory." In *The Oxford Handbook of Memory*, edited by Endel Tulving and Fergus I. M. Craik, 197-211. New York: Oxford University Press, 2000.

Metcalfe, Janet, and David Wiebe. "Intuition in Insight and Noninsight Problem Solving." *Memory and Cognition* 15, no. 3 (1987): 238-46.

Mikulincer, Mario. "Adult Attachment Style and Individual Differences in Functional versus Dysfunctional Experiences of Anger." *Journal of Personality and Social Psychology* 74 (1998): 513-24.

Miller, Fred D. "Rationality and Freedom in Aristotle and Hayek." *Reason Papers* 9 (Winter 1983): 29-36.

Mishler, Elliot G. "The Analysis of Interview-Narratives." In *Narrative Psychology: The Storied Nature of Human Conduct*, edited by Theodore R. Sarbin, 233-55. New York: Praeger Press, 1986.

Mizumori, Sheri J. Y. "Context Prediction Analysis and Episodic Memory." *Frontiers in Behavioral Neuroscience* 7 (October 2013): 1-10.

Moll, Jorge, Ricardo de Oliveira-Souza, and Paul J. Eslinger. "Morals and the Human Brain: A Working Model." *NeuroReport* 14, no. 3 (March 2003): 299-305.

Moll, Jorge, Ricardo de Oliveira-Souza, Paul J. Eslinger, Ivanei E. Bramati, Janaína Mourão-Miranda, Pedro Angelo Andreiuolo, and Luiz Pessoa. "The Neural Correlates of Moral Sensitivity: A Functional Magnetic Resonance Imaging Investigation of Basic and

Moral Emotions." *The Journal of Neuroscience* 22, no. 7 (April 2002): 2730-36.

Moll, Jorge, Roland Zahn, Ricardo de Oliveira-Souza, Frank Krueger, and Jordan Grafman. "The Neural Basis of Human Moral Cognition." *Nature Reviews Neuroscience* 6 (October 2005): 799-809.

Moll, Jorge, Ricardo de Oliveira-Souza, and Roland Zahn. "The Neural Basis of Moral Cognition: Sentiments, Concepts, and Values." *Annals of the New York Academy of Sciences* 1124 (March 2008): 161-80.

Moore, Adam B., Brian A. Clark, and Michael J. Kane. "Who Shalt Not Kill? Individual Differences in Working Memory Capacity, Executive Control, and Moral Judgment." *Psychological Science* 19, no. 6 (2008): 549-57.

Moosa, Imtiaz. "A Critical Examination of Scheler's Justification of the Existence of Values." *The Journal of Value Inquiry* 25, no. 1 (1991): 23-41.

Morein-Zamir, Sharon, and Barbara J. Sahakian. "Pharmaceutical Cognitive Enhancement." In *The Oxford Handbook of Neuroethics*, edited by Judy Illes and Barbara J. Sahakian, 229-44. New York: Oxford University Press, 2011.

Morton, Adam. "Epistemic Emotions." In *The Oxford Handbook of Philosophy of Emotion*, edited by Peter Goldie, 385-99. New York: Oxford University Press, 2010.

Moynihan, Ray, and David Henry. "The Fight Against Disease Mongering: Gathering Knowledge for Action." *PLoS Medicine* 3, no. 4 (2006): 425-28.

Mullen, Mary K. "Earliest Recollections of Childhood: A Demographic Analysis." *Cognition* 52 (1994): 55-79.

Müller, U. J., V. Sturm, J. Voges, H. J. Heinze, I. Galazky, M. Heldmann, H. Scheich, and B. Bogerts. "Successful Treatment of Chronic Resistant Alcoholism by Deep Brain Stimulation of Nucleus Accumbens: First Experience with Three Cases." *Pharmacopsychiatry* 42, no. 6 (November 2009): 288-91.

Müller, Ulf J., Jürgen Voges, Johann Steiner, Imke Galazky, Hans-Jochen Heinze, Michaela Möller, Jared Pisapia, Casey Halpern, Arthur Caplan, Bernhard Bogerts, and Jens Kuhn. "Deep Brain Stimulation of the Nucleus Accumbens for the Treatment of Addiction." *Annals of the New York Academy of Sciences* 1282 (2013): 119-28.

Murray, Thomas H. "What Do We Mean By 'Narrative Ethics'?" In *Stories and Their Limits: Narrative Approaches to Bioethics*, edited by Hilde Lindemann Nelson, 3-17. New York: Routledge, 1997.

Neale, Joanne. *Drug Users in Society.* New York: Palgrave, 2002.

Nelson, Hilde Lindemann, ed. *Stories and Their Limits: Narrative Approaches to Bioethics.* New York: Routledge, 1997.

Nelson, Katherine. "Language and the Self: From the 'Experiencing I' to the 'Continuing Me.'" In *The Self in Time: Developmental Per-

*spectives*, edited by Chris Moore and Karen Lemmon, 15-33. Mahwah, NJ: Lawrence Erlbaum Associates, Inc., Publishers, 2001.

Nelson, Katherine, and Robyn Fivush. "Socialization of Memory." In *The Oxford Handbook of Memory*, edited by Endel Tulving and Fergus I. M. Craik, 283-95. New York: Oxford University Press, 2000.

———. "The Emergence of Autobiographical Memory: A Social Cultural Developmental Theory." *Psychological Review* 111, no. 2 (2004): 486-511.

Newberg, Andrew B. *Principles of Neurotheology*. Routledge, 2010.

Nichols, Shaun, and Ron Mallon. "Moral Dilemmas and Moral Rules." *Cognition* 100, no. 3 (2006): 530-42.

Niedźwieńska, Agnieszka. "Metamemory Knowledge and the Accuracy of Flashbulb Memories." *Memory* 12, no. 5 (2004): 603-13.

Nietzsche, Friedrich. *The Genealogy of Morals*. Translated by Horace B. Samuel. New York: Dover Publications, Inc., 2003.

Nisbett, Richard E., and Yuri Miyamoto. "The Influence of Culture: Holistic versus Analytic Perception." *Trends in Cognitive Sciences* 9, no. 10 (October 2005): 467-73.

Nobler, Mitchell S., and Harold A. Sackeim. "Neurobiological Correlates of the Cognitive Side Effects of Electroconvulsive Therapy." *Journal of ECT* 24, no. 1 (March 2008): 40-45.

Northoff, Georg, Alexander Heinzel, Moritz de Greck, Felix Bermpohl, Henrik Dobrowolny, and Jaak Panksepp. "Self-Referential Processing in Our Brain—A Meta-Analysis of Imaging Studies on the Self." *NeuroImage* 31 (2006): 440-57.

Nussbaum, Martha C. *Love's Knowledge: Essays on Philosophy and Literature*. New York: Oxford University Press, 1990.

Nyberg, Lars, and Roberto Cabeza. "Brain Imaging of Memory." In *The Oxford Handbook of Memory*, edited by Endel Tulving and Fergus I. M. Craik, 501-519. New York: Oxford University Press, 2000.

Ogle, Christin M., David C. Rubin, and Ilene C. Siegler. "Changes in Neuroticism Following Trauma Exposure." *Journal of Personality* 82, no. 2 (April 2014): 93-102.

O'Neill, Onora. "Kant: Rationality as Practical Reason." In *The Oxford Handbook of Rationality*, edited by Alfred R. Mele and Piers Rawling, 93-109. New York: Oxford University Press, 2004.

Osman, Magda. "What are the Essential Cognitive Requirements for Prospection (Thinking About the Future)?" *Frontiers in Psychology* 5 (June 2014): 1-4.

Østby, Ylva, Kristine B. Walhovd, Christian K. Tamnes, Håkon Grydeland, Lars Tjelta Westlye, and Anders M. Fjell. "Mental Time Travel and Default-Mode Network Functional Connectivity in the Developing Brain." *Proceedings of the National Academy of Sciences of the United States of America* 109, no. 42 (2012): 16800-04.

Paharia, Neeru, Karim S. Kassam, Joshua D. Greene, and Max H. Bazerman. "Dirty Work, Clean Hands: The Moral Psychology of

Indirect Agency." *Organizational Behavior and Human Decision Processes* 109, no. 2 (2009): 134-41.

Parens, Erik. "The Ethics of Memory Blunting: Some Initial Thoughts." *Frontiers in Behavioral Neuroscience* 4 (December 2010): 1-2.

Parsons, Talcott. "Definitions of Health and Disease in Light of American Values and Social Structures." In *Patients, Physicians and Illness*, edited by E. Gartley Jaco, 120-44. New York: Free Press, 1979.

Pascal, Blaise. *Pensees*. Edited by Robert Maynard Hutchins. Vol. 33 of *Great Books of the Western World*. Chicago: Encyclopedia Britannica, 1952.

Pasupathi, Monisha. "The Social Construction of the Personal Past and its Implications for Adult Development." *Psychological Bulletin* 127 (2001): 651-72.

———. "Silk from Sows' Ears: Collaborative Construction of Everyday Selves in Everyday Stories." In *Identity and Story: Creating Self in Narrative*, edited by Dan P. McAdams, Ruthellen Josselson, and Amia Lieblich, 129-50. Washington, DC: American Psychological Association, 2006.

Pasupathi, M., E. Mansour, and J. R. Brubaker. "Developing a Life Story: Constructing Relations Between Self and Experience in Autobiographical Narratives." *Human Development* 50 (2007): 85-110.

Pasupathi, Monisha, and Cecilia Wainryb. "Developing Moral Agency through Narrative." *Human Development* 53 (2010): 55-80.

Paxton, Joseph M., and Joshua D. Greene. "Moral Reasoning: Hints and Allegations." *Topics in Cognitive Science* 2, no. 3 (July 2010): 511-27.

Paxton, Joseph M., Leo Ungar, and Joshua D. Greene. "Reflection and Reasoning in Moral Judgment." *Cognitive Science* 36 (2012): 163-77.

Petty, Richard E., and John T. Cacioppo. "The Elaboration Likelihood Model of Persuasion." *Advances in Experimental Social Psychology* 19 (1986): 123-205.

Phelps, Elizabeth A., Kevin S. LaBar, Adam K. Anderson, Kevin J. O'Connor, Robert K. Fulbright, and Dennis D. Spencer. "Specifying the Contributions of the Human Amygdala to Emotional Memory: A Case Study." *Neurocase* 4, no. 6 (1998): 527-40.

Philippe, Frederick L., Richard Koestner, Geneviève Beaulieu-Pelletier, and Serge Lecours. "The Role of Need Satisfaction as a Distinct and Basic Psychological Component of Autobiographical Memories: A Look at Well-Being." *Journal of Personality* 79, no. 5 (October 2011): 905-38.

Philippe, Frederick L., Richard Koestner, Serge Lecours, Geneviève Beaulieu-Pelletier, and Katy Bois. "The Role of Autobiographical Memory Networks in the Experience of Negative Emotions: How

Our Remembered Past Elicits Our Current Feelings." *Emotion* 11, no. 6 (2011): 1279-90.

Pitman, Roger K., Kathy M. Sanders, Randall M. Zusman, Anna R. Healy, Farah Cheema, Natasha B. Lasko, Larry Cahill, and Scott P. Orr. "Pilot Study of Secondary Prevention of Posttraumatic Stress Disorder with Propranolol." *Biological Psychiatry* 51 (2002): 189-92.

Pizarro, David A., and Paul Bloom. "The Intelligence of the Moral Intuitions: Comment on Haidt (2001)." *Psychological Review* 110, no. 1 (2003): 193-96.

Pizarro, David A., Eric Uhlmann, and Paul Bloom. "Causal Deviance and the Attribution of Moral Responsibility." *Journal of Experimental Social Psychology* 39, no. 6 (2003): 653-60.

Polkinghorne, Donald E. *Narrative Knowing and the Human Sciences.* New York: State University of New York Press, 1988.

President's Council on Bioethics. *Beyond Therapy: Biotechnology and the Pursuit of Happiness.* New York: HarperCollins Publishers, 2003.

Price, A. W. "Emotions in Plato and Aristotle." In *The Oxford Handbook of Philosophy of Emotion,* edited by Peter Goldie, 121-42. New York: Oxford University Press, 2010.

Prinz, Jesse J. "The Moral Emotions." In *The Oxford Handbook of Philosophy of Emotion,* edited by Peter Goldie, 519-38. New York: Oxford University Press, 2010.

Przegaliński, Edmund, Małgorzata Filip, Małgorzata Frankowska, Magdalena Zaniewska, and Iwona Papla. "Effects of CP 154,526, A CRF1 Receptor Antagonist, On Behavioral Responses to Cocaine in Rats." *Neuropeptides* 39, no. 5 (October 2005): 525-33.

Qin, Pengmin, and Georg Northoff. "How is Our Self Related to Midline Regions and the Default-Mode Network?" *NeuroImage* 57, no. 3 (August 2011): 1221-33.

Quartz, Steven R. "Reason, Emotion, and Decision Making: Risk and Reward Computation with Feeling." *Trends in Cognitive Sciences* 13, no. 5 (April 2009): 209-15.

Racine, Eric, Aspler, John, eds. *Debates about Neuroethics: Perspectives on its Development, Focus, and Future.* New York: Springer, 2017.

Racine, Eric. *Pragmatic Neuroethics: Improving Treatment and Understanding of the Mind-Brain.* Cambridge, MA: MIT Press, 2010.

Raine, Adrian, and Yaling Yang. "Neural Foundations to Moral Reasoning and Antisocial Behavior." *Social Cognitive and Affective Neuroscience* 1, no. 3 (October 2006): 203-13.

Rasmussen, Anne S., and Dorthe Bersten. "Emotional Valence and the Functions of Autobiographical Memories: Positive and Negative Memories Serve Different Functions." *Memory & Cognition* 37, no. 4 (June 2009): 477-92.

Ratcliffe, Matthew. "The Phenomenology of Mood and the Meaning of Life." In *The Oxford Handbook of Philosophy of Emotion*, edited by Peter Goldie, 349-71. New York: Oxford University Press, 2010.

Rawls, John. "Kantian Constructivism in Moral Theory." *The Journal of Philosophy* 77, no. 9 (September 1980): 515-72.

———. *A Theory of Justice*. Rev. ed. Cambridge, MA: Harvard University Press, 1998.

Reese, Elaine, Catherine A. Haden, and Robyn Fivush. "Mother-Child Conversations About the Past: Relationships of Style and Memory Over Time." *Cognitive Development* 8 (1993): 403-30.

Reichel, Carmela M., Khaled Moussawi, Phong H. Do, Peter W. Kalivas, and Ronald E. See. "Chronic N-Acetylcysteine during Abstinence or Extinction after Cocaine Self-Administration Produces Enduring Reductions in Drug Seeking." *The Journal of Pharmacology and Experimental Therapeutics* 337, no. 2 (2011): 487-93.

Reisenzein, Rainer. "Pleasure-Arousal Theory and the Intensity of Emotions." *Journal of Personality and Social Psychology* 67, no. 3 (1994): 525-39.

Ricoeur, Paul. "L'histoire Comme Récit et Comme Pratique." *Espirit* 54, no. 6 (June 1981): 155-65.

———. *Time and Narrative*. Vols. 1-3. Translated by Kathleen McLaughlin and David Pellauer. Chicago: University of Chicago Press, 1984.

———. "L'identité Narrative." *Esprit* 7-8 (1988): 295-304.

———. "The Creativity of Language." In *A Ricoeur Reader: Reflection and Imagination*, edited by Mario Valéds, 463-81. New York: Harvester Wheatsheaf, 1991.

———. *Oneself as Another*. Translated by Kathleen Blamey. Chicago: The University of Chicago Press, 1992.

Ritchey, Maureen, Florin Dolcos, and Roberto Cabeza. "Role of Amygdala Connectivity in the Persistence of Emotional Memories Over Time: An Event-Related fMRI Investigation." *Cerebral Cortex* 18 (November 2008): 2494-2504.

Rizzo, Matthew, et al., eds. *The Wiley Handbook on the Ageing Mind and Brain*. Wiley-Blackwell, 2018.

Roberts, Robert C. "Emotions and the Canons of Evaluation." In *The Oxford Handbook of Philosophy of Emotion*, ed. Peter Goldie, 561-83. New York: Oxford University Press, 2010.

Robertson, Diana, John Snarey, Opal Ousley, Keith Harenski, F. DuBois Bowman, Rick Gilkey, and Clinton Kilts. "The Neural Processing of Moral Sensitivity to Issues of Justice and Care." *Neuropsychologia* 45 (2007): 755-66.

Robinson, Terry E., and Kent C. Berridge. "Addiction." *Annual Review of Psychology* 54 (2003): 25-53.

Rorty, Richard. *Philosophy and the Mirror of Nature*. Princeton, NJ: Princeton University Press, 1979.

Rosenbaum, R. Shayna, Donald T. Stuss, Brian Levine, and Endel Tulving. "Theory of Mind is Independent of Episodic Memory." *Science* 23 (November 2007): 1257.

Rousseau, Jean-Jacques. *Les Rêveries du Promeneur Solitaire.* In *Oeuvre Complètes.* Vols. 1-5. Paris: Gallimard, 1959.

Rubin, David C. *Memory in Oral Traditions.* New York: Oxford University Press, 1995.

Rubin, David C., Andriel Boals, and Rick H. Hoyle. "Narrative Centrality and Negative Affectivity: Independent and Interactive Contributors to Stress Reactions." *Journal of Experimental Psychology: General* 143, no. 3 (June 2014): 1159-70.

Ryan, Erin. "The Discourse Beneath: Emotional Epistemology in Legal Deliberation and Negotiation." *Harvard Negotiation Law Review* 10, no. 231 (Spring 2005): 231-85.

Samuelson, Paul A. *Foundations of Economic Choice.* Cambridge, MA: Harvard University Press, 1947.

Sarbin, Theodore R. "The Poetics of My Identities." In *Narrative Identities: Psychologists Engaged in Self-Construction,* edited by George Yancy and Susan Hadley, 13-35. London: Jessica Kingsley Publishers, 2003.

Sarlo, Michela, Lorella Lotto, Andrea Manfrinati, Rino Rumiati, Germano Gallicchio, and Daniela Palomba. "Temporal Dynamics of Cognitive-Emotional Interplay in Moral Decision-Making." *Journal of Cognitive Neuroscience* 24, no. 4 (March 2012): 1018-29.

Sartre, Jean-Paul. *Nausea.* Translated by R. Baldick. London: Penguin Books, 1963.

Schacter, Daniel L., Donna Rose Addis, and Randy L. Buckner. "Episodic Simulation of Future Events: Concepts, Data, and Applications." *Annals of the New York Academy of Sciences* 1124 (2008): 39-60.

Schacter, Daniel L., Roland G. Benoit, Felipe De Brigard and Karl K. Szpunar. "Episodic Future Thinking and Episodic Counterfactual Thinking: Intersections Between Memory and Decisions." *Neurobiology of Learning and Memory* 117 (2015): 14-21.

Schaefer, Alexandre, Claire L. Pottage, and Adam J. Rickart. "Electrophysiological Correlates of Remembering Emotional Pictures." *NeuroImage* 54, no. 1 (January 2011): 714-24.

Schafer, Roy. *Language and Insight.* New Haven, CT: Yale University Press, 1978.

Scheler, Max. *Formalism in Ethics and the Non-Formal Ethics of Values: A New Attempt toward the Foundation of an Ethical Personalism.* Translated by Manfred S. Frings and Roger L. Funk. Evanston, IL: Northwestern University Press, 1973.

Schnall, Simone, Jennifer Benton, and Sophie Harvey. "With a Clean Conscience: Cleanliness Reduces the Severity of Moral Judgments." *Psychological Science* 19, no. 12 (December 2008): 1219-22.

Schooler, Jonathan W., and Joseph Melcher. "The Ineffability of Insight." In *The Creative Cognition Approach*, edited by Steven M. Smith, Thomas B. Ward, and Ronald A. Finke, 97-133. Cambridge, MA: The MIT Press, 1995.

Schooler, Jonathan W., Stellan Ohlsson, and Kevin Brooks. "Thoughts Beyond Words: When Language Overshadows Insight." *Journal of Experimental Psychology: General* 122, no. 2 (1993): 166-83.

Schüpbach, M., M. Gargiulo, M. L. Welter, L. Mallet, C. Béhar, J. L. Houeto, D. Maltête, V. Mesnage, and Y. Agid. "Neurosurgery in Parkinson Disease: A Distressed Mind in a Repaired Body?" *Neurology* 66, no. 12 (June 2006): 1811-16.

Scissons, Hannah. "Psychiatrist Studies Treatment for Traumatic Memories." *The StarPhoenix*. 27 January 2010.

Scoboria, Alan, Jennifer M. Talarico, and Lisa Pascal. "Metamemory Appraisals in Autobiographical Event Recall." *Cognition* 136 (2015): 337-49.

Sellnow, Deanna D. *The Rhetorical Power of Popular Culture: Considering Mediated Texts*. Thousand Oaks, CA: SAGE Publications, 2010.

Semkovska, Maria, and Declan M. McLoughlin. "Measuring Retrograde Autobiographical Amnesia Following Electroconvulsive Therapy." *Journal of ECT* 29, no. 2 (June 2013): 127-33.

Sharot, Tali, and Andrew P. Yonelinas. "Differential Time-Dependent Effects of Emotion on Recollective Experience and Memory for Contextual Information." *Cognition* 106 (2008): 538-47.

Sharot, Tali, Mieke Verfaellie, Andrew P. Yonelinas. "How Emotion Strengthens the Recollective Experience: A Time-Dependent Hippocampal Process." *PLoS ONE* 2, no. 10 (2007): e1068.

Sher, Leo. "Memory Creation and the Treatment of Psychiatric Disorders." *Medical Hypotheses* 54, no. 4 (2000): 628-29.

Sherin, Jonathan E., and Charles B. Nemeroff. "Post-Traumatic Stress Disorder: The Neurobiological Impact of Psychological Trauma." *Dialogues in Clinical Neuroscience* 13 (2011): 263-78.

Shoemaker, David. "Moral Responsibility and the Self." In *The Oxford Handbook of The Self*, edited by Shaun Gallagher, 487-518. New York: Oxford University Press, 2013.

Simon, Herbert A. "A Behavioral Model of Rational Choice." *The Quarterly Journal of Economics* 69, no. 1 (February 1955): 99-118.

Singer, Jefferson A. *Personality and Psychotherapy: Treating the Whole Person*. New York: Guilford Press, 2005.

Singer, Jefferson A., Pavel Blagov, Meridith Berry, and Kathryn M. Oost. "Self-Defining Memories, Scripts, and the Life Story: Narrative Identity in Personality and Psychotherapy." *Journal of Personality* 81, no. 6 (December 2013): 569-82.

Singer, Jefferson A., and Peter Salovey. *The Remembered Self: Emotion and Memory in Personality*. New York: Free Press, 1993.

Singer, Peter. "Ethics and Intuitions." *Journal of Ethics* 9 (2005): 331-52.

Slovic, Paul, Melissa Finucane, Ellen Peters, and Donald G. MacGregor. "The Affect Heuristic." In *Heuristics and Biases: The Psychology of Intuitive Judgment,* edited by Thomas Gilovich, Dale Griffin, and Daniel Kahneman, 397-420. New York: Cambridge University Press, 2002.

Solomon, Robert C. *The Passions: Emotions and the Meaning of Life.* Cambridge, UK: Hackett Publishing Company, 1993.

Solymosi, Tibor, Shook, John R., eds. *Neuroscience, Neurophilosophy and Pragmatism: Brains at Work with the World.* Palgrave MacMillan, 2014.

Somers, Margaret R. "The Narrative Constitution of Identity: A Relational and Network Approach." *Theory and Society* 23 (1994): 605-49.

Specht, Jule, Boris Egloff, and Stefan C. Schmukle. "Stability and Change of Personality Across the Life Course: The Impact of Age and Major Life Events on Mean-Level and Rank-Order Stability of the Big Five." *Journal of Personality and Social Psychology* 101, no. 4 (2011): 862-82.

Specio, Sheila E., Sunmee Wee, Laura E. O'Dell, Benjamin Boutrel, Eric P. Zorrilla, and George F. Koob. "CRF1 Receptor Antagonists Attenuate Escalated Cocaine Self-Administration in Rats." *Psychopharmacology (Berl)* 196, no. 3 (February 2008): 473-82.

St. Jacques, Peggy, David C. Rubin, Kevin S. LaBar, and Roberto Cabeza. "The Short and Long of It: Neural Correlates of Temporal-order Memory for Autobiographical Events." *Journal of Cognitive Neuroscience* 20, no. 7 (2008): 1327-41.

Staudinger, Ursula M. "Life Reflection: A Social-Cognitive Analysis of Life Review." *Review of General Psychology* 5, no. 2 (June 2001): 148-60.

Sternberg, Robert J., Fiske, Susan T. *Ethical Challenges in the Behavioral and Brain Sciences. Case Studies and Commentaries.* New York: Cambridge University Press, 2015.

Strange, B. A., R. Hurlemann, and R. J. Dolan. "An Emotion-Induced Retrograde Amnesia in Humans is Amygdala-and β-Adrenergic-Dependent." *Proceedings of the National Academy of Sciences of the United States of America* 100, no. 23 (November 2003): 13626-31.

Talarico, Jennifer M., Dorthe Berntsen, and David C. Rubin. "Positive Emotions Enhance Recall of Peripheral Details." *Cognition and Emotion* 23, no. 2 (February 2009): 380-98.

Tani, Franca, Carole Peterson, and Andrea Smorti. "Empathy and Autobiographical Memory: Are They Linked?" *The Journal of Genetic Psychology* 175, no. 3 (2014): 252-69.

Taylor, Charles. *Philosophical Papers 1: Human Agency and Language.* Cambridge, UK: Cambridge University Press, 1985.

———. *The Ethics of Authenticity.* Cambridge, MA: Harvard University Press, 1991.

———. *Sources of the Self.* Cambridge, MA: Harvard University Press, 1992.

Taylor, Shelley E., and Jonathon D. Brown. "Illusion and Well-Being: A Social Psychological Perspective on Mental Health." *Psychological Bulletin* 103, no. 2 (1988): 193-210.

Terbeck, Sylvia, Guy Kahane, Sarah McTavish, Julian Savulescu, Neil Levy, Miles Hewstone, and Philip J. Cowen. "Beta Adrenergic Blockade Reduces Utilitarian Judgement [sic]." *Biological Psychology* 92 (2013): 323-28.

Thomson, Judith Jarvis. "A Defense of Abortion." *Philosophy and Public Affairs* 1, no. 1 (1971): 67-95.

Thorne, Avril. "Developmental Truths in Memories of Childhood and Adolescence." *Journal of Personality* 63, no. 2 (1995): 139-63.

Thorne, Avril, and Kate C. McLean. "Telling Traumatic Events in Adolescence: A Study of Master Narrative Positioning." In *Connecting Culture and Memory: The Development of an Autobiographical Self,* edited by Robyn Fivush and Catherine A. Haden, 169-85. Mahwah, NJ: Lawrence Erlbaum Associates, 2003.

Todorov, Alexander, and Ingrid R. Olson. "Robust Learning of Affective Trait Associations with Faces When the Hippocampus is Damaged, But Not When the Amygdala and Temporal Pole are Damaged." *Social Cognitive and Affective Neuroscience* 3, no. 3 (September 2008): 195-203.

Tomlinson, Tom. "Perplexed about Narrative Ethics." In *Stories and Their Limits: Narrative Approaches to Bioethics,* edited by Hilde Lindemann Nelson, 123-33. New York: Routledge, 1997.

Torrey, E. Fuller. *Evolving Brains, Emerging Gods. Early Humans and the Origins of Religion.* New York: Columbia University Press, 2017.

Tromp, Shannon, Mary P. Koss, Aurelio Jose Figueredo, and Melinda Tharan. "Are Rape Memories Different? A Comparison of Rape, Other Unpleasant, and Pleasant Memories Among Employed Women." *Journal of Traumatic Stress* 8, no. 4 (October 1995): 607-27.

Tulving, Endel. "Memory and Consciousness." *Canadian Psychology* 26 (1985): 1-12.

Tusche, Anita, Jonathan Smallwood, Boris C. Bernhardt, and Tania Singer. "Classifying the Wandering Mind: Revealing the Affective Content of Thoughts During Task-Free Rest Periods." *NeuroImage* 97 (2014): 107-16.

Unterrainer, Marcus, and Fuat S. Oduncu. "The Ethics of Deep Brain Stimulation (DBS)." *Medicine, Health Care and Philosophy* [Epub ahead of print] (January 2015): 1-11.

Valdesolo, Piercarlo, and David DeSteno. "Moral Hypocrisy: Social Groups and the Flexibility of Virtue." *Psychological Science* 18, no. 8 (2007): 689-90.

van Steenburgh, J. Jason, Jessica I. Fleck, Mark Beeman, and John Kounios. "Insight." In *The Oxford Handbook of Thinking and Reasoning*, edited by Keith J. Holyoak and Robert G. Morrison, 475-91. New York: Oxford University Press, 2012.

van Stegeren, Anda H., Walter Everaerd, and Louis J. Gooren. "The Effect of Beta-Adrenergic Blockade after Encoding on Memory of an Emotional Event." *Psychopharmacology* 163, no. 2 (September 2002): 202-12.

Van Strien, Jan W., Sandra J. E. Langeslag, Nadja J. Strekalova, Liselotte Gootjes, and Ingmar H. A. Franken. "Valence Interacts with the Early ERP Old/New Effect and Arousal with the Sustained ERP Old/New Effect for Affective Pictures." *Brain Research* 1251 (2009): 223-35.

von Herder, Johann Gottfried. *Ideen*. In *Herders Sämtliche Werke*, edited by Bernhard Suphan. Vols. I-XV. Berlin: Weidmann, 1877-1913.

Walker, Margaret Urban. *Moral Understandings: A Feminist Study of Ethics*. New York: Routledge, 1998.

Weiner, Richard D., Helen J. Rogers, Jonathan R. T. Davidson, and Larry R. Squire. "Effects of Stimulus Parameters on Cognitive Side Effects." *Annals of the New York Academy of Sciences* 462 (March 1986): 315-25.

Wheeler, Mark A. "Episodic Memory and Autonoetic Awareness." In *The Oxford Handbook of Memory*, edited by Endel Tulving and Fergus I. M. Craik, 597-608. New York: Oxford University Press, 2000.

Wild, T. Cameron, Jody Wolfe, and Elaine Hyshka. "Consent and Coercion in Addiction Treatment." In *Addiction Neuroethics: The Ethics of Addiction Neuroscience Research and Treatment*, edited by Adrian Carter, Wayne Hall, and Judy Illes, 153-74. San Diego: Academic Press, 2012.

Williams, Bernard. "A Critique of Utilitarianism." In *Utilitarianism: For and Against*, edited by J. J. C. Smart and Bernard Williams, 77-150. New York: Cambridge University Press, 1973.

Williams, J. Mark G., Thorsten Barnhofer, Catherine Crane, Dirk Hermans, Filip Raes, Ed Watkins, and Tim Dalgleish. "Autobiographical Memory Specificity and Emotional Disorder." *Psychological Bulletin* 133, no. 1 (January 2007): 122-48.

Witt, Karsten, Jens Kuhn, Lars Timmermann, Mateusz Zurowski and Christiane Woopen. "Deep Brain Stimulation and the Search for Identity." *Neuroethics* 6 (2013): 499-511.

Wittgenstein, Ludwig. *Tractatus Logico-Philosophicus*. New York: Cosimo, Inc., 2007.

Wolfe, Charles T., ed. *Brain Theory: Essays in Critical Neurophilosophy*. Palgrave MacMillan, 2014.

Wu, Yanhong, Cheng Wang, Xi He, Lihua Mao, and Li Zhang. "Religious Beliefs Influence Neural Substrates of Self-Reflection in Tibetans." *Social Cognitive and Affective Neuroscience* 5, nos. 2-3 (June/September 2010): 324-31.

Yeung, Cecilia Au, Tim Dalgleish, Ann-Marie Golden, and Patricia Schartau. "Reduced Specificity of Autobiographical Memories Following a Negative Mood Induction." *Behaviour Research and Therapy* 44, no. 10 (November 2006): 1481-90.

Yoder, Keith J., and Jean Decety. "Spatiotemporal Neural Dynamics of Moral Judgment: A High-Density ERP Study." *Neuropsychologia* 60 (July 2014): 39-45.

Yonelinas, Andrew P., and Maureen Ritchey. "The Slow Forgetting of Emotional Episodic Memories: An Emotional Binding Account." *Trends in Cognitive Neuroscience* 19, no. 5 (May 2015): 259-67.

Zhou, Hongyu, Jiwen Xu, and Jiyao Jiang. "Deep Brain Stimulation of the Nucleus Accumbens on Heroin-Seeking Behaviors: A Case Report." *Biological Psychiatry* 69, no. 11 (2011): e41-42.

Zhou, Qing, Nancy Eisenberg, Sandra H. Losoya, Richard A. Fabes, Mark Reiser, Ivanna K. Guthrie, Bridget C. Murphy, Amanda J. Cumberland and Stephanie A. Shepard. "The Relations of Parental Warmth and Positive Expressiveness to Children's Empathy-Related Responding and Social Functioning: A Longitudinal Study." *Child Development* 73, no. 3 (May/June 2002): 893-915.

# INDEX

An 'n' following a page number indicates an endnote.

## A

abstract thoughts, 130, 194
ACC (anterior cingulate cortex), 18, 131
accountability, 220. *see also* responsibility
addiction. *see* substance addiction; 12-step recovery programs
Addis, Donna, and colleagues, 63–66
ADHD (attention deficit hyperactivity disorder), 30, 31
Adler, Alfred, 168
adrenaline, 8
affective degeneration. *see* emotional rationality
affective foresight, 79–82
*After Virtue* (MacIntyre), 188
agency, narrative, 196. *see also* autonomy; moral agency; self-agency
"aha" phenomenon, 144
alcoholism, 15, 20, 21, 22, 33, 39n51. *see also* substance addiction
altruism, 160n93
American culture, 178–179, 195, 212–213
amnesia, 9, 10, 20, 21, 65, 74, 89

amygdala
  emotionally rational moral intuition and, 130–131
  emotions and, 8, 67, 69, 70, 82–83
  episodic memory and, 69, 70
  ethical decision making and, 82–83, 84
  moral cognition and, 141
  moral emotion and, 114
  negative contexts and, 71
  post-traumatic stress disorder and, 18, 19
  protected values and, 192
  substance addiction and, 24, 25
anger
  autobiographical narrative and, 72
  body and, 108
  emotional rationality and, 146
  moral agency and, 134
  motivation and, 116
  narrative goals and, 209–210
  practical reason and, 146
  rational evaluative judgment and, 126
  risk-seeking and, 133
  shame and, 108
  trauma and, 17

angular gyrus, 55
Angus, Lynne, 77
anterior cingulate cortex (ACC), 18, 131
anterior temporal cortex, 131, 132
anterior temporal lobes, 193
Arendt, Hannah, 172
Aristotle, 108–109, 110, 146, 173, 174, 204, 215–216
Arras, John, 185–186, 228nn99,103
association cortices, 83, 194
attachments, 73, 176, 210, 211. *see also* parents
attention, 114, 153
attention deficit hyperactivity disorder (ADHD), 30, 31
authenticity, 64, 65–66, 157n29, 204, 207, 211–215, 219, 223, 238
autobiographical (episodic) memory. *see also* affective foresight; casuistry (case-based moral reasoning); coherence; integration; narrative identity; the self
  authenticity and, 64, 65–66, 157n29, 207, 219, 223, 238
  autonoetic awareness and, 49
  centrality of, 76–78
  children and, 56–57
  electroconvulsive therapy and, 11–12
  emotions and, 57, 66–78, 82–85, 92, 98n98
  erasure of, 2, 4n4
  ethical decision making and, 78–91, 78–92, 85–88, 92
  historical views of, 48–49, 61–63
  imaging and socialization of, 54–57
  imprecision of, 200–201
  limited-use memory manipulation and, 36
  memory manipulation and, 2, 4n4, 7–8, 92, 209
  meta-memory and rational prospection of, 63–66, 91–92
  moral judgment and, 48–53, 59, 83–84, 91, 92, 176
  narrative identity and, 179–188
  narrative of emotion and, 66–78
  neurobiology and, 48–49, 50, 53–54, 54–55, 66–67, 69, 79, 82, 94n28, 194
  ontology of, 57–60
  phenomenology and, 61–63
  prospective thinking and, 65–66
  rationality and, 53–66, 61–63, 91–92
  responsibility and, 58–59
  selective loss of, 50
  self and, 48
  semantic memory *versus*, 11
  social and cultural mediation of, 48, 55–60, 200–201, 205
  stories and, 199–200
  subjectivity of, 214–215
  temporal understanding and, 59
autobiographical prospection, 88–91
autonoetic awareness, 49–50, 51
autonomy. *see also* control; desires; free will; goals; moral agency; self-control
  growth and, 48
  limited-use memory manipulation and, 240–242
  myth of global loss of, 32, 36
  narrative self-control and, 215
  reasons and, 203–204
  respect for, 187

Index

substance addiction and, 22–23, 27–29, 32–35, 34
autoregulation, 81

## B

Balfour, David, 33–34
Bartlett, Francis, 55
basal forebrain, 83, 131
basolateral limbic circuit, 98n98
Baumeister, Roy, and colleagues, 63
Bechara, Antoine, and colleagues, 82
behavioral approaches, 77–78
behavioral economics, 108
being in the world, 171–172
beliefs, 64–65, 112, 113, 118, 151, 152, 154–155
Benabid, Alim-Louis, and colleagues, 15
Benjamin, Walter, 186
Benoit, Roland, and colleagues, 79–80
Benveniste, Emile, 196–197
Berns, Gregory, 192
Berntsen, Dorthe, 70
beta-adrenergic receptor-blocking pharmacologicals, 8–10, 17, 35, 36
Billings Jr., F. James, 12
biomedicalization, 29–30, 29–31, 36
bioregulatory states, 142
bipolar disorder, 29–30
Bloom, Paul, 148
Bluck, Susan, and colleagues, 73
the body, 108, 121, 133–134, 142–144, 194, 221. *see also* motor activity
body-loop, 142–143
Boritz, Tali, and colleagues, 76
brain-stem nuclei, 83
Brentano, Franz, 124
Brody, Eugene, 169
Brody, Howard, 170

Bruner, Jerome, 172, 178
Bryant, Richard, 21

## C

Cacioppo, Stephanie, 129
Cahill, Larry, and colleagues, 9, 69
Camille, Nathalie, and colleagues, 90–91
care, 84–85
Carruthers, Mary, 56
Carter, Adrian, 17, 22
Carter, Sid, 133–134
casuistry (case-based moral reasoning), 52–53, 168, 183–184, 227nn84,86, 228n103
categorization, 55, 79, 141, 142
caudate, 88
CDZs (convergence-divergence zones), 194
centro-parietal area, 70
change, 62–63, 77, 130–131, 177
character, 187–188, 204. *see also* personality
Charland, Louis, 28
Charon, Rita, 185–186
children, 56–57, 95n44, 139, 210, 228n107. *see also* parents
Childress, James, 186
choice. *see* autonomy; ethical decision making; rational theory of choice
Christian story, 232n202
cingulate cortex, 18, 48–49, 80, 83, 84, 85, 128
cingulate gyrus, 66–67
circuit remodeling (myelination patterns), 19
class, social, 170, 201
Clifasefi, Seema, and colleagues, 13–14, 39n51
clinical bioethics, 168
clinical reasoning, 168–171
cocaine addiction, 33. *see also* substance addiction

codification of new diseases, 31–32
cognition. *see* being in the world; insight; knowledge; morality and moral judgment; neurobiology; rationality (reason)
cognitive reflection test (CRT), 149
coherence, 178, 179, 189–190, 209, 216–217, 226n61
Cohler, Bertram, 170
commitment, 126–127
community narratives, 196
compassion, 140
conceptual narrative, 180, 182
conscience, 176, 207–208
consciousness, 143, 173, 199. *see also* autonoetic awareness; introspection; noetic consciousness; self-reflection; subjectivity
consistency, 153, 154
consolidation and reconsolidation, 67
contexts. *see also* social and cultural mediation
  consistency and, 153
  emotions and, 68, 69, 70–72
  limited-use memory manipulation and, 36
  narrative identity and, 72, 170, 177, 190–191, 204–205
  neurobiology and, 69, 88–89
  post-traumatic stress disorder and, 18, 19
control, 134, 204. *see also* autonomy; self-control
convergence-divergence zones (CDZs), 194
conversation, 173
Coricelli, Giorgio, and colleagues, 91
cortices, 94n4, 141, 142, 193
corticotropin-releasing factor (CRF), 25–26
cortisol, 18

counterfactual reflection, 88–89, 91
court witnesses, 92n1
CRF (corticotropin-releasing factor), 25–26
crime, 20–22
CRT (cognitive reflection test), 149
cultural narratives, 222, 239. *see also* American culture; narratives, Christian; social and cultural mediation
curiosity, 117–118

# D

Damasio, Antonio, 132, 147, 199
D'Argembeau and colleagues, 54–55, 94n33
DBS (deep brain stimulation), 7, 14–16, 35, 36, 40n63
De Brigard, Felipe, 89
Decety, Jean, 129
declarative memory, 18, 19, 73
deduction, 151
deep brain stimulation (DBS), 7, 14–16, 35, 36, 40n63
default mode network, 193–194
deliberation, 123–124
denial, 217
Dennett, Daniel, 215
depression, 29–30, 31, 72, 76–77, 89, 217
Descartes, René, 63–64, 212, 216
desires, 28–29, 32–34, 33, 108, 113, 116, 126. *see also* autonomy; goals
diagnoses. *see* biomedicalization
Dilthey, Wilhelm, 60
diseases, codification of new, 31–32
disgust, 134
"disorder of choice," 34

dissociation, 217
distancing, 123–124
Disulfiram, 13
*Doctor's Stories* (Hunter), 170
dopamine, 23–24, 25–26, 34–35, 112
Dostoyevsky, Fyodor, 138, 217
drug aversion therapies, 12–13
dual process theory, 148
Duncker, Karl, 149–150

# E

earliest memories, 95n44
East Asian traditions, 195
Ebbinghaus, Hermann, 61
economic models, 152–153
economic value, 87–88
ECT (electroconvulsive therapy), 2, 7, 10–12, 35, 36, 38n31
education, moral, 109, 183–185. *see also* learning
ego depletion, 35
ego identity, 177–178
electric shock (negative stimulus pairing), 39n43
electroconvulsive therapy (ECT), 2, 7, 10–12, 35, 36, 38n31
electrostimulants, 36
electrotherapies, 14–15
Elliot, Carl, 28
emotion, narrative of, 66–72
emotional rationality. *see also* emotions; knowledge; motivation, moral; narrative identity; rationality
  autobiographical memory and, 85–88
  ethical decision making and, 78–79, 88–91, 107–114, 141–156
  evaluative judgment and, 114–127
  the good and, 137, 146
  memory manipulation and, 92, 207

morality and, 127–140, 155
narrative identity and, 179–188
neurobiology and, 128–132
rational affective awareness and, 122–124
emotions. *see also* affective foresight; altruism; depression; emotional rationality; empathic intentionality; feelings; mood; propranolol *and other treatments*
  adrenaline and, 8
  amygdala and, 69, 70
  Aristotle on, 108–109
  autobiographical memory and, 57, 66–78, 82–85, 92, 98n98
  autobiographical moral thought and, 82–85
  described, 114–117
  judgment and, 68
  judgment and decision making and, 111–112
  meaning and, 121–122
  memory manipulation and, 2, 211
  moral, 114, 157n27
  neurobiology and, 69–71, 82, 194
  punishment and, 160n94
  subjectivity and, 157n29
  subthalamic nucleus and, 16
  traumatic events and, 78
empathic intentionality, 73–76, 92, 139
empathy, 131
Engelhardt, H. Tristram, 184, 185, 188
Epictetus, 138
episodic memory. *see* autobiographical (episodic) memory
Erikson, Erik, 177–178
ethical decision making, xvii–xix, 150, 151, 154–156, 218–222. *see also* autobiographical (episodic) memory; emotional rationality;

judgment and decision making; morality and moral judgment; normative ethical decision making (NEDM); reasons, moral
ethics as propositions, 183
event-related potentials (ERPs), 70
event-related potential (ERP) signals, 129–130, 160n104

# F

facial muscles, 134
fairness, 186
false memory creation (FMC), 7, 12–14, 29, 35, 36
familiarity, 80, 122. *see also* recognition memory
fantasy, 74
FDA (Food and Drug Administration), 30
fear
  behavior and, 116
  the body and, 108
  empathy and, 74
  extinction of, 18, 19, 24–25
  Heidegger on, 121–122
  intention and, 121
  irrational, 120
  narrative and, 159
  neurobiology and, 8, 112
  post-traumatic stress disorder and, 17
  practical reason and, 146
  primary inducers and, 83
  without propositional content, 112
  risk and, 113
  value judgments and, 133
feedback, 81
feeling perception, 83
feelings, 116, 120–121, 124–125, 127, 212. *see also* emotions; moral emotion
fiction, 180
Fingarette, Herbert, 33
Fisher, Walter, 179, 180, 226n66
Fivush, Robyn, 63

flashbulb memories, 72
FMC (false memory creation), 7, 12–14, 29, 35, 36
Food and Drug Administration (FDA), 30
forebrain, 23
forecasting, 153–154
forgetting, deliberate, 1–4
*Formalism in Ethics and the Non-Formal Ethics of Values* (Scheler), 124
fornix, 66–67
*Foundations of Bioethics* (Engelhardt), 188
FPC (frontopolar cortex), 131
Frankfurt, Harry, 28
Fraser, Louise, and colleagues, 11
freedom, 212–213, 214, 216, 241–242. *see also* independence *versus* interdependence
free will, 27, 127. *see also* autonomy
Freud, Sigmund, 10, 168, 208
frontal brain areas, 18, 35, 55, 70, 79, 129, 192
frontopolar cortex (FPC), 131
frontotemporal dementia (FTD), 132
FTD (frontotemporal dementia), 132
*Fundamental Principles for the Metaphysics of Morals* (Kant), 187
the future. *see also* autobiographical prospection; autonomy; goals; imagination; narrative identity (life stories); planning; prediction; redemption
  medial left pre-frontal parietal and, 65
  memory manipulation and, 211
  negative events and, 72
  objectivity and, 136
  practical reason and, 146, 155
  the present and, 219

self-consciousness and, 203

## G

GABA (gamma-aminobutyric acid), 18, 25
Gadamer, Hans-Georg, 172
Gage, Phineas, 147
gambling experiments, 82, 90–91
gamma-aminobutyric acid (GABA), 18, 25
gender, 57, 60, 87, 170, 201
*Genealogy of Morals* (Nietzsche), 207–208
gene expression, 19
generativity, 60, 202
Gestalt psychology, 145, 149
Giovanello, Kelly, 89
Glimcher, Paul, 111
glucocorticoids, 11, 19
glutamate hypothesis, 24–25
glutamate transporter 1 (GLT 1), 24–25
Goaillard, Jean-Marc, 81
goals. *see also* autonomy; desires; prospection, rational
 Aristotle on, 109
 autobiographical memory and, 48
 deliberation and, 123
 emotional memory and, 76
 emotional rationality and, 151
 growth and, 216
 memory manipulation and, 209–210
 neurobiology and, 23, 81
 Plato on, 109
 prospective thinking and, 66
 redemption and, 202
 subthalamic nucleus and, 16
the good. *see also* values
 emotions and, 127
 guilt and, 135
 intention and, 139
 knowledge and, 109
 memory manipulation and, xix, 155, 211, 218, 222, 223, 237, 238, 239, 240, 242
 narrative identity and, 177, 181–182, 201–202, 220–222
 neurobiology and, 129, 143, 174
 sociocultural context and, 124, 135, 174, 181–182, 184–185, 201–202, 238
 tradition and, 206
Greenberg, Leslie, 77
Greene, Joshua, 148
Greene, Joshua, and colleagues, 128, 148
Gregg, Gary, 170
grief, 137
growth, personal, 48, 76, 216–218, 223
guess responses, 52–53
Gui, Dan-Yang, and colleagues, 129, 130
guilt, 131, 134–136, 140, 207–208, 210–211, 217

## H

Haidt, Jonathan, 128, 147, 148
Hall, Wayne, 17, 22
Han, Sanghoon, and colleagues, 88–89
happiness, 110, 154, 208, 209–210, 211. *see also* well-being
harm-doing, 175
Harris, Richard, and colleagues, 73
Hart, J.T., 64
Hartmann, Nicolai, 124–125
Hauser, Marc, 130
Hawkins, Anne Hunsaker, 170
Heidegger, Martin, 121–123, 158n54, 172, 203–207
Hemmer, Pernille, 55
Henry, Michael, and colleagues, 20, 22

Herder, Johann Gottfried von, 212–213
Hermans, Hubert, 170
hermeneutic approach, 171–173, 177, 209
Heyman, Gene, 33–34
Higgins, Stephen, and colleagues, 33
Hillman, James, 169
hippocampal formation, 66–67, 79
hippocampus. *see also* parahippocampal gyrus
   autobiographical memory and, 79
   emotion and memory and, 69, 70, 131
   episodic memory and, 48–49, 82
   post-traumatic stress disorder and, 18, 19
   regret and, 91
historical accounts, 16, 59, 239. *see also* narrative identity; tradition
history, 180, 182, 206
homeostatic neural plasticity, 81–82
Homer, 2
HPA (hypothalamic pituitary axis), 25
Hume, David, 112, 137–138
Hunter, Kathryn Montgomery, 168–169, 170
Husserl, Edmund, 124, 125, 173
Hyman, Ira, 12
hypnopsychological memory editing, 36
hypocrisy, 148
hypothalamic pituitary axis (HPA), 25
hypothalamus, 83

# I

the "I," 196, 197, 216. *see also* narrative identity (life stories); the self

identity. *see* ego identity; narrative identity (life stories); self; subjective identity
identity as sameness *versus* as self, 180
*The Illness Narratives* (Kleinman), 170
images, 141, 142, 153, 156, 194
imagination, 63–64, 74–75, 89, 108, 198–199. *see also* affective foresight; autobiographical prospection; fantasy; the future; self-reflection; storytelling
incentives, perverse, 21
incest taboo, 149
independence *versus* interdependence, 195
individualism, 212
inferior frontal gyrus, 55
inferior parietal lobes, 193
informed consent, 31–32
inner voice, 213
insight, 144–145
insula, 18, 112
integration, 194, 199–200, 204, 209, 216–217, 223
   memory manipulation and, 210–211, 223
intention. *see also* empathic intentionality; goals; motivation, moral; planning
   to forget, 208
   morality and, 138–139, 140, 151
   narrative identity and, 201, 219
   rational emotion and, 122, 124, 151
   self-control and, 215
introspection, 51. *see also* consciousness; self-reflection
intuitions, moral, 148, 151–152
Iranians, 195
item memory, 50, 68

## J

James, William, 49, 126
JDM (judgment and decision making), 111–112
Jones, Anne Hudson, 170
Jones, Gregory, 85–87
Jonsen, Albert, 183–184, 186, 227n84
Josselson, Ruthellen, 170
joy, 147
judgment, evaluative. *see also* judgment and decision making; knowledge; morality and moral judgment
  autobiographical memory and, 47–50
  cognition and, 132–133
  competing, 148–149
  emotional rationality and, 124–127, 137–148, 155
  emotions and, 68, 121
  narrative and, 188–191
  neurobiology and, 67, 131
  practical reason and, 147
  rationality and, 137
judgment and decision making (JDM), 111–112
Jung, Carl, 168
justice, 84–85, 134–137, 160n94, 175

## K

Kant, Immanuel, 127, 135, 136, 173, 187
Kaplan, Jonas T., and colleagues, 193, 194–195
Kennett, Jeanette, 215
Kiefer, Markus, and colleagues, 70
Klein, Stanley, and colleagues, 65
Kleinman, Arthur, 170
knowledge. *see also* affective foresight; categorization; judgment, evaluative; rationality (reason); self-knowledge; self-reflection
  background, 151–152, 156
  described, 51–53
  ethical decision making and, 141
  memory judgments *versus*, 104n202
  moral judgment and, 52–53
  Plato on, 109–110
  rational emotion and, 117–120
  self-reflection and, 63–64
Kochanska, Grazyna, 176
Koenigs, Michael, and colleagues, 132
Korsgaard, Christine, 203, 204
Kroes, Marijn, and colleagues, 11
Kuhn, Jens, and colleagues, 15

## L

Lambie, John, 122–123
language, 192, 196–197, 205, 213
late positive potential (LPP), 130, 131
lateral inferior parietal cortices, 194
learning, 19, 142, 183. *see also* education, moral
Leshner, Alan, 28
Levy, Dino, 111
life review *versus* reminiscence, 62
life stories. *see* autobiographical (episodic) memory; narrative identity
life-world (Husserl), 173
limbic system, 66–67, 84, 131. *see also* amygdala *and other structures*
Locke, John, 204, 207, 212
Lombardo, Michael, and colleagues, 74
Lotus-eaters, 2
love, 137
LPP (late positive potential), 130, 131

LUMM (limited-use memory manipulation), 7, 16, 36, 237–243. *see also* PTSD (post-traumatic stress disorder); substance addiction
Lyubomirsky, Sonja, and colleagues, 63

# M

MacIntyre, Alasdair, 184–185, 187–188, 208–209
Mallon, Ron, 148
mammillary bodies, 66–67
mammillothalamic tract, 66–67
Marder, Eve, 81
Martin, Maryanne, 85–87
Masten, Carrie, and colleagues, 75
master narratives, 60, 95n54
Mathur, Vani, and colleagues, 75
McAdams, Dan, 168–169
McCormick, Richard, 238–239, 240–242
McLean, Kate C., 60, 95n54
meaning. *see also* redemption
  emotions and, 121–122
  memory manipulation and, 209, 211
  narrative identity and, 77, 171, 172, 173, 177–178, 207, 208–209
  narrative integration and, 200
  neurobiology and, 174, 194–195
  others and, 214–215
medial frontal gyrus, 131
medicine, 168–171
mediodorsal nucleus, 98n98
medial temporal lobe (MTL), 50, 69–70
Melcher, Joseph, 145
memory. *see* autobiographical (episodic) memory; item memory; propranolol; recognition memory
memory consolidation and reconsolidation, 67
memory manipulation. *see also* autobiographical (episodic) memory; ethical decision making; false memory creation (FMC) *and other manipulations;* the good; memory manipulation, limited-use (LUMM); narrative identity (life stories); neurobiology; neuroscience
  argument against, vii–xix, 1–4
  effects of, 4n3
  enhancement and, 4n3, 5n10, 7, 10, 21
  proportionate reason and, 237–243
  terminology and, 3–4, 4n2
memory manipulation, limited-use (LUMM), 7, 16, 36, 237–243. *see also* PTSD (post-traumatic stress disorder); substance addiction
Menzies, Robin, 9
meta-memory, 63–66
metanarrative, 181, 182
metaontological narratives, 180
Metcalfe, Janet, 145
methylphenidate, 31
midbrain, 131
middle temporal gyrus, 55, 88
Mikulincer, Mario, 210
mimesis, 173
minorities, 201
Mishler, Elliot, 178
Mizumori, Sheri, 81
Mohamed, 138
moods, 71, 121–122. *see also* emotions
Moore, Adam, and colleagues, 148

moral agency, 134, 175–176, 177, 197–198, 207, 223. *see also* agency, narrative; autonomy
moral emotion, 114, 157n27
moral failure, 175
moral identity, 175–176, 206–207
moral intuition, 128–131, 212
morality and moral judgment. *see also* casuistry; conscience; ethical decision making; the good; judgment, evaluative; judgment and decision making; knowledge; motivation, moral; principles and rules; reasons, moral; redemption; responsibility; rules and principles; values
  Aristotle on, 110
  autobiographical memory and, 59, 83–84, 91, 176
  contexts and, 219
  development of, 175–179, 183–185, 228n107
  emotional rationality and, 127–140, 155
  episodic memory and, 48–53
  exceptionless, 238
  factual character of, 222
  memory manipulation and, 242
  modern denial of, 222
  narrative identity and, 175–179, 183–185, 201–202, 222, 228n99, 242–243
  neurobiology and, 82, 83–85, 128–132
  personal identity and, 204
  rational emotion and, 118–119
  substance addiction and, 34
  unity and, 220–221
morphine-induced place preference, 40n63
"most valued object," 87–88

motivation, moral, 66–67, 116, 125–126, 137–140, 178, 210. *see also* intention; reasons, moral
motor activity, 25, 34, 49, 141, 142, 194
MTL (medial temporal lobe), 50, 69–70
Mullen, Mary, 95n44
Müller, Ulf, and colleagues, 15
Murray, Henry, 168
Murray, Thomas, 188
myelination patterns (circuit remodeling), 19

# N

NAC (N-acetylcysteine), 24–25, 43n113
NAc (nucleus accumbens), 15, 23, 24–25, 34, 40n63
narrative as images, 141
*Narrative Based Medicine* (Greenhalgh and Hurwitz), 170
narrative identity. *see also* subjective identity
  relationships and, 221–222, 223
narrative identity (life stories). *see also* affective foresight; autobiographical (episodic) memory; coherence; emotional rationality; self-deception; storytelling; truth
  authenticity and, 217
  autobiographical memory and, 57–58
  autobiographical memory and emotional rationality and, 179–191, 222, 226n66
  autonomy and, 241–242
  case-based moral reasoning and, 228n103
  defined, 176–177
  faulty, 202

limited-use memory manipulation and, 239–240, 241–242
memory manipulation and, 207–223
morality and, 175–179, 183–185, 201–202, 222, 228n99, 242–243
ontology of, 171–174
social and cultural mediation and, 173, 177, 178, 182, 191–192, 198–199, 201–202, 223
theories of, 167–171, 177–178, 199–202
unpredictability of, 191–207, 219–220, 222–223
narratives, Christian, 232n202
narratives, community, 196, 222
narrative self, 195–199
narrative theory, 196
narrative therapies, 77
natural reinforcers, 27
Neale, Joanne, 33
NEDM (normative ethical decision making), 151
negative stimulus pairing, 12, 39nn43,44
Nelson, Hilde Lindemann, 170
Nelson, Katherine, 57–58
nervous system, 173
neurobiology. *see also* dopamine *and other substances;* electroconvulsive therapy (ECT) *and other treatments;* neuroscience; PFC (prefrontal cortex) *and other brain structures*
affective foresight and, 79–81, 82
autobiographical (episodic) memory and, 48–49, 50, 53–54, 54–55, 66–67, 69, 79, 82, 94n28, 98n98, 194
autobiographical prospection and, 88–89

context memory and, 88–89
cultural values and, 192–195
emotionally rational choice and, 88–89, 110–114, 128–132
emotionally rational moral intuition and, 128
emotions and, 69–71, 82, 194
empathic intentionality and, 74–76
ethical decision making and, 82, 141–145, 145–146
insight and, 144–145
language and, 145–146, 192
meaning and, 174, 194–195
moral intuition and, 128–131
moral judgment and, 82, 83–85
narrative identity and, 173–174, 192–193
the past and, 65
post-traumatic stress disorder and, 18–19
propranolol and, 8–9
regret and, 90–91
substance addiction and, 23–26, 34–35
neuroendocrine system, 134
neuropeptides, 18, 19, 25
neuroplasticity, 19
neuroscience, xviii, 3, 35–36. *see also* electroconvulsive therapy (ECT) *and other therapies;* neurobiology; psychoanalytic theory
neuroticism, 78, 208
neurotrophic factors, 19
Nichols, Shaun, 148
nicotine, 39n44. *see also* substance addiction
Nietzsche, Friedrich, 138, 207–208
n-methyl-d-aspartate (NMDA) receptor, 10–11
noetic consciousness, 51

non-conscious influences, 143. *see also* insight
norepinephrine (CHR-NE) cascade, 18–19
normative ethical decision making (NEDM), 151
nucleus accumbens (NAc)
Nussbaum, Martha, 183–185

## O

objectivity, 134–137, 139, 157n29, 203–207, 222. *see also* false memory creation (FMC); truth
Odysseus, 1
OFC (orbitofrontal cortex), 82, 84, 90–91, 111, 112, 131
older adults, 74, 89, 174
*Oneself as Another* (Ricoeur), 172
oral memory, 56
orbitofrontal cortex (OFC), 82, 84, 90–91, 111, 112, 131
orbitofrontal region, 18
originality, 213
Osman, Magda, 66
others. *see also* empathic intentionality; relationships; social and cultural mediation
  narrative identity and, 195–196, 208–209, 213–214, 214–215, 219
  narrative identity of, 220
  neurobiology and, 85
  responsibility and, 218
  self-agency and, 198

## P

palliative sedation, 52–53
"Papez circuit," 98n97
parahippocampal gyrus, 70, 130–131
parents, 33, 56–57, 63, 176, 186, 189–190, 214, 242. *see also* attachment
parietal cortex, 110–111
parietal region, 129
Parkinson's disease, 15–16
Pasqualini, Martha, 133–134
the past, 65, 211, 221. *see also* narrative identity (life stories)
Pasupathi, Monisha, 175, 200
Paxton, Joseph, 148
Paxton, Joseph, and colleagues, 149
PCB (President's Council on Bioethics), 20, 22, 31
PCC (post cingulate cortex), 130–131
periaqueductal gray, 83
perirhinal cortex, 69, 70
Persaud, Kiele, 55
personality. *see also* character; empathic intentionality; narrative identity
  autobiographical memories and, 48, 58, 62, 73, 77–78
  integrated, 204
  memory manipulation and, 4n3
  narrative and, 167, 169–171, 178
  neurobiology and, 193
  script theory of, 168–169
perspective, 74
PFC (prefrontal cortex)
  affective foresight and, 79–80, 82
  autobiographical memory and, 54–55, 94n33
  choice and, 111
  cultural contexts and, 195
  emotionally rational ethical decision making and, 142–143
  emotionally rational moral intuition and, 128, 131
  emotions and, 70
  empathic intentionality and, 75–76
  gambling experiment and, 82
  memory and, 94n29
  moral judgment and, 84–85

post-traumatic stress disorder and, 18, 19
protected values and, 192, 193
regret and, 91
semantic retrieval and, 88
substance addiction and, 23, 24–25
temporal-order memory and, 50
pharmaceutical industry, 22, 31
pharmacologicals, 2, 7, 36, 36n2, 240. *see also* biomedicalization; propranolol *and other pharmacologicals*
Phelps, Elizabeth, and colleagues, 69
phenomenological-hermeneutic approaches, 172
Pizarro, David, 148
planning, 65, 110. *see also* affective foresight; intention
Plato, 108, 109–110, 124–125, 137
political narratives, 180–181
Polkinghorne, Donald, 178
post-central gyrus, 130
post cingulate cortex (PCC), 130–131
posterior brain regions, 54
posterior cortical regions, 80
posterior medial cortices, 193, 194, 195
posterior middle temporal gyrus, 88
post-trauma debriefing, 17, 21
post-traumatic stress disorder. *see* PTSD
practical reason, 146–147, 155
Pratt, Michael, 170
precuneus, 75, 128
prediction, 153–154. *see also* affective foresight; planning

prefrontal brain regions, 49, 53–54, 65. *see also* PFC (prefrontal cortex)
prefrontal cortex. *see* PFC
prefrontal temporal cortex, 132
the present, 219
President's Council on Bioethics (PCB), 20, 22, 31
"prevention," 31
pride, 147, 212
primary inducers, 83
principles and rules, 52–53, 183–188, 190–191, 192, 227n86, 228n103
Prinz, Jesse, 160n94
promises, 207, 208
prophylactics, 31. *see also* propranolol
propranolol
  crime and, 20–22
  informed consent and, 31–32
  memory and, 9–10, 37n11
  neurobiology and, 8–9
  pharmaceutical rebranding and, 30–31
  post-traumatic stress disorder and, 20, 21, 22, 31–32
prospection, rational, 63–66. *see also* goals
protected values, 192–193
Proust, Marcel, 196
pseudo body loop, 142–143, 143–144
psychoanalytic theory, 12, 169
psychopaths, 131
psychotherapy, 202, 240
psychotropic drugs, 36n2. *see also* propranolol *and other psychotropic drugs*
PTSD (post-traumatic stress disorder). *see also* biomedicalization; fear, extinction of
  autobiographical narrative centrality and, 77, 78
  beta-blocking pharmacologicals and, 17

electroconvulsive therapy and, 11
episodic detail and, 89
informed consent and, 31–32
limited-use memory manipulation and, 16–22, 36, 239–240
neurobiology and, 18–19
overview, 17, 19–20
propranolol and, 8–10, 21
repression theory and, 12
punishment, 160n94

## Q

Qur'an, 138

## R

race, 170
radiation problem, 150
Rasmussen, Anne, 70
rational emotion. *see* emotional rationality
rationality (reason). *see also* autobiographical (episodic) memory; cognition; emotional rationality; knowledge; narrative identity (life stories); practical reason
autobiographical, 200–201
bounded, 153
emotionally rational moral intuition and, 131–134
Greek philosophy and, 108–110
memory and, 91–92
normativity and, 107–108, 156n2
practical, 3–4
proportionate, 237–243
self-control and, 215
utilitarian, 128, 132, 149, 152–153
rationalizing, 148
rational prospection, 63–66
rational theory of choice, 152–153

Rawls, John, 134–137
reason. *see* rationality
reasons, moral
autonomy and, 203–204
deliberation and, 123–124
emotional rationality and, 146–147, 155–156, 180
emotions and, 117, 136–137
intentionality and, 138
narrative identity and, 179, 188, 204
rationality and, 180
self-reflection and, 203–204
recall, judge, recognize (RJR) paradigm, 64
recognition memory, 51. *see also* familiarity
redemption, 199–202, 202–203, 211, 212
redemptive life narratives, 178–179
regret, 21, 90–91
regulatory processes, 148
reinforcers, natural, 27
relationships. *see also* attachments; empathic intentionality; others; parents; social and cultural mediation
autobiographical memory and, 48, 178
memory manipulation and, 211
narrative identity and, 221–222, 223
relevant alternatives theory, 118
religion, 195, 206–207
reminiscence, 62, 64–65
remorse, 210–211
representations, 83, 149–150
repression, 12
resilience, 19
responsibility. *see also* accountability
autobiographical memory and, 58–59

emotional rationality and, 140
freedom and, 241–242
for knowledge, 118–119
Lotus-eaters and, 2
memory manipulation and, 210–211, 223
narrative identity and, 189, 208, 217–218
retrieval models, 210
reward magnitude, 111
rhinal and perirhinal cortex, 69, 70
Ricoeur, Paul, 172, 173, 180–181, 208
right-brain hemisphere, 146, 192
right inferior parietal lobule (BA 40), 130
risk, 112–113, 133, 153
RJR (recall, judge, recognize) paradigm, 64
Robertson, Diana, and colleagues, 84–85
romanticists, 205
Rorty, Richard, 195
Rousseau, Jean-Jacques, 212
Rubin, David, 56
rules and principles, 52–53, 183–188, 190–191, 192, 227n86, 228n103

## S

sadness, 209–210
Sales, Jessica, 63
Salovey, Peter, 209–210
Samuelson, Paul, 111
Sarbin, Theodore, 178
Sarlo, Michela, and colleagues, 129
Sartre, Jean-Paul, 122
satiation technique, 12, 39n44
Schafer, Roy, 196
Scheler, Max, 124–125
schizophrenia, 10, 65, 89
Schooler, Jonathan, 145
Schüpbach, M., and colleagues, 15–16

script theory of personality, 168–169
secondary inducers, 83
selective serotonin reuptake inhibitors (SSRIs), 31
the self. *see also* authenticity; autobiographical (episodic) memory; autonomy; the "I"; narrative identity; others
autobiographical memory and, 48
language and, 196–197
narrative and, 180, 195–199
neurobiology and, 75–76, 193
splintering of, 207–211
self-agency, 197–198
self-control, 35, 215–216, 223. *see also* autonomy
self-deception, 217
self-destructive behavior, 62
self-knowledge, 194, 238
self-reflection, 63–64, 140, 188, 190, 203–204, 213, 215. *see also* introspection
self-understanding, 208
semantic integration, 194
semantic memory, 11, 47, 51, 88, 94n4
semantic self-knowledge, 48
sensory cortices, 83, 142, 194
sensory experiences, 130
septal nucleus, 67
serotonin, 18. *see also* SSRIs (selective serotonin reuptake inhibitors)
shame, 108, 131
Sharot, Tali, 67, 68
SIM (social intuitionist model), 128, 148
Singer, Jefferson, 209–210
skin conductance, 133–134
social and cultural mediation. *see also* American culture; independence *versus* interdependence; others; tradition
authenticity and, 220

autobiographical memory
and, 48, 55–60, 200–201,
205
earliest memories and,
95n44
emotional rationality and,
127, 139–140
ethical decision making
and, 221–222
happiness and, 208
narrative identity and, 173,
177, 178, 182, 191–192,
198–199, 201–202, 223
neurobiology and, 193,
194–195
objectivity and, 136, 206
proportionate rationality
and, 27–28
protected values and, 192
Rawls on, 134
self-identities and, 197
social intuitionist model
(SIM), 128, 148
social status, 170
sociological theories, 182
Solomon, Robert, 120–121,
126
somatic marker hypothesis,
82, 141, 143–144, 147
somatosensory structures,
142–143
sorrow, 211
specificity, memory, 76–78
speech, 173
SSRIs (selective serotonin
reuptake inhibitors), 31
STN (subthalamic nucleus
stimulation), 15–16
*Stories and Their Limits* (Nelson, ed.), 170
*Stories of Sickness* (H. Brody),
170
storytelling, 168–171, 193,
199. *see also* narrative
identity (life stories); untold stories
stress, 18, 25–26. *see also*
trauma
STS (superior temporal sulcus), 84, 128, 129

*Studies in Hysteria* (Freud),
10
Studinger, Ursula, 62
subcortical nuclei, 141
subcortical structures, 142,
194
subjective identity, 173, 203–207. *see also* narrative
identity
subjectivity, 157n29, 172. *see
also* consciousness; self-reflection
subject/object, 109–110, 123–124
substance addiction. *see also*
alcoholism; cocaine addiction; desires; drug aversion
therapies
autonomy and, 22–23, 27–29, 32–35, 34
Clifasefi and colleagues'
study and, 39n51
deep brain stimulation
and, 15
limited-use memory manipulation and, 22–23, 26–29, 36, 239–240
neurobiology of, 23–26, 35
subthalamic nucleus stimulation (STN), 15–16
suffering, 211
superior temporal sulcus
(STS), 84, 128, 129
surgeries, 2, 22, 36
synaptic plasticity, 4n4

# T

Taylor, Charles, 170–171, 177,
181, 207, 213
temporal gyrus, 88
temporal-order memory, 50,
55, 59, 61
temporoparietal junction
(TPJ), 128, 192, 193, 195
thalamus, 83, 131
Thematic Apperception Test,
168
Thomson, Judith Jarvis, 184

Thorne, Avril, 60, 95n54, 170, 210
time, 67–68, 69, 78, 81, 87–88, 193. *see also* change; the future; the past; the present
*Time and Narrative* (Ricoeur), 172, 180–181
Tomkins, Silvan, 168–169
Tomlinson, Tom, 189–190
Tormey, Judith, 169
Toulmin, Stephen, 183–184, 186, 227n84
TPJ (temporoparietal junction), 128, 192, 193, 195
tradition, 56, 221. *see also* historical accounts
trauma. *see also* flashbulb memories; PTSD (posttraumatic stress disorder); stress
   autobiographical reason and, 63
   biomedicalization and, 30–31
   informed consent and, 31–32
   narrative identity and, 207
truth, 180, 214–215, 220, 226, 229n114, 242. *see also* objectivity
Tulving, Endel, 51, 52
12-step recovery programs, 243n10

# U

Ulysses, 138
uncertainty, 152–153
untold stories, 198–199
utilitarian reasoning, 128, 132, 149, 152–153

# V

validation, 199
values, 4, 5n14, 124–125, 206, 217. *see also* the good; protected values
ventral striatum, 83, 111
ventral tegmental area (VTA), 23, 25, 26, 34–35
ventromedial cortex, 83
ventromedial sector, 142
verbal overshadowing, 145
Vico, Giambattista, 242
virtue, 187
VTA (ventral tegmental area), 23, 25, 26, 34–35

# W

Wainryb, Cecilia, 175
Walker, Margaret Urban, 188–189
Weber, Max, 3
Weiner, Richard, and colleagues, 11–12
well-being, 60, 63, 76, 154, 170. *see also* happiness
wide reflective equilibrium, 190–191
Wiebe, David, 145
Williams, Bernard, 184
Williams, J. Mark, and colleagues, 76–77
Wittgenstein, Ludwig, 205
worldviews, 126
worry, epistemic, 119

# Y

Yoder, Keith, 129
Yonelinas, Andrew, 67, 68

CPSIA information can be obtained
at www.ICGtesting.com
Printed in the USA
BVHW01*1909280218
509379BV00001B/1/P